History of Rationalities

Luca Sciortino

History of Rationalities

Ways of Thinking from Vico to Hacking and Beyond

palgrave
macmillan

Luca Sciortino
eCampus University
Novedrate, Italy

ISBN 978-3-031-24003-4 ISBN 978-3-031-24004-1 (eBook)
https://doi.org/10.1007/978-3-031-24004-1

This Palgrave Macmillan imprint is published by the registered company Springer Nature Switzerland AG.
The registered company address is: Gewerbestrasse 11, 6330 Cham, Switzerland

To my daughter Patrizia

Preface

This is a book about the idea that in human history it is possible to recognize distinct ways of thinking. It is my contention that, although this idea is dense with important philosophical implications and constitutes a significant ingredient of the philosophies of several noteworthy thinkers, it has not always received the attention it deserves. To think means to direct one's mind towards someone or something in order to accomplish certain tasks, the most interesting of which will, for us, be those concerning the acquisition of knowledge. Thus, before delving deeper into the concept, we can say that a 'way of thinking' is a manner of using one's mind actively for finding out, understanding the world, acquiring knowledge, solving problems, evaluating issues, deciding between various options, analysing ideas or reflecting on one's experience. These are examples of mental actions that can be performed in different ways. For instance, one can understand the world from a certain perspective, make a choice by appealing to a particular value or acquire knowledge by relying on certain methods and preconceptions rather than others.

For the time being, I will consider the characterization above as a working definition of the concept of 'ways of thinking'. I will modify and refine it in the next chapter with additional commentary and nuances, once I have provided the relevant philosophical context for the discussion of why the concept of 'ways of thinking' commands our attention. As to the label I am using, by no means is 'ways of thinking' the only possible choice: 'forms of thought', 'modes of thinking', 'ways of knowing', 'ways of reasoning', 'mental attitudes' and 'worldviews' are only some of the phrases that are used in the same sense and which would have been just as apt.

The concept of 'ways of thinking' has played a crucial role in philosophy at least since the Enlightenment. It appears in different versions, as if they were species of the same genus. In other words, implicitly or not, philosophers have proposed different notions of the concept of 'ways of thinking'. Thinkers such as Giambattista Vico (1668–1744) or Auguste Comte (1798–1857), who have combined the study of history with that of philosophy in an attempt to understand human knowledge, had argued that human beings have thought differently in different historical contexts. However, it was with the emergence of a French tradition in philosophy of science called historical epistemology, which studies the changes and the evolution of epistemological concepts, that the concept of 'ways of thinking' began to acquire considerable weight. Within that tradition, the philosopher Lucien Lévy-Bruhl (1857–1939) was the first to reject the idea of a timeless and universal way of thinking. Based on anthropological studies, he suggested that certain traditional non-literate societies thought in a different way from modern people. His notion of 'primitive mentality', the way of thinking of primitive societies, was of help to other historical epistemologists who reflected on the role in the history of science of concepts such as objectivity, rationality, truth and many others. After him, philosophers and historians of science such as Hélène Metzger (1889–1944), Gaston Bachelard (1884–1962), Ludwik Fleck (1896–1961), Alexandre Koyré (1892–1964) and Michel Foucault (1926–1984) put forward, implicitly or not, their own notions of the concept of 'ways of thinking'. Their ideas were then crucial for the development of other notions, such as that of Thomas Kuhn's 'paradigm' and that of Ian Hacking's 'style of reasoning'. In the 1980s, the latter put forward the idea that there exist distinct styles of reasoning which can be viewed as broad frameworks that govern a certain way of investigating the world. Styles of reasoning have emerged and stabilized at different points within the history of the sciences and involve new types of evidence, questions and methods of inquiry.

This book is motivated by the following thoughts. In its different notions, the concept of 'ways of thinking' is central to historical epistemology. Therefore, a careful analysis of it could provide a new perspective for studying this tradition of thought. Furthermore, a comparative analysis of the different notions of 'ways of thinking' could help us to highlight their merits and defects for the study of the development of human knowledge; and going back in time to the uses of the concept of 'ways of thinking' could allow us to understand more deeply notions that have been put forward more recently, especially when thinking about scientific research. For example, Hacking's project of styles of reasoning can be considered the latest sophisticated attempt to reconstruct the different ways of finding out that can be recognized in the history of science. Among the other things, his original thesis has never been assessed and developed into a comprehensive theory.

Furthermore, the concept of 'ways of thinking' is at the root of two crucial philosophical problems that have always attracted the attention of philosophers: the problem of relativism and the problem of contingency. As for the first problem, one may ask whether or not there are trans-historical and independent criteria for justifying claims made using different ways of thinking. This is an issue of incommensurability between different frameworks: the claim that there might not be common standards of evaluation external to distinct ways of thinking. As for the second problem, in literature it is often asked whether a particular field of science could have taken a different route from the actual one or whether the achievement of certain scientific result has been contingent. However, the achievement of certain scientific results might depend on the emergence of a certain way of thinking. It might therefore be useful to include the concept of 'ways of thinking' in the discussions above regarding the contingency issue.

This book addresses the above problems and, in order to provide a taxonomy of the concept of 'ways of thinking', I shall in the first part discuss what its different notions have in common and how they differ. Since, as I shall show, these notions can be viewed as transient replacements of Kant's a priori, it is possible to say that this book presents and compares different frameworks for the study of how objectivity is possible. I shall also discuss the relativism issue with particular focus on the notion of 'style of reasoning', which I shall develop and present in a more systematic way in the central part of the book. In the final part of the book I shall use Hacking's notion to discuss the contingency issue. The following are some of the questions I shall address: to what extent are the emergence of the styles of reasoning at a certain point of history a contingent circumstance?

Is science bound to converge on a single answer to a given question? Are styles of reasoning long lasting?

To sum up, the concept of 'ways of thinking' is an absolutely fundamental element of most of the philosophies that fall into a tradition of research called 'historical epistemology'. Of this concept. philosophers and historians of science have presented different notions. This book presents, compares and contrasts them. It then focuses on Hacking's notion of style of reasoning in order to assess and develop it into a more systematic theory of scientific thought. Finally, this book argues that Hacking's notion of 'style of reasoning' implies epistemic relativism and discusses its implications for the contingency issue. In the first chapter, I shall outline a history of the concept of 'ways of thinking'. At the end of this presentation, I will better explain the objectives of the book and describe the content of its chapters in more detail.

Most of the reflections that make up this book are the fruit of my doctoral studies at the Department of Philosophy at The Open University and my experience as Research Fellow at the School of Philosophy, Religion and History of Science at the University of Leeds. In those years my work was focused on Hacking's notion of 'style of reasoning'. In the following years, as my research on historical epistemology continued, my project became larger until it became a study on the concept of 'ways of thinking' in historical epistemology. Consequently, this book can be read on three levels: first, as an introduction to the different notions of way of thinking in historical epistemology; second, as an attempt to develop, correct and present in a more systematic way Hacking's notion of 'style of reasoning'; third, as a case-study for a general view on the relativism issue in historical epistemology. There are many people to whom I owe gratitude and appreciation. First of all, I would like to thank Professor Cristina Chimisso for the advice she has provided throughout this project and for all I have learned from her writings. I would also like to thank those members of staff at the Open University, Faculty of Arts and Social Sciences (UK), who have read and provided feedback on the second part of this work when it was still in its infancy, in particular my former second supervisor Professor Sophie Grace Chappell. I am grateful to Professor Gregory Radick for his support and the opportunity to perform research as Visiting Research Fellow in the School of Philosophy, Religion and History of Science, University of Leeds. I should also mention that I really appreciate the valuable suggestions of an anonymous reviewer which helped me to improve the quality of the manuscript. Furthermore, I want to thank the Senior Editor Philip Getz at Palgrave Macmillan for believing in my project and giving me the time to complete it. Finally, I want to say that writing this book would have been all the more difficult were it not for the support provided by my family and my friends. I am indebted to them for their help.

Novedrate, Italy Luca Sciortino

Keywords

Ways of thinking • Historicized Kantianism • French epistemology • Historical epistemology • Thomas Kuhn • Ludwik Fleck • Michel Foucault • Ian Hacking • Styles of reasoning • Objectivity • Relativism • Contingency

About the Book

This book centres on Ian Hacking's proposal that within the history of Western thought a number of distinct 'styles of reasoning' have accumulated since antiquity. The first part of the book places Hacking's proposal within a lineage of thinkers stretching back to Vico who have shared a concern with exploring rationalities, plural. The second part of the book expounds Hacking's proposal more thoroughly and systematically than Hacking himself has done, then examines the implications for debates about relativism and the contingency of knowledge. The idea that human beings have thought, known or found out in totally different ways in the course of history is an absolutely fundamental element of most of the philosophies that fall into a tradition of research called 'historical epistemology'. This is also a book on the concept of 'ways of thinking' and its different notions proposed in philosophy of science. By providing a comparative analysis of these notions, it gives a new perspective for studying historical epistemology. It then focuses on one of them, Hacking's notion of 'style of reasoning', in order to assess and develop it into a more systematic theory of scientific thought. Ultimately, this work can be read on three levels: first, as an introduction to the different notions of the concept of 'ways of thinking' introduced in philosophy of science; second, as an attempt to develop, correct and present in a more systematic way Hacking's notion of 'style of reasoning'; third, as a case-study for a general view on the relativism issue in historical epistemology.

Contents

About the Author

Luca Sciortino received his PhD in Philosophy of Science from the Open University (UK) with a thesis on the project of scientific styles of reasoning put forward by the Canadian philosopher of science Ian Hacking. He has been Armenise-Harvard Science Writer Fellow at the Harvard University and Research Fellow in Philosophy of Science at the University of Leeds, Centre for History and Philosophy of Science (UK). His research focuses on the history and philosophy of scientific thought, especially on 'historical epistemology'. He has written various research articles on the idea that in the history of human thought we can recognize different ways of knowing. Alongside his work as a researcher and adjunct professor at eCampus University (Italy), Luca Sciortino is also a writer and scientific popularizer, author of several books in Italian.

CHAPTER 1

A Brief History of the Concept of 'Ways of Thinking': Introduction and Plan of Work

1.1 The Concept of 'Ways of Thinking': Early Notions

There can be little doubt that in the course of the history of philosophy many thinkers have put forward, implicitly or not, notions of the concept of 'ways of thinking' that are very different from one another. To the best of my knowledge, the Italian Giambattista Vico (1668–1744) was one of the first philosophers to forcefully argue that in different epochs people have thought differently. In his work *The New Science* (1725 [2003]), he presented powerful arguments against the then dominant Cartesian philosophy which viewed human reason as an absolute and unchangeable reality, independent of history. To the Cartesian *cogito ergo sum*, he opposed the *verum ipsum factum* (i.e. what is true is precisely what is made), i.e. the principle that human beings can only understand what they have created. For example, since we have made our own history, we are able to trace its genesis and its laws of development; and, since we have created mathematics, we can understand it. Furthermore, according to Vico, *verum et factum convertuntur* (the true and the made are convertible):[1] what provides 'the norm of truth [is] to have made it' (Vico, 2013 [1710], p. 138) rather

[1] Here and in the following chapters, the translations from texts in languages other than English are mine.

L. Sciortino, *History of Rationalities*,
https://doi.org/10.1007/978-3-031-24004-1_1

than the immediate evidence or the clearness and distinctness of ideas upon which René Descartes (1596–1650) insisted.

In this regard, Vico wrote that 'this civil world was certainly made by men, whence there can be found, because there ought to be found, its principles within the *modification of human mind*' (Vico, 1725 [2003] Book I, Chapter III p. 255 my emphasis). As he remarked in another of his works, *On the Most Ancient Wisdom of the Italians* (1710), 'the Latins called "mens" (i.e., mind) what we call "thought"' (Vico, 2013 [1710] chapter VI, p. 164). Thus, for him, customs, laws and the entire world of human knowledge are historical products of the different ways of thinking that have followed one another throughout human history. However, thinking should not be interpreted solely as reasoning: in Vico, reason cannot produce, alone and autonomously, the whole world of knowledge. What creates the latter is a multiplicity of faculties: sense, memory, imagination, ingenuity or invention and reason or intellect. The faculty of ingenuity [*ingegno*] relates different things and helps us to see an order in them; the faculty of imagination [*fantasia*] is related to that of memory and can create images, the faculty of sense (*senso*) includes emotions, instincts, passions and common sense. A long tradition of thought, which can be traced back to Aristotle (384 BC–322 BC) and Avicenna (980 AD–1037), differentiated these faculties and regarded them as contained in different parts of the brain. Although Vico drew upon this tradition (Tristram, 1988, p. 357), he regarded ingenuity, imagination and memory as belonging to mind but having their roots in the body. For example, he regarded imagination as an extension of bodily experience. For him, in order to produce and assess knowledge, the human mind operates by making use of its faculties. His point was that there are historical variations in the extent to which the mind uses, articulates and applies this collection of faculties. In our terminology, we could say that Vico's 'ways of thinking' are different uses of the faculties of the mind, of which some are rooted in the body.

In any given epoch, it is one of these faculties that prevails over the others, which nevertheless remain active. In particular, Vico argued that human societies pass through ages [corsi] of growths and decay: the ages of gods, heroes and men. In the first two ages, human beings mainly used memory and imagination to comprehend the world as a creation of the gods, while in the third age human beings mainly used reason to create social order. During the first age, that of gods, human beings invoked imagination to understand the world and felt, in the forces of nature that threatened them, the punishment of gods. Vico argued that these first human beings 'were "poets", which in Greek sounds "creators"' (Vico, 1725 [2003], p. 275). For Vico, 'poetry' was a form of knowledge based on the creation of poetical images. He had in mind what we may call a 'metaphorical way of thinking' based on memory and imagination: universal concepts were created through a process of abstraction and conceptualization based on the search for analogies between things very distant from each other. A similar way of thinking was adopted in the age of heroes, during which time human beings followed models of virility and used imagination to create moral values and establish institutions that allowed them to cultivate heroic virtues.

It was in the age of men, when the plebs claimed their equality of nature with the nobles and the popular republics were born, that a new way of thinking emerged. This latter, mainly based on the faculty of reason, replaced the metaphorical way of thinking. As the faculty of imagination weakened and gave way to logic and reason, poetry lost its primary role in the process of acquiring knowledge. The steady ascendance of reason over imagination, which represents a gradual progress of civilization, is not irreversible. Human history is open to uncertainty and the return to the primitive barbarism of the age of the gods is always possible. The historian of ideas Isaiah Berlin (1909–1997) wrote that, by criticizing the notion of fixed laws and highlighting the role of historical and social contexts, Vico laid out the 'whole doctrine of historicism in embryo' (Berlin, 1980, p. 38). As far as we are concerned, to this observation it should be added that Vico also paved the way for later thinkers who regarded knowledge of nature in different times as based on different ways of thinking.

During the unfolding of the Enlightenment various theories of history combined with the belief in the cognitive progress of humankind. However, unlike Vico, their authors assumed that in different epochs human beings had substantially thought in the same way. If there were any differences, they only concerned the correctness of the reasoning. For example, in his *Digression on the Ancients and the Moderns* (Fontenelle, 2019 [1688]), the French thinker Bernard le Bovier de Fontenelle (1657–1757), compared the diversity of plants with that of minds in different climates and epochs. He wondered: 'perhaps the oranges that do not grow as easily here as in Italy indicate there the existence of a particular way of thinking [*certain tour d'esprit*] that cannot be found here [in France] in an identical form' (Fontenelle, 2019 [1688], p. 30). A few pages later he excluded the possibility of the existence of different ways of thinking: 'the centuries make no difference [...] We are all perfectly alike, ancient and modern, Greek, Latin and French' (Fontenelle, 2019 [1688], p. 34). For him, what measured epistemological progress was the 'method of reasoning' [*méthode de raisonner*] employed (Fontenelle, 2019 [1688], p. 44). In particular, he regarded the Cartesian reasoning as an achievement in the history of humankind. If Vico invited us not to project the interpretative schemes of the present into the past, Fontenelle affirmed that 'a precision and exactness, hitherto unknown, reign today not only in good works of physics and metaphysics, but also in those of religion, morality and ethics' (Fontenelle, 2019 [1688], p. 44).

Later, the economist Robert Turgot (1727–1781) explained what he regarded as a linear and cumulative progress of humanity by invoking the concept of culture, the idea that knowledge is conceptualized, communicated in symbolic form and passed on from generation to generation. In his words, 'the arbitrary symbols of speech and writing have created a common treasure-house of all individual's knowledge, which one generation contributes to the next, a heritage constantly increased by the discoveries of each age' (Turgot, 1913–1923, p. 215). In agreement with his mentor and friend Turgot, the philosopher and mathematician Nicolas de Condorcet (1743–1794) stressed the importance of passing the intellectual achievements of important individuals on to the entire community and to subsequent generations. Thus, progress was guaranteed to happen thanks to the role of education and the efforts of single scientists, whose cognitive capabilities were the same globally. The linear accumulation of knowledge could be inhibited by political and religious factors: 'the triumph of Christianity was thus the signal of the entire decline both of the sciences and philosophy', wrote Condorcet, and 'had the art of printing been known, the sciences would have been able to preserve their ground' (Condorcet, 1795, p. 128).

1.2 The Concept of 'Ways of Thinking' from Kant to Hegel and Comte

Vico's claim that human beings can only know what they have made implies that the cognitive object is always somehow 'constructed' by the knower as a condition of knowledge. It is precisely on this idea that Kant's transcendental argument rests: we cannot know the thing-in-itself but only the object as it appears to us through our forms of intuition. For Vico as for Immanuel Kant (1724–1804), human beings do not know the physical world as it is but as it appears to them. However, unlike Vico, Kant proposed a *non-historical* epistemological constructivism[2] based on an abstract notion of the subject, which he conceived as independent of the social context. One of the problems Kant addressed was to understand how judgements become possible: to determine what he called the 'transcendental conditions', i.e. the 'conditions of possibility' of our concepts. He believed he could state them independently of history or of any necessary constraints imposed on us from outside. Indeed, the Kantian transcendental conditions are related only to the structure of human thought—it is reason that imposes them on the raw sense data by making a set of sensations into objective experience. In particular, the spatial and temporal forms of intuition represent the conditions within which we perceive objects as existing in space and time and within which the propositions describing these objects are *objective*, i.e. they have a truth-value. Since transcendental conditions are prior to experience, Kant referred to them as a priori conditions. As Michael Friedman wrote, 'what is objective [became] that to which all human beings—in virtue of their rational faculties of sensibility and understanding—must necessarily agree' (Friedman, 2011).

It must be noted that Kant was engaged in a particular stage of scientific process, Newtonian science,[3] which for him explained the world of appearances. In order to justify the kind of necessary universal laws proposed by Isaac Newton (1642–1726/27) he defended the intellectual structures underlying Newtonian mechanics: Euclidean geometry, continuous classical time, absolute space and the concepts of causality and substance. Kant did not see the possibility that the transcendental conditions could change in the course of history. He took the constitutive principles of Newtonian physics as forming fixed and timeless categories of the human mind. Only when it became clear that the categories of thought have a history was it possible to reimagine the existence of different ways of thinking. The latter could be defined, as we shall see better in the next chapter, in terms of the different categories of thought employed in the production of knowledge.

[2] For an outline of the constructivist approach to knowledge of Vico and Kant see (Rockmore, 2005, 2016).

[3] For a comprehensive account of Kant's scientific sources see (Calinger, 1979).

It was the new historical consciousness of the nineteenth century, which had emerged at the dawn of the Enlightenment that spawned the first philosophies at odds with the view of an universal rationality governing all human knowledge at all times and places. It is noteworthy in this regard that the philosopher George Wilhelm Friedrich Hegel (1770–1831), one of the most important figures of German idealism, criticized Kant's belief that categories are timeless and universal and allowed them to change through experience. His *Science of Logic* (Hegel, 2010 [1816]) has been interpreted by some scholars[4] as a category theory, i.e. a dialectic deduction of the categories of thought and a study of their interrelations and changes. If this interpretation is correct, one might say that Hegel reconstructed the genesis and the changes of the rules for thinking, which were represented by the different Hegelian 'moments'. In this sense, as Daniel Dahlstrom put it, 'Hegel's *Science of Logic* is an analysis of the presuppositions of various ways of thinking' (Dahlstrom, 2008, p. 109).

The new historical consciousness of the nineteenth century also favoured the birth of notions of 'ways of thinking' that had a more direct link to Enlightenment views of progress. In this regard, one cannot fail to mention the founder of positivism August Comte (1798–1857). The influence of the ideas of Turgot and Condorcet on Comte's theory of progress has been noted by many scholars (see for example Belrose, 1892; Morgenthaler, 2007; Vinogradova, 2020). As far as we are concerned, in his *Cours de philosophie positive* (1830–1842) Comte argued that humanity had developed through three stages, the theological, the metaphysical and the positive age, in each of which a particular way of thinking was predominant. Indeed, he wrote that 'the human mind [*esprit*], by its nature, employs successively in each of its inquiries three methods of philosophizing [*méthodes de philosopher*], the character of which is essentially different and even radically opposed' (Comte, 1968 [1830–1842], p. 11). He called these 'methods of philosophizing' the 'theological method', the 'metaphysical method', and the 'positive method' and defined them as 'general systems of conceptions on all phenomena that are mutually exclusive: the first is the necessary starting point of human intelligence; the third, its fixed and final state; the second is only intended to serve as a transition' (p. 11). Thinking theologically for Comte meant that 'the human mind [...] represents the natural phenomena as produced by the

[4] See for example (Pinkard, 1988; Burbridge, 1993; Di Giovanni, 2010).

direct and continuous action of supernatural agents, more or less numerous. whose arbitrary intervention explains all the apparent anomalies of the universe' (p. 11). If thinking in the metaphysical way was not so different from thinking in the theological way ('in the metaphysical age the supernatural agents are replaced by abstract forces' (p. 11)), thinking in the positive way meant 'to give up knowing the inner causes of phenomena and focus solely on discovering, through the well combined use of reasoning and observation, their effective laws, that is to say their invariable relations of succession and similitude' (p. 11). The kind of knowledge gained by thinking in this way amounts to 'the connection established between the various particular phenomena and some general facts, of which the progress of science tends more and more to diminish the number' (p. 11).

Both Vico and Comte believed that human beings were able to think in different ways. Although the relative importance of these ways of thinking had varied in time, they all were rooted in human capabilities that had always been the same in all times and places. However, while for Vico progress was not inevitable and a return to primitive conditions was possible, for Comte, being in the positive age was for humanity a permanent and definitive condition. For the latter, humanity's tendency towards a positive stage of reason was the result of a logical instinct that leads human beings to establish relationships between phenomena and seek unity (Comte, 1968 [1830–1842], p. 20). According to Comte, it was with Descartes that human thinking reached the positive stage, an idea that he had inherited from Enlightenment thinkers such as Turgot and Condorcet.

Comte's positivist doctrine had a series of consequences, including that of dealing a fatal blow to German idealism and bringing the very role of philosophy towards a re-conception. As Peter Dews wrote, 'Comte, [...] in the first lesson of the *Cours de philosophie positive*, debunk[ed] the "psychological method" ("the so-called study of the mind by the mind is a pure illusion") and argue[d] that the task of positive philosophy is to trace "the course actually followed by the human mind in action"' (Dews, 1992, p. 348). Indeed, once Comte had put aside transcendental logic and considered the age of metaphysics as a lower stage of human development, philosophy was left with the task of classifying the different sciences, describing what they do and searching for the rules of the formulation and corroboration of scientific theories.

1.3 Historicized Kantianism and the Concept of 'Ways of Thinking'

At the beginning of the twentieth century some experimental observations challenged the classical conception of the physical world and, finally, the theory of relativity and the theory of quanta resulted in a non-Newtonian conception of space, time and motion. Relativity theory revealed that the Euclidean-Newtonian picture of space and time—according to Kant's synthetic a priori constitutive principles—were not a necessary and unrevisable constituent of our empirical knowledge.[5] Indeed, in the context of general relativity, Euclidean geometry is constitutive a priori only in a small enough region of space-time but not globally, in a large region. On its part, quantum theory brought into question the realist view of the world in which individual objects and their properties have an unequivocal existence independent of any act of measurement. The uncertainty principle stated the impossibility of separating the behaviour of atomic objects and their measurement apparatuses—the properties of objects depend on their context. In other words, propositions in quantum theory do not possess an objective truth-value that exists regardless of our means of warranting its assignment. Ultimately, the upheavals of quantum theory and the theory of relativity called into question Kant's claim that the conditions of experience can be found in the structure of human mind and the very basic assumption of classical epistemology that the knower has direct access to nature. As Friedman put it 'What we end[ed] up with is thus a relativised and dynamical conception of a priori mathematical-physical principles, which change and develop along with the development of the mathematical and physical sciences themselves' (Friedman, 2001, p. 31).

Although Comte had convincingly argued that knowledge could only be studied in its historical development, his philosophy did not have anything to say about the discontinuities of the scientific process and the changes of the norms of science. His notion of way of thinking ('way of philosophizing') was not adequate to describe the profound changes in the categories and principles imposed by the development of new

[5] The last chapter of *Einstein's Theory of Relativity* (Born, 1962) gives a brief account of the failure of Euclidean geometry and the first chapters of *Causality and Chance in Modern Physics* (Bohm, 1984 [1957]) provide a clear exposition of why quantum mechanics brought about a renunciation of causality.

empirical theories. Thus, the philosophical schools of the first half of the twentieth century sought to resolve the question of how knowledge was possible by appealing to replacements for the Kantian a priori.

For its part, logical positivism, arising in Vienna in the 1920s, proposed a solution that could be synthesized by these words of Hans Reichenbach (1891–1953), one of its proponents: 'We wish to add some general remarks to the resolution of Kant's doctrine of the a priori. It seems to have been Kant's mistake that he who had discovered the essence of epistemology in his critical question confused two aims in his answers to this question. If he searched for the conditions of knowledge, he should have analysed *knowledge*; but what he analysed was *reason*. He should have searched for *axioms* instead of *categories*' (Reichenbach, 1965 [1920], p. 72). Thus, Reichenbach replaced Kant's a priori principles with the 'principles of coordination' by which he mainly referred to general rules, e.g. theoretical principles expressed by mathematical equations, which order the content of what is perceived, and 'define the individual elements of reality and in this sense constitute the real object' (Reichenbach, 1965 [1920], p. 53). He also introduced the distinction between the principles of coordination and the 'axioms of connection' which interconnect specific variables. For example, according to Reichenbach, that events are governed by a probability distribution is a principle that cannot be empirically established and that he calls the principle of probability. The latter acts as a principle of coordination in that an experimenter has to assume a distribution of experimental errors in order to interpret a set of numbers as the result of different measurements of a given object. On the other hand, the specific distribution, e.g. the Gaussian distribution, interconnects different variables that describe that object. In this sense, the principle of probability is an a priori, precondition of empirical knowledge. Rudolf Carnap (1891–1970), to mention one of the most important advocates of logical positivism, introduced a distinction between logical rules (L-rules) and physical rules (P-rules) of a linguistic framework that parallels Reichenbach's distinction between principles of coordination and principles of connection. The logical rules of a linguistic framework establish the standards of the correctness of propositions; the physical rules are empirical laws such as Einstein's equations of gravitation or even observation sentences. A linguistic framework and its logical rules provide the conditions of objectivity for new kinds of entities, concepts and propositions (Carnap, 1950, p. 21). The move of the logical positivists—to search for the conditions of knowledge by analysing not reason but knowledge itself—cut off any transcendental inquiry into the conditions of possible knowledge and, with it, any inquiry into possible differences in the way human beings think.

From 1870 until the end of the nineteenth century, the two German schools of Neo-Kantianism—the Marburg School and the Southwest School—had tried to strengthen Kant's position. When, at the beginning of the twentieth century, physics had evolved to the point that classical physics could not cope with some inadequacies that had begun to emerge, the philosophers of the Marburg School realized that the categories and principles can change as scientists develop new empirical theories. Thus, they gave up in their attempt to wed transcendental philosophy to Euclidean geometry and Newtonian mechanics and wondered what concepts or laws within some parts of science were necessary for its objective validity, i.e. for its truth and its independence from individual subjects. For Ernst Cassirer (1874–1945), the last Neo-Kantian philosopher of the twentieth century, what guaranteed the objectivity of knowledge was not the constitutive role of logic, as Carnap maintained, but rather the existence of symbolic forms of representation such as myth, art, science, history and language. Cassirer had at heart the search for the conditions of the possibility of the cultural fact and his point was that the symbolic forms are codes through which thought is objectified, concept and object are mediated and the human thought is expressed. As Donald Verene put it, 'Cassirer view[ed] the later two [Kant's] *Critiques* as attempts to account for other realms of objectivity. Cassirer's philosophy of symbolic forms is an attempt to continue this tendency in Kantian thought by regarding each major area of culture as having its own sort of objectivity' (Verene, 1966).

According to Cassirer thought is structured by means of categories such as space, time and causality which are present in all symbolic forms but take different configurations from time to time. For example, scientific space and mythical space differ: the latter is not homogeneous in that particular places are endowed with special meanings. Thus, the concept of symbolic form can be regarded as the particular notion of way of thinking put forward by Cassirer: a particular mental scheme to represent reality that makes use of particular configurations of categories of thought. In the course of history, argued Cassirer, human beings have adopted ever new symbolic forms, of which the mathematical formalism of modern physics represented the most advanced stage. In our terminology, these ever new symbolic forms gave rise to a rich variety of different ways of thinking about the same object.

1.4 Sociology of Knowledge, Artistic Styles and the Concept of 'Ways of Thinking'

Sociologists also joined the project of expanding Kant's *Critique of Pure Reason* into a *Critique of Historical Reason.* Like the German thinker George Simmel (1858–1918), the Austro-Hungarian sociologists and philosophers György Lukács (1885–1971) and Karl Mannheim (1893–1947) worked in a neo-Kantian perspective: "they [...] searched for a priori *conditions of possibility* in the context of which various forms of knowledge, in the broadest possible sense of the term, could have emerged' (Demeter, 2012, p. 40). In pursuing this goal, they all rejected positivist approaches that adapted the methods of natural sciences to social inquiry. In his work *Ideology and Utopia* (1979 [1936]), Mannheim tried to identify the roots of human thought within the sociological context that produced it. One of the conclusions that he reached was precisely the existence of different ways of thinking, or *styles of thought* in his terminology, which he characterized in a way that is reminiscent, as we shall see, of Kantian forms of human experience. To understand better Mannheim's notion of way of thinking and its importance, it is better first of all to illustrate some ideas that were a crucial element of reflection for its elaboration.

To begin with, as Jeremy Tanner (2019) has documented, the critical tradition of German art history played an important role in the development of Mannheim's thought. Between 1912 and 1915, when he was an undergraduate in Budapest, Mannheim joined the so-called Sunday Circle, which discussed philosophy, art and literature under the leadership of Lukács. The members of the circle shared a strong enthusiasm for the work of the Austrian art historian Alois Riegl (1858–1905) (Tanner, 2019, p. 756) who, in his book *Problems of Style: Foundations for a History of Ornament* (1992 [1893]), argued that the evolution of the ornamental style does not depend so much on the technical and material development but is shaped by the 'creative intention' or the 'artistic will' [*kunstwollen*] of the artist. What Riegl wanted to disprove was a misinterpretation of the thesis of the art critic Gottfried Semper (1803–1879): 'whereas Semper did suggest that material and technique play a role in the genesis of art forms', wrote Riegl, 'the Semperians jumped to the conclusion that all art forms were always the direct product of materials and techniques [...] They were, of course, not acting in the spirit of Gottfried Semper, who would never have agreed to exchanging free and creative artistic impulse [*Kunstwollen*] for an essentially mechanical and materialist drive to imitate' (Alois Riegl, 1992 [1893], p. 4). Later, in his book *The Late Roman Art Industry* (Alois Riegl, 1985 [1901]), he developed his ideas by examining ancient Egyptian, classical Greek and late Roman art and described the ***kunstwollen*** as the way in which each culture perceives and interprets the world in relation to a worldview [*Weltanschauung*][6] (see for example Alois Riegl, 1985 [1901] chapter v).

Riegl regarded the transition from a style to another as a 'necessary transition[s] made by the human thought' (1985 [1901], p. 232), a change in the creative intention of the artist as a consequence of a mutation of the worldview of an epoch. Accordingly, he rejected narratives of decline in art history by arguing, in a specific example, that the fourth century reliefs of the Arch of Constantine marked a new way of perceiving space and symmetry rather than an epoch of regress. By contrast, as Jaś Elsner illustrated, in describing the arch of Constantine both the Italian painter Raffaello Sanzio [Raphael] (1483–1520) and the art historian Giorgio Vasari (1511–1574) 'use[d] stylistic analysis to create a historical picture of decline' (Elsner, 2003, p. 101). Later, even the English historian Edward Gibbon (1737–1794) incorporated the arch into *The History of the Decline of the Roman Empire* (1776) as 'a melancholy proof of the decline of the arts and a singular testimony of the meanest vanity' (Elsner, 2003, p. 102).

It is important to highlight that, in a letter to Pope Leo X that described the sculptures of the Arch of Constantine as a sign of decline, Raphael used the term 'style' [*stile*].[7] Indeed, as Elsner noted, 'the art-historical concept [of "style"] was born from a vocabulary generated by Renaissance

[6] The German word *Weltanschauung* is generally translated as 'worldview'. 'Welt' means 'world' while 'Anschauung' means 'view' or 'perception'.

[7] The document, datable to the last months of 1519, concerned the protection of the vestiges of ancient Rome. It was written by the Renaissance author Baldassarre Castiglione (1478–1529) with the contribution of Raphael and was addressed to Pope Leo X. In the first part, the authors described the Gothic *style* as a sign of degradation of the German art in the Middle Ages. Furthermore, regarding the Arch of Constantine they wrote: 'Although literature, sculpture, painting, and almost all the other arts had long been declining and had grown worse and worse until the time of the last emperors, yet architecture was still studied and practiced according to the good rules and buildings were erected in the same *style* as before […] Of this there are many evidences: among others the Arch of Constantine, which is well designed and well built as far as architecture is concerned. But the sculptures of the same arch are very feeble and destitute of all art and good design. Those, however, that come from the spoils of Trajan and Antoninus Pius are extremely fine and done in perfect *style*' (Sanzio, 2020 my emphasis).

artists. Their interest la[id] in defining good and bad practice in relation to canons of contemporary production (hence the significance of style to the thematic of "decline")' (2003, p. 108). These artists were the first to use the concept of style consciously, in the way a modern art historian uses it: 'in medieval or archaic art the presence of "style" (whether "period style" or "personal style") is largely unselfconscious; the style is made explicit through art-historical analysis and hence made conscious to the modern consumer [...]. By contrast, post-Renaissance and especially modern art practice [...] is wholly self-conscious and deliberate about its artistic evocations' (Elsner, 2003, p. 108).

With Mannheim, the concept of 'style', born in art-historical discourse, was adapted to the study of the history of human thought[8] becoming the notion of 'thought style' or 'style of thought' [*denkstil*].[9] For him, 'what is true of art also holds *mutatis mutandis* good for knowledge' (Mannheim, 1979 [1936], pp. 243–244; 1997 [1922]):

> [...] art does in fact develop in 'styles', and [...] within the styles there is a gradual change from one phase to another which makes it possible to 'place' an unknown work of art. Now it is our contention that human thought also develops in 'styles', and that there are different schools of thought distinguishable by the different ways in which they use different thought patterns and categories. Thus it should be just as possible to 'place' an anonymous piece of writing as an anonymous work of art, if we only took the trouble to reconstruct the different styles of a given epoch and their variations from individual to individual. (Mannheim, 1971a [1927], p. 261)

Ultimately, Mannheim claimed that in the history of thought it is possible to distinguish different styles of thinking characterized by the use of different categories and patterns of reasoning. He assumed that, just as it is possible to identify an artistic style from an analysis of a work of art, so it is possible to identify a style of thought from its cultural objectifications: for example, writings, speeches, political choices. The task of what he calls

[8] Lukács had already introduced the concept of 'style' in the study of the history of literature in 1910 in an article first published in Hungarian (Lukács, 1910) and later in German under the title *Bemerkungen zu einer Theorie der Literaturgeschichte* (i.e. 'Remarks on a theory of literary history') (Lukács, 1969 [1910]).

[9] In *On the Interpretation of Weltanschauung* (Mannheim, 1971b [1952]), Mannheim mentioned Riegl several times explicitly praising him for providing an exact description of the style in art. He also wrote that 'the basic intention behind each style of thought is in many ways similar to that of Riegl's art motive' (1971b [1952], p. 264).

'sociology of knowledge' is to reconstruct a style of thought from the social background in which it is rooted and from its cultural objectifications:

> There are two main ways of writing the history of thought. On the one hand there is what might be called the narrative way [...], on the other hand there is [...] the recently developed sociology of knowledge. At the heart of this method is the concept of a style of thought. The history of thought from this point of view is no mere history of ideas, but an analysis of different styles of thought as they grow and develop, fuse and disappear; and the key to the understanding of changes in ideas is to be found in the changing social background, mainly in the fate of the social groups or classes which are the 'carriers' of these styles of thought. (Mannheim, 1971a [1927], p. 260)

In 1922, with the essay *The Distinctive-Character of Cultural-Sociological Knowledge*, published for the first time in German in 1980 and then in English in 1982 in the volume *Structures of Thinking* [*Strukturen des Denkens*] (1997 [1922]), Mannheim had already considered the possibility of using the notion of 'style' in sociology of knowledge. In that article, Mannheim followed Lukács (1969 [1910]) in considering the notion of 'style' as both an aesthetic and a sociological category. In other words, according to him the term 'style' points towards two different directions. In the first case, 'style' designates a formal, compositional element, which appears repeatedly in many works of art of the same trend. To this 'immanent' sense of the term 'style', i.e. concerning the works of art as such, Mannheim contrasts a sociological-functional sense: one that alludes to principles and norms related to representation common within a social group. Thus, an immanent approach towards the works of the Gothic period allows us to grasp the compositional principles and describe their development; on the other hand, a sociological approach allows us to focus on the common definition, within a social group, of the problems concerning representation and its solutions.

Mannheim's notion of 'style of thought' [*denkstil*] is systematically discussed by Mannheim in *Conservative Thought* (1971a [1927]) but also appears in *Ideologie und Utopia* and in the essay *On the Interpretation of Weltanschauung* (Mannheim, 1971b [1952]). In *Conservative Thought* Mannheim described conservatism in the first half of the nineteenth century as a political force which 'transcends the political sphere proper and comes to imply a particular form of experience and thought' (1971a

[1927], p. 286). In his terminology 'conservative' and 'progressive' are adjectives that allude not only to two distinct political aims but also to two different modes of thought. Mannheim's analysis of the conservative style was different from a mere history of ideas since he wanted to know how individuals take patterns of thought and categories over from their social group.

What he investigated in particular was the historical and social configuration which brought forth the conservative style of thought. Indeed, he considered social groups or classes as the 'carriers' of these political styles of thinking (1971a [1927], p. 260). From his analysis, it emerged that the progressive and conservative styles are characterized by specific categories of thought and therefore by different ways of experiencing things, conflicts, institutions and so on. A style of thought establishes the conditions within which the experience and the concepts of a certain epoch are possible or impossible. Thus, a style of thought in the sense of Mannheim can be considered a replacement of Kantian a priori.

To give an example, he argued that different pairs of contrasts are relevant to the conflict between the conservative and the progressive styles of thought. One of them is 'concreteness' versus 'abstractness', which can be seen in the conservative and progressive different concepts of property (1971a [1927], p. 290). Another is 'norm' versus 'germ' (understood as an initial stage):

> The peculiarity of the conservative way of putting things into a wider context is that it approaches them in some way from behind, from their past. For progressive thought, everything derives its meaning in the last analysis from something either above or beyond itself from a future utopia or from its relation to a transcendent norm. The conservative, however, sees all the significance of a thing in what lies behind it, either its temporal past or its evolutionary germ. Where the progressive uses the future to interpret things, the conservative uses the past; the progressive thinks in terms of norms, the conservative in terms of germs. (p. 296)

Not only different categories lead to different conceptualizations, e.g. that of 'liberty' (1971a [1927], p. 291) but also to different ways of experiencing the world:

> The conservative picture of things as a whole is like the inclusive sort of picture of a house which one might get by looking at it from all possible sides, a concrete picture of the house in all its detail from every angle. But

> the progressive is not interested in all this detail; he makes straight for the ground plan of the house and his picture is suitable for rational analysis rather than for intuitive representation. And within this difference of ways of fitting individual things into their wider context, there lies a further radical difference between progressive and conservative patterns of experience—this time a difference in the way of experiencing time. (1971a [1927], p. 297)

Since a style of thought is rooted in a particular social configuration, as opposed to the 'subjectivity' of the isolated individual, it cannot maintain a definite static form—it grows and develops because it is historically and socially conditioned. To reconstruct a style of thought means to trace it back to the social forces that make it possible and to the *Weltanschauung* of which it is expression (Mannheim, 1979 [1936], p. 276).

In summary, I have presented Mannheim's version of the concept of 'ways of thinking', i.e. the notion of 'style of thought' [denkstil]. First of all, we have seen that a concept of 'artistic style' emerged in the writings of Renaissance artists such as Raphael and Vasari. This concept played a crucial role in the elaboration of Riegl's work, who conceived a style as an expression of the artist's way of perceiving the world. This same concept was adapted by Mannheim to sociology of knowledge. For him, a style of thought is rooted in a certain historical and social context, changes over time, is characterized by very specific categories of thought and makes a certain way of experiencing the world possible or impossible. As we shall see, the term 'thought style' [*denkstil*] will make again its appearance in the paper 'Some Specific Features of the Medical Way of Thinking' (1986 [1927], pp. 39–46) and in the book *Genesis and Development of a Scientific Fact* (1935) by the microbiologist and philosopher Ludwik Fleck (1896–1961).

1.5 Phenomenology and the Concept of 'Ways of Thinking'

It may be said that the phenomenology movement, originating in the first half of the twentieth century, sought a substitute of the Kantian a priori in the structures of consciousness which condition the possibility of concrete lived experience. Edmund Husserl (1859–1938) was one of the first philosophers who drew attention to the subject's intentional process of consciousness. He invited his readers to focus on the description of phenomena as they appear to the subjective point of view without worrying about the question of whether the natural world exists. Ultimately, his method, called 'transcendental idealism' like Kant's account of objectivity, aimed to identify the conditions of the possibility of conscious experience.

For Husserl, as Friedman noted, 'the subject of "pure consciousness" is thoroughly ahistorical. [He was] interested in precisely the essential structures of consciousness eternally present in all historical epochs whatsoever, for it is only so, in fact, that phenomenologically philosophy can be a strict science as opposed to a mere historically relative *Weltanschauung*' (Friedman, 2000, p. 47). However, in the latter part of his life, historicity became an important element of his reflections. He described the aim of his last work *The Crisis of European Sciences and Transcendental Phenomenology* (Husserl, 1970 [1936]), as an 'attempt, by way of a teleological-historical reflection upon the origins of our critical scientific and philosophical situation, to establish the unavoidable necessity of a transcendental-phenomenological reorientation of philosophy' (Husserl, 1970 [1936], p. XXIV). To identify the origins of the crisis of the European culture, Husserl embarked on a long analysis of Greek mathematics and the rise of modern science. He did not want to condemn the idea of science in general but just stigmatize its anti-humanistic use which reduces rationality to naturalism and objectivism, excluding subjectivity. In his analysis he used the term 'style' several times in referring to certain ways of acquiring knowledge and in highlighting radical changes in the development of scientific thinking in the European tradition, although he never gave a definition or presented a systematic account of the corresponding concept. For example, he pointed out that 'the positive sciences, including pure mathematics and the exact natural sciences [...] have proved to be changeable in the total *style* of their systematic theory-building and methodology' (Husserl, 1970 [1936], p. 4 my emphasis). Accordingly, he referred to classical physics as the 'confirmed *style* for centuries' (1970 [1936], p. 4 my emphasis) and to Greek mathematics and natural science as 'a *style* which was *new in principle*, unknown to the ancients' (1970 [1936], p. 21 my emphasis). Regarding this latter style, he drew attention to 'the immense change in meaning whereby *universal* tasks were set, primarily for mathematics [...] the idealization of empirical numbers, units of measurement, empirical figures in space, points, lines and the impressive idea of a systematically coherent deductive theory' (1970 [1936], p. 21).

In this regard, it is worth mentioning that in a communication presented at the New York Academy of Sciences in 1976, the Nobel laureate in physics Steven Weinberg (1933–2021) said: 'we have all been working in what Husserl called the "Galilean style"; that is, we have all been making abstract models of the universe to which at least the physicists give a higher degree of reality than they accord the ordinary world of sensation' (Weinberg, 1976, p. 28). Perhaps Husserl did not exactly use the phrase 'Galilean style' but he surely spoke of a 'Galilean attitude of idealizing and mathematizing' (Husserl, 1970 [1936], p. 41). In 1982 the philosopher Ian Hacking picked up Weinberg's observation and mentioned Husserl as an example of a philosopher who used the term 'style' to indicate a scientific way of thinking[10] (Hacking, 1982, p. 324).

The main difference between the phenomenology of Husserl and Martin Heidegger (1889–1976), his assistant during 1920–1923, is that the latter did not ask how the phenomena constitute themselves in our consciousness but how the human being is subjectively disposed towards the world when the world constitutes itself. Heidegger did not want to analyse consciousness but what he called 'Dasein'. This latter is the 'there being', the human being inasmuch as the human being is a being thrown into the world, subjected to its several different limitations but also able to transcend it and project herself into the future with an act of freedom. 'The "essence" of Dasein lies in its existence', wrote Heidegger in *Being and Time* (1962b [1927], p. 67 Sect. 9), and its existence is an unfolding in the temporal dimension, the past, the present and the future.

Dasein is also that privileged entity, the human being, which can question itself posing 'the problem of Being': what does it mean to be? According to Heidegger, this question had originally been formulated by the early Greeks but had been increasingly relegated to oblivion as 'metaphysics' because with Plato and Aristotle the distinction between beings [*das Seiende*] and Being [*Sein*] began to blur. For Heidegger there is an *ontological difference* between these two concepts: beings are entities such as tables and pens; while the term 'Being' refers to the meaning of being,

[10] Together with Husserl, Hacking cited the English philosopher Robin George Collingwood (1889–1943). As far as I know, the latter mentioned the term 'style' only in reference to the field of art history (Collingwood, 1945, 1979 [1939]). However, in *The Idea of Nature* (Collingwood, 1945), Collingwood claimed that 'in the history of European thought there have been three periods of constructive cosmological thinking [i.e. the "Greek view of nature", the "Renaissance view of nature", "the Modern view of nature"], three periods, that is to say, when the idea of nature has come onto the focus of thought, become the subject of intense and protracted reflection, and consequently acquired new characteristics which in their turn have given a new aspect to the detailed science of nature that has been based upon it' (Collingwood, 1945, p. 1).

i.e. to how beings are intelligible as entities. Being is always the Being of a being (which is an entity as such).

In the years after *Being and Time* (1962b [1927]) Heidegger focused on the idea that the concept of Being of beings has been understood in different ways throughout the history of the Western thought. *The Age of the World Picture*[11] (2002 [1938]), a brief essay written on the occasion of a conference held in Freiburg in 1938, can be regarded as a history of the understanding of Being in the West. Heidegger wondered what characterizes modernity and distinguishes it from previous eras. His starting idea was that the foundation of the thought of a certain epoch is constituted by a given metaphysical conception, meaning by that a conception of the Being of beings and a conception of truth. In his words:

> In metaphysics, reflection on the essence of beings and a decision concerning the essence of truth is accomplished. Metaphysics grounds an age in that, through a particular interpretation of beings and through a particular comprehension of truth, it provides that age with the ground of its essential shape. (2002 [1938], p. 57)

Further on in the text, in note 8, Heidegger added that the entire metaphysical conception of an epoch involves not only an understanding of Being and truth, but also the fundamental question of Dasein, i.e. 'the manner and way in which man is man, that is, himself: the essential nature of selfhood which by no means coincides with I-ness, but is rather determined by the relationship' (2002 [1938], p. 79).

Based on these thoughts, Heidegger focused on the modern epoch in order to identify its metaphysical foundations. For him, it is science the essential phenomenon of modernity, an era which retains its essential unity in spite of the changes that occurred over four centuries. Science, he claimed, reduces the mode of Being of beings to the being of independent objects removed their every-day context and viewed as a magnitude in the space-time: 'we first arrive at science as research when the being of beings

[11] *The age of the world picture* is a 1938 lecture by Heidegger and one of the six essays contained in *Off the Beaten Track*, the translation of the essay collection *Holzwege*, published in 1950. Heidegger himself explains the meaning of the title: 'In the wood there are paths, mostly overgrown, that come to an abrupt stop where the wood is untrodden. They are called *Holzwege*. Each goes its separate way, though within the same forest. It often appears as if one is identical to another. But it only appears so. Woodcutters and forest keepers know these paths. They know what it means to be on a *Holzwege*' (Heidegger, 2002 [1938], p. 0).

is sought in such objectness' (2002 [1938], p. 66). Likewise, truth is reduced to correspondence-to: 'science as a research arrives when, and only when, truth has transformed itself into the certainty of representation' (2002 [1938], p. 66). This metaphysical conception of beings is inseparable from a specific mode of Being of Dasein:

> [...] in order fully to [sic] grasp the modern essence of representedness we must scent out the original naming power of that worn-out word and concept "to represent": to put forth ad relate to oneself. It is through this that the being comes to stand as an object and so first receives the seal of being. That the world becomes picture is one and the same process whereby, in the midst of beings, man becomes subject. (2002 [1938], p. 69)

In other words, the Being of beings is reduced to being and the Being of Dasein to a subject engaged in the explanation and prediction of objectified beings. Ultimately, the metaphysics at the root of the modern era consists in an interpretation of beings as objectuality, of the human being as subject and of truth as certainty. This metaphysical conception underlies a way of knowing that Heidegger described in this way:

> [...] knowing establishes itself as a procedure within some realm of beings in nature or history. Procedure, here, does not just mean methodology, how things are done. For every procedure requires, in advance, an open region within which it operates. But precisely the opening up of such a region constitutes the fundamental occurrence in research. This is accomplished through the projection, within some region of (for example, natural) beings, of a ground-plan [*Grundriss*] of natural processes. Such a projection maps out in advance the way in which the procedure of knowing is to bind itself to the region that is opened up. This commitment [*Bindung*] is the rigor of research. Through the projection of the ground-plan and the prescribing of rigor, procedure secures for itself, within the realm of being, its sphere of objects. (2002 [1938], p. 59)

For example, in classical physics the states of beings are represented as points in the phase space and their motions as paths, i.e. the phase space trajectories. In this specific case, the phase space is the 'open region within which [research] operates' and its points are objectified natural entities. The rigour of research is the commitment to the mathematical rules of the phase space and the differential equations which govern the dynamical systems. Through the phase space and its mathematical rules the procedure 'secures for itself, within the realm of being, its sphere of objects': objectified beings represented as the bearers of two measurable properties, position and momentum. The differential equations express rules or laws which can be used to make predictions. In turn, these laws can be tested through experiment. Heidegger, remarked that:

> Natural science does not first become research through experiment. It is rather the other way round: experiment is only possible where knowledge of nature has already transformed itself into research. It is only because contemporary physics is a physics that is essentially mathematical that it is capable of being experimental. (2002 [1938], p. 61)

By 'research' Heidegger meant a way of knowing characterized by 'the projection within some region of (for example, natural) beings, of a ground-plan [*Grundriss*] of natural processes' (2002 [1938], p. 59), by the identification of stable and permanent elements within these represented processes and by the existence of specific institutions which facilitate the communication of results and the consolidation of methods. Heidegger's point is that the habit of experimentation and observation, important feature of the way of knowing of modernity, hinges on a metaphysical conception of being which decontextualizes natural entities by transforming them into mathematical abstractions. It is the specific understanding of Being of modernity that makes possible the experimental way of knowing.

In addition to the modern age, Heidegger mentioned two other epochs, each characterized by its own specific metaphysical conception of beings: the Middle Ages and Antiquity. One important point to bear in mind is that, according to him, 'that beings acquire being in and through representedness makes the age in which this occurs [i.e. modernity] a new age, distinct from its predecessors [...]' (2002 [1938], p. 68). In neither of the two epochs just mentioned does the being consist in its representation. 'In the Middle Ages [...] the being is the *ens creatum*, that which is created by the personal creator-God who is considered to be the highest cause' (2002 [1938], p. 68). Furthermore, 'genuine possession of the truth has, through Christianity, been transferred to faith—to the truth preserved in the written word' (2002 [1938], p. 62). Certain ways of acquiring knowledge in the Middle Ages were deeply rooted in these two assumptions. To begin with, 'knowledge [was] not research but rather right understanding of the normative word and of the authorities who proclaim it' (2002 [1938], p. 62). Furthermore, on the basis of the belief that the created things possessed some characteristics of the cause that created them, the Thomistic principle of the *analogia entis* asserted that by examining the attributes of beings it was possible to know something about the characteristics of God.

On the other hand, the fundamental metaphysical position of the era of Antiquity consisted in the Greek 'interpretation of being as presence and truth as unconcealment—an unconcealment which remains itself an essential determination of being [...]' (2002 [1938], p. 68). By that Heidegger meant that the early Greeks did not reduce Being to being by objectifying it, rather they experienced beings as what is present, what shows itself, what unveils itself and becomes manifest. Instead of transforming beings into objects for the thinking subject, instead of viewing all there is to Being as a conceptualizable and manipulable being, the early Greeks let beings unfold and reveal themselves, let them be as they were. In this regard, in his *Introduction to Metaphysics* (2000 [1953]) Heidegger wrote that among the early Greeks 'beings were called *phusis*' (2000 [1953], p. 14) and when this word was translated into the Latin 'natura' 'the authentic philosophical naming force of the Greek word [was] destroyed' (2000 [1953], p. 14). What the word 'phusis' suggested to the early Greeks was: 'what emerges from itself (for example, the emergence, the blossoming of a rose), the unfolding that opens itself up, the coming-into-appearance [...]. Phusis is being itself, by virtue of which beings first become and remain observable' (2000 [1953], p. 15). The conception of truth of the early Greeks is consonant with their way of relating to Being: according to Heidegger, truth was conceived as 'unconcealment', which is the literal meaning of aletheia, i.e. 'what comes out of oblivion (lethe)' and lets itself to be seen, after being hidden. According to Heidegger, this conception of truth started to dissolve with Plato and Aristotle: from that moment, the history of metaphysics was characterized by the belief that truth is a correspondence between thought, language and being. Heidegger expressed the need to return to the original pre-Socratic conception of truth and open up to the moments in which the being lets itself be glimpsed.

Neither in the antiquity nor in the Middle Ages was experiment a common practice for acquiring knowledge, although observation directed towards knowledge became important from Aristotle onwards. When Heidegger emphasized this point, he remarked that experiment was incompatible with the metaphysical presuppositions of Antiquity and the

Middle Ages: 'for what is completely absent here is what is decisive about the experiment' (Heidegger, 2002 [1938], p. 61). As we have seen, for him it is a conception of truth as certainty and of beings as objectuality that makes possible experiment.

In conclusion, the point I want to make is that Heidegger implicitly put forward a notion of way of thinking. This was not the outcome of an inquiry into the forms of knowledge practised in different periods. Rather, Heidegger's first move was to replace Kant's (and Husserl's) question 'what does it mean to *know*?' with the question 'what does it mean to *be*?'. He then answered this latter question with regard to three different historical periods. As we have seen, once he identified the interpretation of Being and truth in a given era, he showed that it made possible (or impossible) a certain way of acquiring knowledge. For example, the interpretation of beings in terms of objectuality in the modern era made possible a way of knowing based on experiment, which was impossible in the era of Antiquity when Being was essentially interpreted as 'unconcealment'. In his words:[12] 'Any *form of thought* [*Denkart*], however, is always the actualization and the consequence of a certain form of historical existence, of a certain fundamental position regarding being in general and truth, that is, of the way in which the being as such appears' (Heidegger, 1962a, p. 74 my emphasis). Thus, a way of thinking in the sense of Heidegger is essentially a way of acquiring knowledge which is rooted in a certain 'form of historical existence', which includes a particular conception of truth. Ultimately, for Heidegger what really changes from an epoch to another is our sense of reality. A new way of acquiring knowledge comes about when a new understanding of Being emerges and takes shape at a certain point in time. It is this changing sense of what-is, that makes it impossible to compare different systems of knowledge:

> As we use the word science these days, it means something essentially different from the *doctrina* and *scientia* of the Middle Ages, different, too, from the Greek episteme. Greek science was never exact precisely because, according to its essence, it neither could be, nor needed to be, exact. Hence, it makes no sense at all to assert that contemporary science is more exact than

[12] The original passage, taken from the 1935/36 lecture course *Die Frage nach dem Ding*, is: 'Wir sagten, es sei ein Grundzug des neuzeitlichen Denkens. Jede Denkart ist aber immer nur der Vollzug und die Folge einer jeweiligen Art des geschichtlichen Dasei (Thomson, 2011) ns, der jeweiligen rundstellung zum ein überhaupt und zu der Weise, wie das Seiende als ein solches offenbar ist, d.h. zur Wahrheit' (Heidegger, 1962a, p. 74).

> the science of antiquity. Neither can one say that Galileo's doctrine of free-falling bodies is true while Aristotle's teaching that light bodies strive upwards is false. *For the Greek understanding of the nature of body and place and of the relation between them rests on a different interpretation of beings. It determines, therefore, a correspondingly different way of seeing and questioning natural occurrences.* No one would presume to say that Shakespeare's poetry is more advanced than that of Aeschylus. It is even more impossible to say that the contemporary understanding of beings is more correct than that of the Greeks. If, then, we wish to grasp the essence of contemporary science we must free ourselves of the habit of comparing modern with older science—from the perspective of progress—merely in terms of degree. (2002 [1938], p. 58 my emphasis)

Some scholars have noted substantial convergences between Heidegger's and Kuhn's views on the history of scientific thought (see for example Belgrano, 2021). Others have claimed that 'Foucault adopt[ed] Heidegger's epochs in his investigation of the different occidental *epistemes*' (Thomson, 2011, p. 107). As a matter of fact, Heidegger was an 'essential philosopher' for Foucault who, in his last interview in 1984 said: 'my entire philosophical development was determined by my reading of Heidegger' (Kritzman, 1988, p. 250). However, another important point to notice, for what will concern us here, is that Husserl had pointed out that the Greek idealization of figures in space, points, lines was part of a new 'style' in the study of nature. From Heidegger's perspective, we may say Greek geometry marks the beginning of the reduction of Being to its representation. It is no coincidence that Heidegger argued that Plato represents the beginning of the history of the oblivion of Being. We will see that Hacking claimed that the discovery of demonstrative proof in ancient Greece marks the emergence of a new way of thinking which, among other things, introduces new types of abstract objects, such as figures in space, points, lines and new criteria for the truth or falsehood of statements about those objects.

1.6 Classical Historical Epistemology and the Concept of 'Ways of Thinking'

The concept of 'ways of thinking' is an absolutely fundamental element of most of the philosophies that fall into a tradition of research called 'historical epistemology', which can be regarded, as we shall see, as a collective effort to historicize the philosophy of science. From the time when this tradition fully emerged in France at the beginning of the twentieth century to its current approaches, several of its practitioners have inextricably linked their philosophies to the idea that ways of thinking have been different in different epochs, civilizations and, sometimes, social or geographical contexts. However, their versions of this concept were different, shaped for their specific goals and applications. The label 'historical epistemology' [*épistémologie historique*] was employed as early as 1907 by the French philosopher Abel Rey (1873–1940) in the introduction to his doctoral thesis *La théorie de la physique chez les physiciens contemporains* (1907) (Rey, 1907, p. 13), as was pointed out by (Braunstein, 2006), (Chimisso, 2008) and (Brenner, 2016). His declared purpose was to identify the 'relationship of the logic of science with the history of the scientific thought' and promote a 'history of the scientific thought' from a study of 'the general mode of thought [*esprit*] of physical science' (Rey, 1907, p. 13). Anastasios Brenner noted that, since Rey was born about ten years before Moritz Schlick, Otto Neurath and Philipp Frank, some of the first exponents of the Vienna Circle, his early use of the expression 'historical epistemology' 'shows that the latter 'is not an afterthought reaction to the analytical philosophy of science, but arises at the same time. If their methods diverge, they originate in the same context, that of the crisis of science at the end of the 19th century' (Brenner, 2016, p. 1).[13]

As a young scholar, Rey joined the conventionalist movement initiated by the theoretical physicist and philosopher of science Henri Poincaré. In 1919, he succeeded Gaston Milhaud (1858–1918) as professor of 'history of philosophy in its relationship with the sciences' at the Sorbonne and was the doctoral supervisor of Gaston Bachelard (1884–1962). Although the latter, together with Georges Canguilhem (1904–1995), is generally considered the acme of historical epistemology, other French scholars, such as Lucien Lévy-Bruhl (1857–1939), Léon Brunschvicg (1869–1944), Hélène Metzger (1889–1944), Alexandre Koyré (1892–1964), Michel Foucault (1926–1984) and many

[13] In the previous sections I have provided some reasons for the origins of the historicization of epistemology. However, the latter is an issue still under debate. Brenner (2003, pp. 1–9) identified as a pivotal point the work of scholars such as Henri Poincaré (1854–1912), Pierre Duhem (1861–1916) and Gaston Milhaud (1858–1918) at the turn of the twentieth century. Rheinberger (2010) remarked that the historicization of epistemology was the solution to a crisis that emerged in the late nineteenth century. Bordoni (2017) argued that its beginnings can be found in the debates between different scientists in France in the 1860s.

others, have provided equally important contributions to the same tradition of thought. As Cristina Chimisso (2008) has convincingly shown, all these thinkers worked, discussed, lectured and organized talks within the same rich and varied intellectual milieu[14] and their philosophies form a coherent enterprise. They opposed the idea of a fixed set of categories to grasp the nature of science and their common assumption was that reason is different in distinct historical and social contexts and that, therefore, to study a priori its interaction with the world is impossible.

The thesis that there are different ways of thinking, which is central to their philosophies, as we shall see more clearly in the next chapter, marks an important difference from the thinkers of French epistemology who preceded them (e.g. Émile Boutroux (1845–1921), Pierre Duhem (1861–1916), Henri Poincaré (1854–1912), Émile Meyerson (1859–1933) and others) and who, for this reason, will be excluded from my analysis. Today historical epistemology has expanded beyond the studies of these French thinkers and the label 'historical epistemology' indicates both their tradition of thought and some more recent variety of projects inspired by that tradition and which I will discuss in the next section. To avoid confusion, I shall follow (Chimisso, 2011) who called the former 'classical historical epistemology' in order to distinguish it from the more recent versions of 'historical epistemology', an expression by which I will refer to the whole tradition, classical and modern.

In various ways, the early historical epistemologists called into question the basic assumption of classical epistemology that the knower has direct access to nature. In their writings, they complicated the subject–object relationship: phenomena were no longer 'out there', waiting to be discovered, but they had to be purified, isolated and investigated through instruments calibrated according to certain rules, which were in turn the result of negotiations. Consequently, objectivity ceased to be something given once and for all: its rules had to be established in the course of a dialectics between subject, world and context, i.e. an 'objectification process' in which certain ways of thinking, methods, standards of evidence came to be shared by a community of people. For example, Gaston Bachelard (1884–1962) drew attention to the role of scientific instruments in the

[14] The most important academic and intellectual spaces in which the thinkers of classical historical epistemology discussed their ideas are: the Sorbonne (where in the early 1930s Rey established the 'Institut d'histoire et de philosophie des sciences et des techniques' to promote a philosophical history of science), the Collège de France, the Centre international de synthèse, the Société française de philosophie and the conferences organized by the members of the academic journal *Revue de métaphysique et de morale*.

production of scientific knowledge as key to the concept that objectivity is variable in the course of history. For him, 'the knowledge of reality [...] is never direct [...] everything is constructed' (Bachelard, 2011 [1938], pp. 14–15), i.e., mediated by human values, unconscious ideas and normative rules established within the *cité scientifique*. Ultimately, objectivity became the result of a dialectics between a certain way of thinking, world, technology and social structure, so that the question as to how it becomes possible in different contexts came to be crucial.

It is certainly true that the beginnings of the historicization of Kant, i.e. 'the assumption that knowledge can be adequately understood only if it is studied in its historical development, as a situated dialectic of theory and experience, rather than being considered as the product of an encounter between a pregiven empirical reality and certain immutable faculties of mind [...] can be traced back to Auguste Comte' (Dews, 1992, p. 40). However, the tradition of historical epistemology understood history not so much as a method for finding out specific details about theories, but rather as a field of study where it is possible to find evidence of the existence of different ways of thinking. As I shall argue in the next chapter, what most of the historical epistemologists ultimately helped to clarify was how facts, propositions, concepts and other epistemological items become objective; and in their explanations the different versions of the concept of 'ways of thinking' played the role of a replacement of the Kantian a priori. To give an account of objectivity was also the goal of neo-Kantians such as Cassirer, as I have explained above. Indeed, the latter's writings were present to the mind of the historical epistemologists.[15] It was under the influence of neo-Kantianism that Lévy-Bruhl and Brunschvicg, the 'fathers' of historical epistemology, began the historicization of Kant which characterizes this tradition of thought. Brunschvicg, struck by the philosophical implications of the relativity theory, argued that Kant had in fact presented a particular way of thinking, of which Newton's science was the product; Lévy-Bruhl argued that the categories of thought change in different cultures. Other historical epistemologists inherited from Lévy-Bruhl and Brunschvicg the idea that human way of thinking changes across different times or cultures and pursued the objective of historicizing Kant, as we shall see in the next chapter.

[15] On the influence of Kantianism and Neo-Kantianism on the tradition of historical epistemology see for example (Goyard-Fabre, 2005; Sinaceur, 2006; Ferrari, 2021).

Classical historical epistemologists's view that the analysis of history of thought reveals ways of reasoning very different from one another was in contrast with the ideas of the logical positivists who, impressed by the exciting scientific advances of their time, considered science as a model of rationality and its growth as cumulative. The belief that categories and certain laws and values are timeless and universal had already been the subject of Hegel's criticism. However, Brunschvicg, like other French historians of philosophy at the Sorbonne, rejected any idea of laws of development in the history of thought. Although the classical historical epistemologists absorbed the lesson of Comte that the study of history is crucial for understanding knowledge, they 'were on the whole opposed to laws of progress, and regarded intellectual history as an open-ended process' (Chimisso, 2008, p. 5). In this regard, it must be borne in mind that, for Comte, humanity had reached the final and definitive stage of its development, something to which his way of thinking was now fixed forever. Classical historical epistemologists, on the other hand, never denied the possibility that the changes of historical-social contexts could lead to the birth of new ways of thinking. There was no end of history in sight.

What 'a way of thinking' is and what characteristics it has were questions whose answers differed from thinker to thinker. As we shall see in the next chapter, Lévy-Bruhl put forward the concept of 'mentality', according to which certain traditional non-literate societies adopted a form of explanation different from that of modern people, based on supernatural causes, i.e. on primary rather than secondary causes. Brunschvicg, Metzger and Bachelard kept Lévy-Bruhl's notion of mentality uppermost in their mind, although they presented their own versions and applications. Brunschvicg, struck by the philosophical implications of the relativity theory, maintained that Kant had in fact presented a particular way of reasoning, of which Newton's science was the product. Metzger argued that in different epochs human beings have particular worldviews, which appear irrational to us, dominated by specific assumptions, concepts and practices. Gaston Bachelard (1884–1962) thought that scientific knowledge and common sense differ radically (just as primitive and modern mentalities differed for Lévy-Bruhl). According to Bachelard, in order for scientific knowledge to emerge, it is necessary to have an 'epistemological break' with immediate experience that is rooted in imagination and emotions. Within scientific thought there have also been epistemological breaks, which have produced new ways of thinking.

The Polish philosopher Ludwik Fleck (1896–1961) is another thinker who, in the 1930s, rejected the 'outside time and history' investigation of scientific activities conducted by the Vienna Circle (Fleck, 1979 [1935], p. 50) and claimed that scientists' practices do not follow a timeless rule. Although Fleck was probably unfamiliar with many works of French historical epistemology,[16] his thought had 'a surprising number of points in common' with Gaston Bachelard (Rheinberger, 2010, p. 19). It was in his paper 'Some Specific Features of the Medical Way of Thinking' (1986 [1927] [1927], pp. 39–46) and then in his book *Genesis and Development of a Scientific Fact* (1935) that Fleck introduced the notion of *thought style*, the distinctive mode of thinking of a community in which knowledge is produced (a 'thought collective').[17] As we shall see, this notion implies an idea central to historical epistemology: objectivity is not the result of the personal ideas and ambitions of singular individuals but of a dialectics between a certain way of thinking of an entire community and the world.

Koyré's academic career took place not only in Paris but also in Germany and in the United States. For this reason, he is rarely remembered as a historical epistemologist. Yet, a common link with classical historical epistemology is precisely the fact that he regarded human thought as subject to change. In particular, he questioned how some concepts appear absurd in certain epochs and plausible in others and, in answering this question, he spoke of different 'types of thinking' (Koyré, 1943, p. 406) or 'worldviews', of which examples are the Aristotelian physics or Galilean physics. According to him, what made the principle of inertia possible was the radical change from different European worldviews. On his part, Foucault was a student of Canguilhem and, unlike Fleck, belonged to the genealogy of the French historical epistemologists. His notion of *episteme* essentially highlighted the structures common to the practices of a historical period

[16] This is not surprising since in the Eastern block there was a strong sociological approach to science. See for instance the papers of the Soviet delegation at the international congress of history of science, especially Nikolaj Bukharin's and Boris Hessen's (*Science at the Cross Roads. Papers from The Second International Congress of the History of Science and Technology 1931*, 2014).

[17] We have seen the Mannheim had already introduced a notion named 'style of thought' [*Denkstil*] at least five years before Fleck. Although the latter mentioned sociologists of knowledge such as Durkheim, Lévy-Bruhl, Jerusalem and Simmel, he did not refer to Mannheim. Sady (2021) claimed that Fleck did not know works of the 'founders of the sociology of knowledge': Max Scheler and Karl Mannheim while Fuller, in note 41 of his book on Kuhn (Fuller, 2000) provided reasons to explain why Fleck was better disposed to Lévy-Bruhl's mentalities approach to the sociology of knowledge than to Mannheim's.

and was proposed in *The Order of Things* (1966), a book in which Foucault wanted 'to reveal a ***positive unconscious*** of knowledge: a level that eludes the consciousness of the scientist and yet is part of scientific discourse' (Foucault, 1994 [1966], p. xi).

It must be added that *The Structure of Scientific Revolutions* (Kuhn, 1996 [1962]) is widely presented as a work that has substantive continuities with classical historical epistemology. Indeed, Gary Gutting (2003) compared Thomas Kuhn (1922–1996) with Bachelard regarding their discontinuist reading of the history of science and Chimisso (2019) documented the immense influence of the entire tradition of classical historical epistemology on Kuhn's work, in particular that of scholars such as Alexandre Koyré and Hélène Metzger. Kuhn himself recognized as sources of inspiration Metzger and Koyré (Kuhn, 1996 [1962], p. VII) (Fleck, 1979 [1935], p. VIII) in addition to scholars such as the French philosopher Émile Meyerson (1859–1933) and the German historian of science Anneliese Maier (1905–1971). As we shall see in the next chapter, his concept of ***paradigm***, which he introduced as a general framework with components such as exemplars, ontological assumptions and values, can be viewed as a particular notion of way of thinking.

It was during the 1970s that the philosopher Paul Feyerabend (1924–1994) became a critic of Karl Popper (1902–1994) but some years earlier he had already begun his historical turn in philosophy of science. Like Kuhn, Feyerabend belonged to an intellectual milieu quite different from that of classical historical epistemology. However, as I shall explain in the next chapter, especially in the last part of his philosophical production, he defended the thesis that Kant's categories are not historically invariant and that there exist different ways of thinking. As we shall see, his account of the transition from the Homeric to the Pre-Socratic worldview shows that he conceived of ways of thinking essentially as ways of perceiving the world involving specific questions, goals and standards of truth.

1.7 Modern Approaches of Historical Epistemology and the Concept of 'Ways of Thinking'

The label 'historical epistemology' [*Historische Epistemologie*] reappeared in a correspondence about the establishment of the Max Planck-Institute in Berlin in the first years of 1990s (Hacking, 1999, p. 54) and has been used over the past few years for a wide variety of programmes (for an overview see Feest & Sturm, 2011). Among them it should be mentioned the work of Lorraine Daston and Peter Galison, who characterized epistemological categories out of knowledge practices. Daston defined historical epistemology as 'the history of the categories that structure our thought, pattern our arguments and proofs, and certify our standards of explanation' (Daston, 1994, p. 282) and cited the philosophers Arnold Davidson and Ian Hacking as its leading practitioners (Daston, 1994, p. 283). The former conducted investigations into the conceptual formation of new disciplines and mentioned historical epistemology in relation to a kind of study that attempts to show how novel forms of experience are linked to the emergence of new structures of knowledge (Davidson, 2001, p. XIII). The latter, instead of 'historical epistemology', preferred the expression 'historical meta-epistemology' to mark a distinction with the previous use. Indeed, for him the label was too full of reminiscences of a book by the French philosopher Dominique Lecourt (1944–2022) on Gaston Bachelard (1884–1962) entitled *L'Epistémologie historique de Gaston Bachelard* (Lecourt, 1969). He meant that the expression 'historical epistemology' already indicated a tradition in philosophy of science whose main protagonists were Gaston Bachelard and Georges Canguilhem (1904–1995) (Hacking, 1999, p. 54). The prefix 'meta' he added wanted to mark a distinction with that prior use: 'for Bachelard objects are the sciences in their historical development [...] our objects of study are different, not knowledges, but ideas about knowledge' (Hacking, 1999, pp. 54–55). Hacking described 'historical meta-epistemology' as the study of *organizing concepts*, i.e. as I shall explain below, epistemological concepts such as probability, evidence, reason and objectivity that shape our ideas and practices of knowledge (Hacking, 1999, p. 53) (see also Sciortino, 2017). He added that 'where Bachelard insisted that historical considerations are essential for the practice of epistemology, the historical meta-epistemologist examines the trajectories of the objects that play certain roles in thinking about knowledge and belief' (2002a, p. 9). For him, historical meta-epistemology falls under the concept of 'historical ontology', the study of the historical trajectories of 'not only "material" objects but also classes, kinds of people and ideas' (2002a, p. 2). It is a fact that today historical epistemologists have expanded their field of research, among other things, from the study of organizing concepts to that of the historicity of objects of scientific practice and of epistemic values and beyond.[18]

[18] The introduction to the conference entitled 'What (good) is historical epistemology?' held at the Max Plank Institute for the History of Science in Berlin, provides further details about aims and programmes of historical epistemology (Sturm, 2008).

In two papers published in 1982 and 1992 (Hacking, 2002b [1982], 2002c [1992]), Hacking outlined the 'project of scientific styles of thinking' (Hacking, 2012b), for short the 'styles project'. The label 'style of reasoning' came from the historian Alistair Crombie (Crombie, 1994), who had specified six methods of scientific enquiry central to the history of Western thought: the style of geometry, the experimental style, the style of hypothetical modelling, the taxonomic style, the statistical style and the historico-genetic style. Hacking noted that each of these styles involves new types of objects, standards of evidence, laws and true-or-false sentences. He also suggested that styles have sharp beginnings and are self-authenticating, i.e. they do not answer to any higher standard of truth. In his later writings (e.g. Hacking, 2009), he insisted that styles are not only ways of thinking but also ways of *doing*: humans are embodied creatures that use their minds and bodies to think and act in the world (Hacking, 2012a, p. 2). Hacking's main objective was to investigate the philosophical significance of the thesis that there exist distinct scientific styles of reasoning which have emerged and stabilized at different points of the history of the sciences. Although over the last 40 years Hacking has added to his original nucleus of ideas several constructive suggestions, his original thesis has never been fully developed into a comprehensive theory. This book, especially Chaps. 3, 5, 6, 7 and 8, presents an attempt to develop, correct and present in a more systematic way the styles project so to make of it something coherent and complete, what might be called the 'theory of scientific styles of thinking and doing'. Since to infer philosophical implications from Hacking's project and analyse them in order to pursue these aims has been central to my methodology, this book is also an analytical account of Hacking's styles project and its philosophical implications.

It will be one of the main points of the third chapter that the notion of 'style of reasoning', just like the other notions of 'ways of thinking' proposed by the historical epistemologists and mentioned above, implies the existence of different 'objectivities', conceptual spaces in which certain facts, sentences or other epistemological items are candidates for truth or falsehood. For example, Hacking's styles are what settles what it is to be objective because, when a style comes about, it brings into being a set of novelties, for instance, a new kind of evidence. This new form of evidence provides the new criteria by which new sentences become candidates for truth or falsehood and eventually *objective*. Thus, Lorraine Daston and Peter Galison's book *Objectivity* (Daston & Galison, 2007), which charts the history of the nature of new conceptions of objectivity that emerged in the last three centuries, cannot fail to fit into this line of thought. According to them, at a certain moment in history one or more conceptions of objectivity came to represent the governing principles according to which we plan our experiments, decide which kinds of instruments to use and which ones to discard, train our senses to make subjective decisions and establish which details of a phenomenon to omit. For them, each conception of objectivity is rooted in an *epistemic virtue*, as they call the scientific ideal to which scientists are committed in a particular period or circumstance. As we shall see, epistemic virtues can be viewed as particular ways of thinking, i.e. ways of looking at the portion of the world under study, which shed light on particular aspects of it and put others in the shade.

Given their importance in the context of historical epistemology, it is better to go back to Hacking to clarify what *organizing* concepts are. A good start for a discussion might be to remember that Kant's categories (and the two innate intuitions of space and time) lie at the top level of our conceptual organization. To such a degree that according to the *Critique of Pure Reason* (Kant, 2007 [1781]), there could not be *empirical* concepts without *organizing* concepts such as a priori categories: the former are forged from the raw material of experience by applying the categories to intuitions. Like the Kantian categories, the organizing concepts studied by historical (e.g. probability and objectivity, mentioned in the passage above) 'govern and control ground level concepts' (Hacking, 1999, p. 65); however, unlike Kantian categories, organizing concepts are categories of thought that change or come into being through specific historical processes. When they emerge, they come to shape our practices of knowledge, direct us in what to do, fashion our concepts, structure the ways in which the ideas that permeate our world are connected. In this sense, organizing concepts are the conditions of the possibility of our experience. As such, although they are changeable throughout history, they are indispensable for knowing: 'they are [...] essential to the very functioning of our society. We are stuck with them, which is not to say [...] that they are not changing as I speak' (Hacking, 1999, p. 65).

By saying that organizing concepts are 'essential' to the very functioning of our society Hacking means that 'they produce a feeling of inevitability' (Hacking, 2002a, p. 21): when we think of an organizing concept we find it difficult to conceive of a way of experiencing our world and ordering our society which does not rely on it. In this sense they are 'inevitable'. To sum up, an organizing concept is an epistemological concept that: (1) structures our thought about the world; (2) has emerged in the course of history or has changed over time; (3) is inevitable. As an example of *non*-organizing concept Hacking considers the concept of horse: it is not essential to our thinking about the world and it is a concept that 'we can understand totally outside the history of any idea' (Hacking, 1999, p. 59). I am going to clarify points *(1)*...*(3)* above by discussing two examples of organizing concepts: probability and objectivity.

The concept of probability, which for Hacking fully emerged in the seventeenth century has come to structure our experience (point *1* above) in many ways. First of all, it shapes the way we know: to prove a statistical sentence involves, for all intents and purposes, a whole body of practices, from selecting samples to measuring, classifying, dealing with errors or assessing the difference between experimental distributions and theoretical ones. Secondly, concepts such as sample, population, distribution and various criteria of assessing statistical statements are constituted within a 'space of possibilities' opened by the concept of probability. That is, the emergence of the latter has allowed the coming to the fore of a domain of scientific discourse, with its conceptual material, its practices and its theoretical structures. Other concepts such as variability, correlation and uncertainty, which occupy a pivotal role in our ways of thinking about our experience, are couched and conceived of in terms of the concept of probability. The latter gives rules for the usage of these concepts: for example, uncertainty is thought of in terms of standard deviation, which in turn is defined in terms of a probability distribution. Through the concept of probability other concepts, such as 'life expectancy', are connected with the concept of mean and statistical distribution. We can say that the *organizing* concept of probability governs and controls the way *ground level* concepts such as variability, correlation, uncertainty, expectancy, mean, distribution, population are thought of or related to one another. Finally, the 'feeling of inevitability' (point *3* above) mentioned by Hacking is provided by the fact that we now cannot conceive of understanding the world without the concept of probability: it has come to determine the way we organize our thinking and how strongly we ought to believe that something will happen—so much so that 'the mean and dispersion are now thought of as objective properties of some part of the world, as "out there" as the location of a planet' (Hacking, 1992, p. 148). The same considerations apply to the concept of objectivity. Outside science, in their expression of imagination and creative skill, people may well persistently ignore objectivity. However, in order to do their best to disregard objectivity they must have the idea of what to be objective means. Even though objectivity may not be the organizing principle in the arts, it is essential to the functioning of our society and our daily life. It is present in our minds when we discuss the election results, when we apply the law, when we try to eliminate subjective valuations in our relations with other human beings.

Among other things, Hacking says that historical epistemology is not a theory *of* knowledge. Indeed, traditional epistemology studies the conditions, the limits and the sources of knowledge and enquires into what makes knowledge scientific. It considers concepts such as proof, rationality, objectivity, evidence and the like as permanent and, to understand them, it analyses the timeless structure of logic and consults human understanding. Historical epistemology starts from an opposite presupposition: even though organizing concepts are necessary for our thinking, they get their meaning from the uses we make of them in a certain period of history and the way in which they are connected with other concepts. They are 'situated': they 'have no constitution other than tradition and use' (Hacking, 1999, p. 56). Furthermore, historical epistemology cannot be assimilated to the history of ideas either: the latter is generally centred on the subject and tells how key concepts are transmitted from a thinker to another; the former, on the other hand, is an inquiry into the very general structures in which human beings think. In conclusion, unlike traditional epistemology, historical epistemology does not formulate or defend theories of knowledge; rather, it reflects on the historical conditions under which we know, that is on organizing concepts. *In nuce*, historical epistemologists make a double claim: (1) *at a particular stage* of history there is a set of organizing concepts that play a role similar to Kantian pure concepts by allowing us to make judgements; (2) organizing concepts are not permanent. By bearing in mind these two points, historical epistemology can be viewed as the historicization of Kant.

It is important to note that, as the next chapter will show, there is much more continuity between modern and classical epistemology than Hacking and other modern practitioners of modern epistemology are willing to admit. Suffice to say that the history of the formation and use of concepts has been central to classical historical epistemology as well, and not only to modern historical epistemology, as several critics have pointed out: for example, David M. Peña-Guzmán has characterized the former in terms of two features: 'its claim that scientific discourse is the object of epistemology and (ii) its claim that scientific concepts are the building blocks of scientific discourse' (Peña-Guzmán, 2020, p. 1). This characterization is particularly fitting in the case of Jean Cavaillès (1903–1944) or of Georges Canguilhem who, in *Le normal et le pathologique* (1966), examined the role of the concept of normality in medicine and pathology and its dependence on the experience of the individual. As Dews pointed out, Canguilhem 'was far more interested in the social relativity and normative foundations of certain basic concepts, and even allowed that the life sciences may be permanently dependent on certain figurative modes of expression, often of political or ideological origin' (Dews, 1992, p. 39).

However, it is wrong to reduce historical epistemology to the study of concepts and overlook the role played by the different notions of way of thinking put forward by several historical epistemologists. As it will be clearer in the next chapter, these notions should be viewed as frameworks that help historical epistemologists to identify and describe the relationships between concepts that are important in the production of knowledge. For instance, Hacking's notion of style of thinking helps to understand how probability, evidence and objectivity have interacted with each other throughout the history of science; Daston and Galison's notion of 'epistemic virtue' helps to clarify the relations between values, research practices and standards of truth. In addition to that, as we shall see, the different notions of the concept of 'ways of thinking' can be viewed as frameworks also in the sense that they consist of sets of hypotheses and presuppositions that enable us to emphasize certain aspects of the past and construct historical narratives. One may protest that these notions have been introduced for the purpose of philosophical research rather than for understanding and narrating the history of science, but these aims align: Hacking describes the notion of style as a 'philosophical tool' and adds: 'if the tool does not provide a coherent and enlightening ordering of the record, then it has no more place in sound philosophy than would any other fantasy' (Hacking, 2002c, p. 178). Indeed, when using history to address philosophical questions, historical epistemologists must decide which happenings are relevant to the answer, how to interpret them and how they can be used to produce explanatory narratives. Thus, any notion employed must also help to order the historical record—it must be a framework for highlighting, omitting or unifying episodes. In the following chapters, I will alternatively use both the term 'framework' and 'notion' to refer to the different versions of the concept of 'ways of thinking' proposed by scholars.

1.8 Main Questions Addressed in This Work

When we look at the entire history of the concept of 'ways of thinking' that I have outlined above, we note that the specific social entities on which philosophers concentrated their attention were often not the same: societies, cultures, scientific communities, to mention just a few. Some of these philosophers saw changes in thought in certain epochs, others in different cultures, still others in research communities or even professional and semi-professional groups. There were also differences between philosophers as to which components of human thought were to be considered variable throughout history, e.g. reasoning, logic, imagination, thought categories, values. Furthermore, some of them placed less importance on the idea that human beings have thought in distinct ways through the course of history and rather emphasized the changes of various elements crucial in the process of acquiring knowledge, e.g. methods and practices. Finally, of the different terms philosophers used to label their own articulations of the concept of 'ways of thinking', some do not even directly refer to the human thought, e.g. 'world-view', 'episteme', 'paradigm', while others make the idea of thinking explicit, e.g. 'thought style', 'type of thinking', 'style of reasoning'.

Such a rich variety of notions implying that human thought has a history is not surprising, given the differences in the historical contexts or intellectual milieux of these philosophers. However, in Chaps. 2 and 3 I shall show that, within the tradition of historical epistemology, all the different notions of the concept of 'ways of thinking' share a distinctive trait. The different notions put forward by the historical epistemologists, such as 'mentality', 'episteme', 'paradigm' and 'style of reasoning', will all be viewed as replacements for the Kantian a priori or, *rectius*, its historicizations. They will all be part of the same project of historicizing the categories of scientific thought, which Kant had conceived as unchanging. In other words, there will be every reason to use the biological metaphor and claim that in the 'conceptual ecosystem' of historical epistemology there exist different 'species' of the 'genus ways of thinking', the genus of the historicizations of Kant's a priori. The latter will include different philosophical characterizations provided by the historical epistemologists of what makes objectivity possible in different historical and social contexts. To put it another way, I shall argue that there is a basic concept, which I shall call 'ways of thinking' (or, within the biological metaphor, the 'genus ways of thinking'), of which historical epistemologists have presented different versions (species), which is central to historical epistemology.

The analysis of the different articulations of the concept of 'ways of thinking' I will have provided in the first three chapters will help to shed light not only on the fundamental presuppositions of historical epistemology but also on important questions in philosophy of science. One of them concerns the consequences of the historicization of human thought: does a given notion of the concept of ways of thinking imply a form of relativism? Are ways of thinking incommensurable? These questions are particularly relevant today in relation to historical epistemology: whether relativism is a philosophical consequence of the historicization of epistemology is a general worry, even among historical epistemologists, as I shall explain in the fourth chapter. What Peña-Guzmán says about Bachelard and Canguilhem is true for much of historical epistemology: 'Bachelard and Canguilhem have to fight an epistemological battle on two philosophical fronts at once. To one side, they fight against externalist interpretations of the a priori conditions of rational knowledge. To the other, they fight against relativistic theories of scientific history that would rather do without the a priori altogether' (Peña-Guzmán, 2016, p. 179).

Another important issue has to do with the fact that all the different notions of the concept of 'ways of thinking' that have been proposed within historical epistemology are at odds with an image of science as exhibiting progress and methodological unity. Thus, in this respect one may ask whether a given notion of 'ways of thinking' implies that science is bound to converge on a single answer to a given question or that science itself converges on truth over time. A further question concerns the role of contingency in the history of science and, in particular, to what extent the emergence of ways of thinking (in its different conceptions) is a contingent circumstance or whether or not they are long-lasting. This is another crucial issue in the philosophy of science in that the emergence of a way of thinking may make it possible or impossible the emergence of certain questions so as to influence the development of knowledge.

Obviously, the answer to these questions depends on the features we attribute to a given notion of 'ways of thinking'. Thus, my strategy of addressing them will be: (1) to investigate how the different versions of the concept of 'ways of thinking' differ from each other; (2) to provide an answer for the case of a particular concept of 'ways of thinking', that of 'style of reasoning' introduced by Hacking; (3) finally, to ask whether the same answer is valid for other notions by considering how they differ from

Hacking's notion. I shall explain better in due time my reasons for choosing the latter as the notion of 'ways of thinking' that deserves closer examination. Here, I can anticipate that the question whether or not the concept of 'ways of thinking' implies relativism is particularly interesting in the case of Hacking's notion. The latter inherits various features from notions put forward by previous thinkers and it is better defined than most of the other notions of 'ways of thinking'. Furthermore, Hacking's notion represents the latest version of a long series of notions of the concept of 'ways of thinking'. His notion is also a useful basis for comparisons: other notions can be assessed by noting whether or not they possess some of its features. Furthermore, what makes the study of the notion of style of reasoning so interesting is Hacking's anti-relativistic stance: any relativistic implication of his notion would be at odds with part of his philosophy of science, in particular his realism about theoretical entities. Finally, the fact that the notion of 'style of reasoning' plays a more fundamental role in revealing fundamental aspects of the process by which objectivity becomes possible, as I shall show in the last chapters, makes it particularly suitable for the study of the contingency issue.

1.9 Plan of the Work

The next two chapters of this book will describe the different notions of the concept of 'ways of thinking' that have been introduced in historical epistemology. I do not claim this to be complete and exhaustive. What I want to do is to trace a guiding thread running through historical epistemology in an attempt to unify the researches of some important authors. This guiding thread is mainly constituted by the project of historicizing Kant's a priori by replacing it with a given notion of 'ways of thinking'. Since it is this common thread that I want to follow, my schematic account of the development of the concept of 'ways of thinking' will make other ideas central to historical epistemology less prominent. What Chaps. 2 and 3 will show is that all the different versions of the concept of 'ways of thinking' introduced by many historical epistemologists can be viewed as tools for understanding how one of the most important organizing concepts—objectivity—emerges, changes or endures. Also, these notions act as a framework that sheds light on how the concepts studied by historical epistemology interact in the production of knowledge.

In the fourth chapter I shall compare and contrast the distinct species of the genus 'ways of thinking' presented in the previous chapters in order to trace their common characters and provide a conceptual taxonomy. For example, I shall compare Hacking's styles project with Daston and Galison's study on objectivity as well as with other projects such as Fleck's, Kuhn's and Foucault's. Thanks to these comparative analyses, I shall identify points where all these projects support one another. In particular, I shall argue that different notions of the concept of 'ways of thinking' can be used together as a toolkit for studying objectivity and that styles of reasoning and epistemes are the most appropriate notions to highlight fundamental factors crucial for its emergence. An important point will be that Hacking's notion, together with Foucault's, illuminates conditions of possibility that lie at a deeper level than that of other notions of the concept of 'ways of thinking'. Furthermore, as we shall see, it accounts not only for the changes of objectivity but also for its endurance in certain historical periods. These features will make Hacking's notion more suitable than others for the study of contingency in science in the last chapter (Metzger, 1969 [1918]).

Finally, I shall present and explain the importance of two main issues I want to address. The first amounts to the question as to whether or not a notion of 'ways of thinking' implies a form of relativism; the second concerns the role of contingency in the history of science. In order to address these questions in the case of the notion of 'style of reasoning', in the fifth chapter I shall develop Hacking's styles project by characterizing four more styles of reasoning, of which Hacking has not provided a systematic characterization, in addition to the 'statistical' and 'laboratory style of reasoning' described in the third chapter, which Hacking has described in some detail. I shall present the relativism issue in the sixth chapter: in brief, the issue will amount to the question of whether there is an independent and atemporal justification for every claim made within a given style of reasoning. My answer will be in the negative and my argument will be developed in Chap. 7: I shall argue that the claim of entity realism made by Hacking is a case of a claim in a style of reasoning that has no independent and atemporal justification outside it. At a more general level, my point will imply that Hacking's experimental realism—the claim that we can know about unobservable entities—is not consistent with the epistemic relativistic implications of his styles project. I shall also mention a more general argument for the thesis of incommensurability that concerns all of Hacking's styles of reasoning, i.e. an argument that is not specific to the styles of reasoning discussed in the previous chapter.

In the last chapter, I shall examine the implications of historical epistemology for the contingency issue: (1) I shall give a picture of the evolution of sciences in terms of the emergence and stabilization of the styles of reasoning; (2) I shall discuss how both inevitability and contingency play a major role in the emergence and endurance of styles of reasoning; (3) I shall point out that each style is grounded in certain objective features of part of the world it studies (4) I shall argue that there are good reasons for seeing as inevitable the convergence of science on correct answers to single questions; although, as Hacking maintains, it is contingent that those questions have been asked.. To finish, on the basis of the features of the styles of reasoning I shall conclude that the idea of the convergence of science in the long run to a single true, coherent and complete description does not jibe with the claim that the styles of reasoning continually enrich our present ontology.

Bibliography

Bachelard, G. (2011 [1938]). *La formation de l'esprit scientifique*. Librarie philosophique J. Vrin.

Belgrano, M. (2021). World and Paradigm in Heidegger and Kuhn. *Franciscanum. Revista de las Ciencias del Espíritu, 63*, 2–2.

Belrose, L. (1892). Comte and Turgot. *The Monist, 3*(1), 118–122.

Berlin, I. (1980). *Vico and Herder: Two studies in the history of ideas*. Chatto and Windus.

Bohm, D. (1984 [1957]). *Causality and change in modern physics*. Routledge.

Bordoni, S. (2017). *When historiography met epistemology*. Brill.

Born, M. (1962). *Einstein's theory of relativity*. Dover Publication.

Braunstein, J.-F. (2006). Abel Rey et les débuts de l'Institut d'histoire des science et techniques (1932–1940). In M. Bitbol & J. Gayon (Eds.), *L'épistémologie française, 1830–1970*. Presses Universitaires de France.

Brenner, A. (2003). *Les origines françaises de la philosophie des sciences*. Presses Universitaires de France.

Brenner, A. (2016). L'épistémologie historique d'Abel Rey. [Abel Rey's historical epistemology]. *Revue de Métaphysique et de Morale, 90*(2), 159–176. https://doi.org/10.3917/rmm.162.0159

Burbridge, J. (1993). Hegel's conception of logic. In F. C. Beiser (Ed.), *The Cambridge companion to Hegel* (pp. 86–101). Cambridge University Press.

Calinger, R. (1979). Kant and Newtonian science: The pre-critical period. *Isis, 70*(3), 349–362.

Canguilhem, G. (1966). *Le normal et le pathologique*. PUF/Quadrige.

Carnap, R. (1950). Empiricism, semantics, and ontology. *Revue Internationale de Philosophie, 4*(11), 20–40.
Chimisso, C. (2008). *Writing the history of mind.* Ashgate.
Chimisso, C. (2011). *The diverse lives of concepts in Canguilhem's historical epistemology.* The Open University.
Chimisso, C. (2019). *Hélène Metzger, historian and historiographer of the sciences.* Routledge.
Collingwood, R. (1945). *The idea of nature.* Clarendon Press.
Collingwood, R. (1979 [1939]). *An Autobiography.* Oxford: Oxford University Press.
Comte, A. (1968 [1830–1842]). *August Comte. Textes choisis et présentés par Pierre Arnaud*, Paris.
Condorcet, J.-A.-N.d. C. (1795). *Outlines of an historical view of the progress of the human mind: Being a posthumous work of the late M. de Condorcet.* Printed for J. Johnson.
Crombie, A. (1994). *Styles of scientific thinking in the European tradition* (Vol. 1,2,3). Duckworth.
Dahlstrom, D. O. (2008). *Philosophical legacies: Essays on the thought of Kant, Hegel, and their contemporaries (Studies in Philosophy and the History of Philosophy, Volume 50).* Catholic University of America Press.
Daston, L. (1994). Historical epistemology. In J. Chandler, D. Arnold, & D. H. Harry (Eds.), *Questions of evidence: Proof, practice, and persuasion across the disciplines* (pp. 282–289). University of Chicago Press.
Daston, L., & Galison, P. (2007). *Objectivity.* Zone Books.
Davidson, A. I. (2001). *The emergence of sexuality: Historical epistemology and the formation of concepts.* Harvard University Press.
Demeter, T. (2012). "Weltanschauung" as a priori: Sociology of knowledge from a 'romantic' stance. *Studies in East European Thought, 64*(1/2), 39–52. http://www.jstor.org/stable/41477748
Dews, P. (1992). Foucault and the French tradition of historical epistemology. *History of European Ideas, 14*(3), 347–363. https://doi.org/10.1016/0191-6599(92)90213-V
Di Giovanni, G. (2010). Introduction to G.W.F. Hegel, the science of logic. In *The Science of Logic.* Cambridge University Press.
Elsner, J. (2003). Style. In R. Nelson & R. Shiff (Eds.), *Critical terms for art history* (p. 98–109). The University of Chicago Press.
Feest, U., & Sturm, T. (2011). What (good) is historical epistemology? *Erkenntnis, 75*(3), 285–302. https://doi.org/10.1007/s10670-011-9345-4
Ferrari, M. (2021). Ernst Cassirer's legacy: History of philosophy and history of science. *Journal of Transcendental Philosophy, 2*(1), 85–109. https://doi.org/10.1515/jtph-2021-0008
Fleck, L. (1979 [1935]). *Genesis and development of a scientific fact.* The University of Chicago Press.

Fleck, L. (1986 [1927]). Some specific features of the medical way of thinking. In R. S. Cohen & T. Schnelle (Eds.), *Cognition and fact – Materials on Ludwik Fleck*. Springer.

Fontenelle, B. L. B.d. (2019 [1688]). *Digression Sur Les Anciens Et Les Modernes* (S. Audidière, ed.). Edizioni ETS.

Foucault, M. (1994 [1966]). *The order of things*. Vintage Books Edition.

Friedman, M. (2000). *A parting of the ways. Carnap, Cassirer and Heidegger*. Open Court.

Friedman, M. (2001). *Dynamics of reason*. Stanford: The 1999 Kant Lectures at Stanford University.

Friedman, M. (2011). Extending the dynamics of reason. *Erkenntnis, 75*(3), 431–444.

Fuller, S. (2000). *Thomas Kuhn: A philosophical history of our times*. The University of Chicago Press.

Goyard-Fabre, S. (2005). Léon Brunschvicg et l'héritage kantien. In J. e. a. Ferrari (Ed.), *Kant et la France – Kant und Frenkreich* (pp. 261–269). George Holms.

Gutting, G. (2003). Thomas Kuhn and French philosophy of science. In T. Nickels (Ed.), *Thomas Kuhn* (pp. 45–64). Cambridge University Press.

Hacking, I. (1982). Language, truth and reason. In M. Hollis & S. Lukes (Eds.), *Rationality and relativism* (pp. 48–66). [Reprinted in Hacking 2002])). Blackwell.

Hacking, I. (1992). Statistical language, statistical truth and statistical reason: The self-authentication of a style of scientific reason. In E. McMullin (Ed.), *The social dimension of science*. University of Notre Dame Press.

Hacking, I. (1999). Historical meta-epistemology. In W. Carl & L. Daston (Eds.), *Wahrheit und Geschichte* (pp. 53–77). Vandenhoeck & Ruprecht.

Hacking, I. (2002a). *Historical ontology*. Harvard University Press.

Hacking, I. (2002b [1982]). Language, truth and reason. In *Historical ontology* (pp. 159–177). Harvard University Press.

Hacking, I. (2002c [1992]). 'Style' for historians and philosophers. In *Historical ontology* (pp. 178–199). Harvard University Press.

Hacking, I. (2009). *Scientific reason*. NTU Press.

Hacking, I. (2012a). 'Language, Truth and Reason' 30 years later. *Studies in History and Philosophy of Science Part A, 43*(4), 599–609.

Hacking, I. (2012b). 'Language, Truth and Reason' 30 years later. *Studies in History and Philosophy of Science Part A, In Press [Available online 28 July 2012]*. https://doi.org/10.1016/j.shpsa.2012.07.002

Hegel, G. W. F. (2010 [1816]). *Science of logic*. Cambridge University Press.

Heidegger, M. (1962a). *Die Frage Nach Dem Ding. Zu Kants Lehre Von Den Transzendentalen Grundsätzen*. Max Niemeyer Verlag.

Heidegger, M. (1962b [1927]). *Being and time*. Blackwell.

Heidegger, M. (2000 [1953]). *Introduction to metaphysics*. Yale Nota Bene.

Heidegger, M. (2002 [1938]). *Off the Beaten track*. Cambridge University Press.
Husserl, E. (1970 [1936]). *The crisis of European sciences and transcendental philosophy*. Northwestern University Press.
Kant, I. (2007 [1781]). *Critique of pure reason*. Penguin.
Koyré, A. (1943). Galileo and Plato. *Journal of the History of Ideas, 4*(4), 400–428. https://doi.org/10.2307/2707166
Kritzman, L. (Ed.). (1988). *Michel Foucault, politics, philosophy, culture: Interviews and other writings 1977–1984*. Routledge.
Kuhn. (1996 [1962]). *The structure of scientific revolutions*. The University of Chicago Press.
Lecourt, D. (1969). *L'épistémologie historique de Gaston Bachelard*. Librairie Philosophique Vrin.
Lukács, G. r. (1910). Megjegyzések az irodalomtörténet elméletéhez. In *Dolgozatok a modern filozófia köréből: Emlékkönyv Alexander Bernát hatvanadik születése napjára* (pp. 388–421). Franklin.
Lukács, G. r. (1969 [1910]). Megjegyzések az irodalomtörténet elméletéhez [Bemerkungen zu einer Theorie der Literaturgeschichte]. In *Ders.: Művészet és társadalom. Válogatott esztétikai tanulmányok [Kunst und Gesellschaft. Ausgewählte ästhetische Aufsätze]* (pp. 34–65). 2., erweiterte Ausgabe.
Mannheim, K. (1971a [1927]). Conservative thought. In K. H. Wolff (Ed.), *From Karl Mannheim*. Routledge.
Mannheim, K. (1971b [1952]). On the interpretation of Weltanschauung. In K. H. Wolff (Ed.), *From Karl Mannheim*. Routledge.
Mannheim, K. (1979 [1936]). *Ideology and Utopia*. Routledge.
Mannheim, K. (1997 [1922]). *Structures of thinking*. Routledge.
Metzger, H. (1969 [1918]). *La genèse de la science des cristaux*. Albert Blanchard.
Morgenthaler, J.-L. (2007). Condorcet, Sieyès, Saint-Simon et Comte. *Socio-logos*. https://doi.org/10.4000/socio-logos.373
Peña-Guzmán, D. M. (2016). The nineteenth century in Ruins: A genealogy of French historical epistemology. *Foucault Studies, 21*, 159–183.
Peña-Guzmán, D. M. (2020). French historical epistemology: Discourse, concepts, and the norms of rationality. *Studies in History and Philosophy of Science Part A, 79*, 68–76. https://doi.org/10.1016/j.shpsa.2019.01.006
Pinkard, T. (1988). *Hegel's dialectic: The explanation of possibility*. Temple University Press.
Reichenbach, H. (1965 [1920]). *The theory of relativity and a priori knowledge*. University of California Press.
Rey, A. (1907). *La théorie de la physique chez les physiciens contemporains*. Alcan.
Rheinberger, H.-J. (2010). *On historicizing epistemology*. Stanford University Press.
Riegl, A. (1985 [1901]). *Late roman art industry* (R. Winkes, Trans.). Bretschneider.
Riegl, A. (1992 [1893]). *Problems of style: Foundations for a history of ornament*. Princeton University Press.

Rockmore, T. (2005). *On constructivist epistemology*. Rowman & Littlefield Publishers.

Rockmore, T. (2016). *German idealism as constructivism*. The University of Chicago Press.

Sady, W. (2021). Ludwik Fleck. In E. N. Zalta (Ed.), *The Stanford encyclopedia of philosophy*.

Sanzio, R. (2020). *Lettera a Papa Leone X*. Skira.

Science at the Cross Roads. Papers from The Second International Congress of the History of Science and Technology 1931. (2014). (N. I. Bukharin, ed.). Routledge.

Sciortino, L. (2017). On Ian Hacking's notion of style of reasoning. *Erkenntnis*, 1–22. https://doi.org/10.1007/s10670-016-9815-9

Sinaceur, H. B. (2006). From Kant to Hilbert: French philosophy of concepts in the beginning of the twentieth century. In J. Ferreiròs & G. Jeremy (Eds.), *The architecture of modern mathematics* (pp. 349–376). Oxford University Press.

Sturm, T. (2008). What (good) is historical epistemology? http://www.mpiwg-berlin.mpg.de/en/news/features/feature1

Tanner, J. (2019). Karl Mannheim and Alois Riegl: From art history to the sociology of culture. In *Art history* 32, 4 (pp. 755–784).

Thomson, I. (2011). Ontotheology. In D. O. Dahlstrom (Ed.), *Interpreting Heidegger: Critical essays* (pp. 174–196). Cambridge University Press.

Tristram, R. J. (1988). Vico on the production and assessment of knowledge. *Philosophy and Phenomenological Research, 48*(3), 355–388. https://doi.org/10.2307/2107469

Turgot, A. R. J. (1913–1923). *Oeuvres de Turgot et Documents Le Concernant avec Biographies et Notes* (Vol. 1). Librarie Felix Alcan.

Verene, D. (1966). Cassirer's view of myth and symbol. *The Monist, 50*(4), 553–564. http://www.jstor.org/stable/27901663

Vico, G. (1725 [2003]). La scienza nuova In N. Abbagnano (ed.), *La scienza nuova e altri scritti*. Utet.

Vico, G. (2013 [1710]). L'antichissima sapienza degli Italici. Da estrarsi dalle origini della lingua italiana. In N. Abbagnano (ed.), *La Scienza Nuova e altri scritti*. Utet.

Vinogradova, O. A. (2020). Reception of Nicolas de Condorcet's idea of the progress of the human mind. *Russian Journal of Philosophical Sciences, 63*(9), 90–114. https://doi.org/10.30727/0235-1188-2020-63-9-90-114

Weinberg, S. (1976). The forces of nature. *Bulletin of the American Society of Arts and Sciences, 29*(4), 28–29.

CHAPTER 2

Notions of the Concept of 'Ways of Thinking': From Classical Historical Epistemology to Kuhn

2.1 Introduction

The purpose of the next two chapters is to briefly present the main notions of 'ways of thinking' that have been conceived in the wake of the tradition of historical epistemology and highlight their common features. Their proponents form anything but a homogeneous group. On the one hand, Lévy-Bruhl, Brunschvicg, Metzger, Bachelard and Koyré were exponents of the so-called classical historical epistemology, an approach to the history of the sciences adopted by some French scholars who, at the beginning of the twentieth century, developed their philosophies and discussed their ideas within the same intellectual milieu. On the other hand, Fleck was not attached to a traditional school of philosophy of science although his works, rooted in his particular experience as a scientist, had strong affinities with Bachelard's. Foucault was influenced by different philosophical schools, and not only by classical historical epistemology, to whose institutional and philosophical lineage he belonged. Like Feyerabend, Kuhn has been described as a representative of post-positivism, that line of thought in the philosophy of science that was developed on the ruins of logical positivism, but certainly he was one of the heirs of classical historical epistemology. Hacking combined the tools of analytic philosophy with those of historical research and, together with scholars such as Daston and Galison, is one of the practitioners of modern historical epistemology.

L. Sciortino, *History of Rationalities*,
https://doi.org/10.1007/978-3-031-24004-1_2

Despite the difficulties in pinpointing the common beliefs of the thinkers mentioned above, there is no doubt that they share a common approach to the study of the production of knowledge that closely connects history and philosophy. They stand in a row of thinkers that have produced a decisive transformation in the predominant image of science as exhibiting progress and methodological unity. Already in 1968, Canguilhem had considered Kuhn's work as one of historical epistemology (Canguilhem, 1994 [1968]). More recently, different critics have included not only Kuhn (Vernant, 1990; Brenner, 2006) but also Feyerabend (Brenner, 2006) among the historical epistemologists and Jean-François Braunstein, pointed out that Fleck's, Crombie's and Hacking's philosophies share the traits of the French style in history and the philosophy of science (Jean-François Braunstein, 2008).

More importantly for our purposes, all of them have proposed notions of 'ways of thinking'. These notions have been compared with each other since at least 1968, when Jean Piaget (1896–1980) (2015, p. 132) observed that 'Foucault's epistemes are reminiscent of Kuhn's paradigms' and discussed their differences from a structuralist perspective. In more recent times, Mößner (2011) pointed out important differences between the concept of 'thought style' and that of 'paradigm'; Elwick (2012) depicted Hacking's styles of reasoning as conditions of possibility and likened them to Foucault's epistemes; Winther (2012) investigated the simultaneous application of the notions of 'paradigm' and 'style of reasoning' to the history of science; Sciortino (2021) proposed a layered diagram that allows the differences between the notions of Fleck, Foucault, Kuhn and Hacking to be mapped, while acknowledging their similarities. Less attention has been given to the different articulations and applications of the concept of 'ways of thinking' in classical historical epistemology.

The fact that most of these notions are often compared to each other conceals the admission that they belong to the same genus. However, the general physiognomy of this genus has not been described yet: are these notions tools for achieving overlapping objectives? How do they reflect the different perspectives of their proponents? Clarifying these questions will help me to make explicit certain conceptual links between different authors so as to justify the choice of considering them part of historical epistemology, i.e. a single movement, however broad and varied, that consists of classical historical epistemology and its modern developments.

Viewing these notions as frameworks central to historical and philosophical enquiry provides a perspective from which they can be studied. I believe that for characterizing a framework, it is necessary, firstly, to specify its *explanatory goal*: why has it been put forward? What is it meant to explain? Secondly, as Currie and Walsh (2019, p. 7) have noticed, it is necessary to pinpoint its underlying *explanatory expectations*, the preconceptions of what a good explanation should look like. For instance, Crombie's explanatory expectations were those of a continuist and internalist historian. These two features of frameworks (explanatory goal and explanatory expectations) act as principles of selection: they shape the explanatory narrative by foregrounding events and selecting its central subject (what the narrative is about) and its set of contrasts and comparisons (events or ideas compared across time).

My main point will be that, despite the fact that these thinkers disagree on certain assumptions, their frameworks have the same explanatory goal—to understand how objectivity is possible. I shall present this goal as a necessary element of a common project—that of historicizing Kant's a priori. Furthermore, I shall argue that these notions help to illuminate different conditions of possibility of the emergence of the concepts important for knowledge such as objectivity, evidence, probability, proof, detachment, etc., and help to understand how these epistemological concepts interact in the production of knowledge. Ultimately, in Chap. 4, I shall conclude that there exists a 'genus ways of thinking' whose main features amount to possessing the same explanatory goals—to understand how certain concepts important for the production of knowledge, in particular objectivity, emerge, change and interact with each other.

Since the thesis I am going to argue for is that the main explanatory goal of these frameworks is to explain how objectivity is possible, it is better to clarify first in which sense the term 'objectivity' must be understood. Douglas (2004, p. 4) has argued that there are 'different types of processes we can examine in practice when determining whether to describe the product of that process as objective'. We can focus on: (1) 'processes where humans attempt to interact with the world'; (2) 'an individual's thought processes'; and (3) 'the process used to reach an agreement' (2004, p. 4). Once we focus on a particular kind of process, there are multiple senses of objectivity. In case (3), the outcome of a process can be considered objective in three senses: when 'the same outcome is always produced, regardless of who is performing the process' ('procedural objectivity'); when the members of a community agree on the outcome ('concordant objectivity'); and when they have 'argued with each other to ferret out the sources of their disagreements', in particular, when scientific data, theories, methods, etc. have been shared or discussed ('interactive objectivity') (Douglas, 2004, pp. 12–15).

When I say that the aim of all the notions of historical epistemology is to explain how objectivity emerges, I mainly allude to a form of interactive and concordant objectivity. In other words, to explain how a proposition (or a theory, a scientific 'discovery' or other epistemological items) becomes objective essentially will mean to explain how a community of people come to share and discuss that very proposition as a candidate for truth or falsehood (interactive objectivity), and to agree on its truth (or falsehood) (concordant objectivity). Interactive and concordant objectivity implies that the members of that community agree (tacitly or not) on methods, standards of evidence, techniques, practices, mental operations and schemas of perception that are to be used to prove or refute that proposition. As we shall see, for the authors under study, the elements (methods, standards of evidence, etc.) on which agreement is obtained are historically variable. Therefore, objectivity itself comes into being through specific historical processes. Although other senses of objectivity are important too, those I have just mentioned are of particular importance for understanding the explanatory goals of their frameworks.

2.2 Lévy-Bruhl's 'Mentalities'

'Mentality' is a particular notion of the concept of 'ways of thinking' introduced by Lévy-Bruhl, as shall be explained in this section. The early works of this philosopher reflect an interest in German thought, which he shared with other French philosophers such as Boutroux and Victor Delbos (1862–1916). In 1902, at a meeting of the Société française de philosophie, he defended Comtism as a living philosophy that, as Warren Schmaus (1996, p. 430) pointed out, he wanted to develop in the Kantian direction of an investigation into the conditions of knowledge. In 1903, Lévy-Bruhl published *La morale et la science des moeurs* [*Ethics and Moral Science*] (1903), whose sociological approach testified to the influence of Émile Durkheim (1858–1917), and afterwards he developed an interest in the study of primitive societies. His purpose of finding ethnological data for a philosophical study of human nature dates back to the years in which he became Professor of Modern Philosophy at the Sorbonne (in 1904) and wrote *Les Fonctions mentales dans les sociétés primitives* [*How Natives Think*] (2015 [1910]).

As he recounted in (Lévy-Bruhl, 1923), the translation into French of a book received from a Chinese historian made him wonder whether 'the logic of the Chinese coincided with ours' (Lévy-Bruhl, 1923, p. 634). Further study of Chinese philosophy convinced him that human thought changes across different cultures (Mousalimas, 1990, p. 34). Thus, his previous intention to investigate the conditions of human knowledge in a 'Kantian direction', combined with this new belief, resulted in an effort to understand how the categories of thought change in different cultures, i.e. in an attempt to historicize Kant. This research objective is clearly expressed in the introduction of *Les fonctions mentales dans les sociétés primitives* in which Lévy-Bruhl rejects beforehand any idea of reducing mental operations of different societies to a single type (2015 [1910] p. 28) and asserts that their study 'can throw some light upon the genesis of our categories and our logical principles' (2015 [1910] p. 14). Indeed, according to him, Durkheim had already shown that a *comparative* study of the various types of mental operations of different societies led to a theory of knowledge (2015 [1910] p. 14).

His attempt to understand how the categories of thought and mental operations change in different cultures was closely connected to that of understanding how certain epistemological items become objective for a community of people. In a terminology borrowed from Durkheim, he expressed this point by claiming that a community elaborates certain *collective representations*: concepts, ideas, values or mystical images of objects 'common to the members of a given social group [...] transmitted from a generation to another [...]', which 'awaken in them sentiments of respect, fear, adoration' and whose 'existence does not depend upon the individual' (Lévy-Bruhl, 2015 [1910], p. 14). He was interested in understanding how these collective representations, emerging in so-called primitive societies,[1] differ from those of more advanced societies and are connected with each other so to form other collective representations such as ideas concerning spirits, witchcraft and the power of objects.

The concept of 'collective representation' was the cornerstone of Durkheim's social theory, while the term 'representation' was widely current among neo-Kantians of the time such as Charles Renouvier (1815–1903) and Octave Hamelin (1856–1907) (cited by Durkheim in

[1] Lévy-Bruhl wrote in a note that his use of the term 'primitive', in the sense of 'undeveloped', was 'rendered almost indispensable through common usage' (Lévy-Bruhl, 2015 [1910], p. 13).

(2008 [1912], p. 12) (see also Paoletti, 2004; Falasca-Zamponi, 2014; Garlitz, 2020)). In *Les formes élémentaires de la vie religieuse* [*The Elementary Forms of Religious Life*] (2008 [1912]), which has been viewed as a work of neo-Kantianism (see for example Edwards (2002) and Garlitz (2020)), by 'collective representations' Durkheim mainly referred to religious beliefs and thought categories that are socially generated and cannot be reduced to the features of single individuals. He believed that categories are concepts which are subject to time and change: 'categories are [...] artful instruments of thought that human groups have laboured to forge over the centuries' (2008 [1912], p. 21), he wrote. For him, Kant was right in thinking that human reason operated with categories but wrong in believing that these categories are inherent in humans.

As far as we are concerned, there were two points of disagreement between Lévy-Bruhl and Durkheim.[2] To begin with, while Durkheim considered primitive collective representations as concepts, Lévy-Bruhl pointed out that they were concepts that differed profoundly from ours: 'the mental activity [of a primitive] is too little differentiated for it to be possible to consider ideas or images of objects by themselves apart from the emotions and passions which evoke those ideas or are evoked by them' (Lévy-Bruhl, 2015 [1910], p. 36). Furthermore, while for Durkheim there was continuity between the mental activity of primitive individuals and that of modern societies, for Lévy-Bruhl there was a radical discontinuity. Scientific thought, according to the latter, was not a later stage of a growing sophistication of the mental processes of human beings.

Lévy-Bruhl focused on the difference between our collective representations and those of primitive societies and on the difference between our rules of logic and the rules that govern collective representations in the primitive societies. He wanted to understand how individuals in different societies use, apply and connect their collective representations to each other in order to make sense of the world and interpret social relationships. Ultimately, he wanted to reveal how people in different societies think. In the final chapter of *Les Fonctions mentales dans les sociétés primitives*, he concluded that his analysis of the ethnological data borne out two theses, which I am going to explain in detail:

[2] For a discussion of the differences between Lévy-Bruhl's and Durkheim's theses (see Merllié, 2012).

1. 'The institutions, practices, and beliefs of "primitives" imply a prelogical and mystical *mentality*, oriented otherwise than our own' (Lévy-Bruhl, 2015 [1910], p. 361 my emphasis)
2. 'The collective representations and the connections among these representations that constitute this mentality are governed by the *law of participation*, and, as such, indifferent to the logical law of contradiction' (Lévy-Bruhl, 2015 [1910], p. 361 my emphasis).

To make these two statements clearer, I will begin to clarify the meaning of the term 'mentality' and then move on to that of 'law of participation' and that of 'prelogical'. Although the term 'mentality' is never defined in Lévy-Bruhl's work, its meaning can be gleaned from its usage. In *Les Fonctions mentales dans les sociétés primitives*, Lévy-Bruhl wrote that 'different mentalities will correspond with different social types, and all the more because institutions and customs themselves are at bottom only a certain aspect of collective representations' (Lévy-Bruhl, 2015 [1910], p. 27). Stanislas Deprez noted that this passage seems to suggest that 'a mentality is nothing other than a series of collective representations' (Deprez, 2009, p. 10) but he rightly pointed out that this definition would not be satisfactory. Certainly, the particular collective representations of a society as well as its institutions, conventions and customs are a characterizing element of what Lévy-Bruhl calls a mentality. However, the crucial element to which the notion of mentality alludes is the way the members of a community connect their representations. Indeed, in the second thesis above, Lévy-Bruhl says that 'the collective representations *and the connections* among these representations [...] *constitute* this [primitive] mentality' (my emphasis). Thus, as Gaston Bouthoul (1961, p. 30) put it, Lévy-Bruhl's mentalities are 'sets of concepts and intellectual dispositions integrated in the same individual, linked together by logical relationships and belief relationships'.

So, how do primitive societies connect their representations? What are the laws governing the connections between primitive collective representations? What characterizes the mentality of the primitives? From Lévy-Bruhl's comparative study between different societies (Lévy-Bruhl, 2015 [1910], p. 28), it emerges that the collective representations that constitute the primitive mentality 'are not connected with each other as ours would be' but are governed by the law of participation (thesis two), according to which things can be at the same time themselves and something else or can emit or receive mystic powers that can be felt elsewhere:

> In the collective representations of primitive mentality, objects, beings, phenomena can be, though in a way incomprehensible to us, both themselves and something other than themselves. In a fashion no less incomprehensible, they give forth and they receive mystic powers, virtues, qualities, influences, which make themselves felt outside, without ceasing to remain where they are. (Lévy-Bruhl, 2015 [1910], p. 76)

Thus, the law of participation allows multilocation, consubstantiality and multinumeration and therefore allows violations of the principle of non-contradiction. For example, the Huichols, an indigenous people of Mexico, affirm the identity of corn, deer, plumes and the Hikuli dance (Lévy-Bruhl, 2015 [1910], p. 128) and the Bororo, indigenous people of Brazil, believe that they are red parakeets (Lévy-Bruhl, 2015 [1910], p. 77). It is not a relation between collective representations of objects that they claim but a mystical participation. 'Mystical' is the adjective Lévy-Bruhl used in thesis one above to point out that the collective representations of primitives differ profoundly from our ideas and concepts: 'primitive representations express, or rather imply, not only that the primitive actually has an image of the object in his mind, and thinks it real, but also that he has some hope or fear connected with it, that some definite influence emanates from it, or is exercised upon it' (Lévy-Bruhl, 2015 [1910], pp. 37–38). The 'law of participation' plays the role of an organizing concept which governs the connections among different collective representations within the mentality of primitive peoples.

Lévy-Bruhl did not deny that the primitive mentality made use of concepts but insisted that, since their abstraction was the result of the use of the law of participation, for the most part they differed from our own (Lévy-Bruhl, 2015 [1910], p. 116). He gave as examples symbolic objects that in different circumstances have different meanings (Lévy-Bruhl, 2015 [1910], p. 117) and asserted that they were clear instances of *mystic abstractions*: attention is not paid to certain features considered important to the subject, as it happens in the process of abstraction, but to the features that establish relations between the visible and some mysterious force such as a spirit or a soul.

To explain the other adjective, 'prelogical', which qualifies the primitive mentality, Lévy-Bruhl wrote: 'I do not mean to assert that such a mentality constitutes a type of antecedent stage, in point of time, to the birth of logical thought [...] I merely wish to state that it does not bind itself down, as our thought does, to avoiding contradiction' (Lévy-Bruhl, 2015

[1910], p. 78). What accounted for the connections between primitive collective representations were principles wholly different from the principles of logic such as the law of participation. The latter is indifferent to the rule of non-contradiction and allows the fallacy of ***post hoc, ergo propter hoc*** to which, according to Lévy-Bruhl, primitives were undoubtedly prone (Lévy-Bruhl, 2015 [1910], pp. 73–74). Indeed, the connection between the antecedent and the consequent consisted in a mystical relation between them. Thus, the primitives adopted a form of explanation based on supernatural causes, i.e. on primary rather than secondary causes. However, it is important to bear in mind that for Lévy-Bruhl, depending on the circumstances, the same individual could think in two different ways: in accordance with the mentality of the primitive society in which he lived or as *we* think: 'considered as an individual, the primitive, in so far as he thinks and acts independently of these collective representations where possible, will usually feel, argue and act as we should expect him to do' (Lévy-Bruhl, 2015 [1910], pp. 78–79).

The features of a mentality—its collective representations and the way individuals connect them—throw light on how objectivity emerges and changes from one society to another. For example, as a consequence of the law of participation, certain propositions (e.g. 'animals are angry with the hunters', 'fishes have their regular paths through the waters', 'if a man uses a bone against an enemy, the water becomes infectious') are perceived as real and objective by the primitives. While in primitive societies certain objective propositions are made possible by the use of the law of participation, i.e. a sort of law of communion between an object and other entities, on the contrary, in advanced societies objectivity is the result of an inverse process, one in which the object to be known is separated as much as possible from any external influence to eliminate any element of subjectivity. For those who adopt a scientific mentality, wrote Lévy-Bruhl, 'to objectify is to project beyond oneself, as something which is foreign to oneself, that which is to be known. How intimate, on the contrary, is the communion between entities participating of each other, which the collective representations of prelogical mentality assures!' (Lévy-Bruhl, 2015 [1910], p. 384).

When individuals of a primitive society, influenced by the law of participation affirm the identity of objects or the common presence of a mystical power in each of them, a type of classification is established. This classification includes a complexity of participations with some of them serving as a nucleus around which other participations cluster. For Lévy-Bruhl these 'primitive classifications' to a certain extent occupy the position held by the Kantian categories in logical thought (2015 [1910], p. 364). However, the way objectivity emerges through primitive classifications does not resemble the way objectivity emerges through our logical classifications. 'These [latter] involve a series of concepts whose extent and connotation are definite, and they constitute an ascending scale the degrees of which reflection has tested. The prelogical mind does not objectify nature thus. It lives it rather, by feeling itself participate in it, and feeling these participations everywhere' (Lévy-Bruhl, 2015 [1910], pp. 129–130).

Another important point is that in the mentality of a primitive individual, there is hardly any room for questions such as 'how?' and 'why?'. Indeed, 'the ensemble of collective representations which master him and excite in him an intensity of feeling which we cannot even imagine, is hardly compatible with that disinterested contemplation of a matter which a purely intellectual desire to probe into its cause would demand' (Lévy-Bruhl, 2015 [1910], p. 25). Thus, on the one hand the primitive mentality brings to the fore questions related to the alleged existence of powers, spirits and signs, which have no place in the modern mentality; on the other hand, it overshadows all those questions generated by a curiosity seeking rationally for causes. Furthermore, what counts for the primitive mind are not the attributes of which our senses inform us but certain 'signs' or certain mystical properties of things. In other words, a mentality is also a way of perceiving the world.

Ultimately, the concept of mentalities is a notion of the concept of 'ways of thinking' introduced by Lévy-Bruhl: a particular mode of operating connections between certain concepts. He conceived this notion by examining the difference between our own way of thinking and that of primitive societies and, in particular, by relying on ethnological data. The concept of mentality can also be viewed as a framework whose goal is to explain how objectivity emerges in primitive and modern societies. For Lévy-Bruhl, the core of the mentality of primitive peoples is the law of participation, a high-level concept that governs how collective representations are connected to each other in the mind in order to make sense of the world. In this sense, a mentality substitutes Kant's a priori. In a primitive mentality what makes certain sentences objective or certain questions possible or impossible is the specific way concepts are connected rather than the existence of given methods and criteria of evidence. The explanatory expectations of Lévy-Bruhl, i.e. the preconceptions of what a good explanation should look like, mainly consisted in the belief that the mental operations of different societies could not be reduced to a single type and that a *comparative* study of the various types of mental operations of different societies led to a theory of knowledge. What his framework provided was an explanatory narrative of how objectivity changes from primitive to modern societies.

2.3 Brunschvicg's 'Ages of Intelligence'

Brunschvicg's philosophy marks another stage in the evolution of the concept of 'ways of thinking'. Professor at the Sorbonne for 30 years, starting in 1909, Brunschvicg was a central figure in twentieth-century French philosophy. He believed that the task of the philosopher was to reflect on the history of science, as the development of philosophy was not autonomous and could not lead to a truth higher than that of science. In particular, he focused on the history of mathematics convinced that the mathematization of nature was the only truly reliable method for progressing in the production of knowledge. Alongside the three main works on the history of mathematics and physics, *Les Etapes de la philosophie mathématique* (1912), *L'expérience humaine et la causalité physique* (1922) and *La Physique au vingtième siècle* (1939), Brunschvicg wrote *Les âges de l'intelligence* (1934), a book on the history of science and philosophy built around the lectures given at the Sorbonne during the winter of 1932–1933. In this latter, he used the concept of 'age of intelligence' as a framework for narrating a history of scientific thought from the early humans to Albert Einstein (1879–1955). An age of intelligence was a particular stage of human history characterized by a precise way of thinking, open to change and with no definite purpose in sight. As I shall argue, this is a particular notion of 'ways of thinking' which, for simplicity, we can continue to call 'age of intelligence'.

The philosophy of Brunschvicg complicated the subject–object relationship in two ways. On the one hand, he believed that, by mathematizing nature, the subject had the power of enriching experience and therefore of 'creating the object'. He described this view as an 'idealism that corresponds to the expression of authentic scientific activity, which has always consisted in applying to the raw data of physical perception the mathematical connections due to the legislative power of the mind' (Brunschvicg, 1934, p. 144). On the other hand, as Metzger put it,

> [Brunschvicg] did not postulate as self-evident the eternal fixity of our reason; on the contrary, he tried to demonstrate that this reason, far from being forever permanent in fixed frames from the dawn of civilization, and assisting as an impassive spectator in the development of our knowledge, that this reason, I say, is also likely to progress; that in contact with the experience, which it wants to bring under her laws, human reason becomes more flexible, it discovers new presences thanks to which little by little its field of penetration increases. (Metzger, 1914–1939, p. 171)

Here, by 'fixed frames', Metzger alluded to Kant's immutable a priori categories. What the historical evidence showed to Brunschvicg was that the categories of thought were not given once for all. In the introduction to *L'expérience humaine et la causalité physique* he wrote that 'it is turning one's back on the good philosophical method of wanting, in the name of an a priori concept of causality, to prescribe to experience what it must be' (Brunschvicg, 1922, p. 13). On the contrary, it was necessary 'to consult experience as it is, asking it to guide us through the diversity of conceptions that successive generations have made of causality' (Brunschvicg, 1922, p. VII). Thus, since for Brunschvicg subject and object, i.e. reason and reality, have a history his problem was to understand how objectivity emerges from their dialectic:

> The role of mathematical philosophy is in no way [...] to enclose the truth within the framework of a logical artifice. [...] It is necessary, on the contrary, to reach the source where the original contact of intelligence with things would be manifested [...] The aim of mathematical philosophy would no longer be to add a system to the systems which have already taken place in history; it turns to history itself, it seeks the convergence and coordination of the results which have been obtained in different periods, and it registers them as the positive marks of objectivity. (Brunschvicg, 1912, pp. 462–463)

Of the two elements, subject and object, Brunschvicg focused on the former because, as Cristina Chimisso pointed out, 'he did not believe that a mind-independent reality—the thing-in-itself, in Kantians terms—existed; and even if it did, he argued, we could not know it, so it would be of no interest to us' (2019b, p. 27). Indeed, Brunschvicg wrote that 'our task is not to know how the nature of things is made, but to say how the mind of man is made' (Brunschvicg, 1922, p. XII). This meant one needed to understand how people thought in different 'ages of intelligence', as he called them, and in particular to know how the categories of thought changed. Although the phrase 'ages of intelligence' has vague Comtean connotations, Brunschvicg's narrative in his homonym book (Brunschvicg, 1934) did not follow the 'law of three stages' or any other law. What characterized human thought in Brunschvicg's account is the power to go beyond any given and to restructure it according to a norm that is never definitive. 'At the basis of science there is no rigid framework—the Kantian "waffle mold" that Fichte mocked—there is the absolute spontaneity of the mind. There is nothing beyond freedom', wrote Maurice Loi (1984). Subject and object, reason and reality transform each other incessantly, giving shape from time to time to what is objective with no definite end in sight. An age of intelligence is a particular moment in this open-ended dialectic characterized by a particular mode of thought: 'In every epoch, no doubt, a kind of synthesis is established which will seem to fix the laws of thought in the way in which the various idioms seem to have stabilized their vocabulary and their syntax forever' (Brunschvicg, 1934, p. 129). In particular, an age of intelligence is made up of two elements, which Brunschvicg calls 'substratum' and 'norm': the former is 'the background of mental habits which constitutes the biological and social infrastructure of the spirit' and the latter is 'the impetus of reason, freeing itself from the realism of the imagination to create combinations of relations in increasingly close connection with the results of the experimental technique' (Brunschvicg, 1934, p. 129).

These passages of Brunschvicg can be interpreted by saying that our categories of thought, mental dispositions, values, presuppositions and practices constitute the substratum, i.e. the infrastructure in which human beings think. On the other hand, the norms are what governs the way in which concepts, evidence and methods are connected to each other in the production of scientific knowledge. Indeed, Brunschvicg continued the idioms metaphor above by suggesting that in the evolution of Italian, French, Spanish and Romanian 'we can see the influence of two factors: the ideal standard of Latin, and the substrate which varied from one province to another' (Brunschvicg, 1934, p. 129). Just as syntax governs the grammatical relationships between words and phrases that make up linguistic expressions, so the norms govern the connections between concepts. In each age of intelligence the equilibrium between substratum and norm is always precarious and 'doomed to remain unstable and ruinous, unless the proper nature and the respective roles of the two factors have been fully illuminated and radically discerned' (Brunschvicg, 1934, p. 129).

Brunschvicg used the notion of age of intelligence as a framework that shaped the history of Western thought and helped to compare different periods of the history of science. In *Les âges de l'intelligence*, in order to account for the thought of the early humans he relied on the ethnological data already examined by Lévy-Bruhl and did not hesitate to define it as *primitive mentality*:[3]

> The thought of the uncivilized, like the thought of the child, must be considered, not in terms of our own, but for itself in its intrinsic behaviour. That is to say that the primitive mentality will designate only the field beyond which our means of investigation do not allow us to penetrate [...]; we will not say anything absolute when we talk about the first age of intelligence [...]. And immediately we are in the presence of the expression that M. Lévy-Bruhl must have chosen as the most suitable [...]: the primitive mentality is prelogical. (Brunschvicg, 1934, p. 18)

Lévy-Bruhl had introduced the notion of mentality in the context of anthropological studies. Brunschvicg adapted it to the early history of the science of human thought by viewing it as a first age of intelligence. It is legitimate to say that the norm of the first age of intelligence consisted of the law of participation, while the substratum consisted of collective representations, practices, beliefs in mystical powers, virtues, qualities, influences and so on.

According to Brunschvicg, it is around the sixth century BC that the following age of intelligence arose, when human thought purified itself from mystical beliefs. For the first time, 'the internal properties of numbers, their combinations, as they were represented in spatial figures or as they reflected the structure of celestial constellations, or even as they corresponded to the harmony of the sounds of the scale, became the subject of a disinterested and systematic study' (Brunschvicg, 1934, p. 41). This is the moment in which 'thought became aware of its vocation and its dignity by releasing a form of judgment which owed only to oneself the *norm* of its certainty' (Brunschvicg, 1934, p. 41 my emphasis).

As a final example of the age of intelligence we cannot fail to mention the one in which Einstein formulated the theory of relativity. Brunschvicg,

[3] The notion of mentality, which had quickly spread to other disciplines soon after its introduction by Lévy-Bruhl, has also been a point of reference for Abel Rey, Gaston Bachelard and Hélène Metzger.

struck by the philosophical implications of the relativity theory, argued that Kant had in fact presented a particular way of thinking, of which Newton's science was the product (Brunschvicg, 1922, p. 458). The age of Einstein replaced the substratum of the age of Kant, in which absolute space and time were considered the containers of our experiences, with another substratum in which there was no abstract space and no abstract time which could be thought, respectively, as containers of objects and experiences.

To sum up, Brunschvicg introduced his own version of the concept of 'ways of thinking'. For him, in certain stages of history human beings have thought in different ways, in the sense that their efforts to know were governed by certain norms and particular sets of categories of thought, presuppositions and values (substrata). Brunschvicg called these stages of human thought 'ages of intelligence' and used this notion to provide an explanatory narrative of how in different moments objectivity is possible. The substrata that characterize an age of intelligence can be viewed as replacements of the Kantian a priori. The norms are the means by which Brunschvicg's notion of the concept of 'ways of thinking' governs the connections of epistemological items important in the production of knowledge. It is from the dialectics between substrata, norms and reality that objectivity emerges over time.

2.4 Metzger's 'A Priori'

The French historian and philosopher of science Hélène Metzger wrote her first monography *La génèse de la science des cristaux* [*The Emergence of the Science of Crystals*] (1969 [1918]) during the First World War and starting from 1921 she collaborated with the historian of science George Sarton (1884–1956) on the magazine 'Isis'. Member of the Centre international de Synthèse, she published a philosophical volume, *Les concepts scientifiques* [*The scientific concepts*] (1926), in addition to books on the history of chemistry in the seventeenth and eighteenth century,[4] a popular

[4] Aside from *La génèse de la science des cristaux*, Metzger published these works on the history of chemistry: *Les doctrines chimiques en France du début du XVIIe à la fin du XVIIIe siècle* [*Chemical Theories in France from the Beginning of the 17th to the End of the 18th Century*] (1923), *Newton, Stahl, Boerhaave et la doctrine chimique* [*Newton, Stahl, Boerhaave and Chemical Theory*], (1930), *La Philosophie de la matière chez Lavoisier* [*Lavoisier's Philosophy of Matter*] (1935).

history of chemistry[5] and a book on the history of physics.[6] What I would like to discuss here is Metzger's version of the concept of 'ways of thinking', whose features emerge from her reflections on the history of chemistry and, in particular, from her book *Les concepts scientifiques*. The research on the 'mentality of the primitives' conducted by her uncle Lévy-Bruhl and the historicized neo-Kantian works of Brunschvicg were decisive elements for the development of her notion of the concept of 'ways of thinking'. As we have seen, both these thinkers had concluded that the categories of thought were not timeless and universal. Not only did Metzger share this conclusion but, like Brunschvicg, she believed that in the process of knowledge thought and reality interacted perpetually, modifying each other. She wrote that '[…] this reason […] in contact with the experience that it wants to bring under its laws becomes more flexible, it discovers new processes thanks to which little by little its field of penetration increases' (Metzger, 1914–1939, p. 171). On the other hand, reality changes continuously as we look at it from ever new perspectives: 'our mind at work is much more concerned with moving forward, with constantly discovering new points of view on the real, than with putting order in our vision of the world' (Metzger, 1914–1939, p. 119).

The transformation in the concepts of measure, time, space and causality brought about by quantum mechanics and the theory of relativity is an example of how our categories of thought change in the process of knowledge: 'causality, which once seemed on a plane prior to that of experience, in an a priori intuition which tried to assimilate a nature external to our thought, no longer appears under the same aspect […]. Experience and causality intertwine and form an indissoluble whole in the mathematical language of modern science' (Metzger, 1914–1939, p. 171).

What Metzger wanted to do was to study the entire process by which a theory becomes objective without limiting itself to a pure and simple interpretation of the texts in question: 'we must distinguish the theory as it arises in the mind of the one who thinks it, from the thought which has become a thing, an object, and which is found stabilized […]. The intelligence of the historian must not begin and end with the reading and the objective interpretation of the text' (Metzger, 1914–1939, p. 65). In

[5] *La chimie* [*Chemistry*] (1930).

[6] *Attraction universelle et religion naturelle chez quelques commentateurs anglais de Newton* [*Universal Attraction and Natural Religion According to Some English Commentators on Newton*] (1938).

Metzger, the Kantian problem of identifying the forms of intuition that structure our experience took the form of this question: 'By what various procedures does our mind, spontaneously and without a preconceived will, manage to classify in the same classes things which present superficial resemblances or deep analogies?' (Metzger, 1926, p. 1). She had in mind the classifications of chemical elements, the great variations of which she had become aware of in her studies of the history of chemistry. According to her, to address this question meant not to discover a logic of classification but to identify particular organizing concepts which allow us to consider certain objects as similar in some respect. These organizing concepts can be viewed as the categories of thought peculiar to a certain epoch. Thus, for her, one of the great tasks confronting the historian of science was to detect the conceptual system of a certain era that allowed people to classify things and organize their experience. Furthermore, in her view, in order to understand the process of the production of knowledge it was also important to study the interplay between organizing concepts, what she called intellectual tendencies (emotions, esthetical considerations, mental predispositions etc.) and metaphysical views.

An example of 'concept', which for Metzger is 'a supposition concerning the nature of things' (Metzger, 1926, p. 12), is that of 'virtual analogy' [*analogie virtuelle*]. The latter is the hypothesis that some objects share certain properties based on the observation of particular similarities: in the seventeenth century, minerals such as arsenic and mercury were considered as having the same nature as poisons and were included in the same category (Metzger, 1926, p. 19). Another organizing concept was that of 'active analogy' [*analogie agissante*], when it was believed that two similar objects, since they resembled each other, acted on each other. Active analogy helped seventeenth-century chemists to establish correspondences between different sets of objects, e.g. where different classes of similar objects attracted or repelled each other. Other important organizing concepts identified by Metzger concerned the idea of evolution. Among them, that of teleological evolution was used by the alchemists to order the metals found in nature: each metal corresponded to a particular stage of an evolution whose culminant point was gold (Metzger, 1926, p. 93).

It is important to note that, by making these points, Metzger did not mean that in a certain historical period all the scholars agreed in considering certain elements in the same class. For example, as she remarked, in the seventeenth century there was disagreement about the virtual analogies to be attributed to arsenic, mercury and lead (Metzger, 1926, p. 19). But the key point I wish to emphasize is that the concepts of virtual or active analogy were 'objective', i.e. shared by people of a certain epoch who used them to discuss, classify and formulate theories or assertions. In turn, these theories and assertions were made objective by the use of the concepts of analogy: they became candidates for truth or falsehood.

In Metzger, the study of organizing concepts cannot be separated from that of passions, emotions and intuitions. For her, these non-rational factors play a role in the choice of organizing concepts or in the way the latter are used and applied to the world. For example, she remarked that alchemists employed the concept of analogy, in its different forms, guided by a 'fundamental tendence' to think of the world as a homogenous whole. She thought that what Lévy-Bruhl called participation was no more and no less than the concept of active analogy that in primitive peoples was inspired by mystical feelings. Indeed, according to her, the concept of active analogy was not the exclusive prerogative of primitive thought but of all scientific thought at all times in its nascent state, which she called 'spontaneous thought' (Metzger, 1914–1939, pp. 122–123) (see also Chimisso, 2019a, pp. 119–120). Furthermore, for Metzger, in different epochs the same concept can be used in different ways: for example, alchemists used it to organize their chemical experiments while primitive people to decide a strategy of hunting. Thus, the specificity of an organizing concept is also the way it is used for thinking and to *doing*, i.e. to intervene in the world.

Ultimately, from Metzger's thought emerges a unitary vision of human nature that overcomes the mind–body dualism: her vision is reminiscent of that of Vico who, as we have seen in the previous chapter, was opposed to Descartes' rational mind with a view of human nature in which the sensitive faculties of memory, imagination and ingenuity were fundamental components. With Metzger, the poetic horizon, linked to corporeality and sensitivity, regained a crucial role in the process of production of knowledge. Like Vico, Metzger explicitly criticized the Cartesian view: for example, she emphasized the role of imagination in the work of so-called Cartesian chemists (Metzger, 1914–1939, p. 243).

In conclusion, what is a 'way of thinking' for Metzger? We can answer that for a given epoch or society, to think in a certain way means to adopt a particular system of organizing concepts that in the process of production of knowledge interact in specific ways with certain presuppositions, intellectual tendencies, mental attitudes or emotional considerations. Metzger called this system of organizing concepts and the intellectual tendences that inspire them 'a priori':

> [...] [T]he a priori will not just represent all the concepts in place prior to experience and on which experience relies. The a priori will also represent the fundamental tendencies that produce these concepts; or, to put it in scholastic language, to the a priori in actuality, which is made a reality by the first notions, we shall add the underlying a priori in potentiality which, on contact with the experience of life (and not only scientific experience), effectively becomes a priori in actuality.

And she added that:

> The same general concept, that of metal for example, can either inspire heterogeneous hypotheses or, conversely, be inspired by these hypotheses. Undoubtedly, the most diverse assumptions react on the concept, transform them more or less by modifying the limits of the genre which they serve to define; so that the theoretical data, from which the work that each researcher provides, is not absolutely stable and sometimes changes so quickly that it becomes unrecognizable to the next generation of scientists. (Metzger, 1914–1939, p. 13)

Thus, what characterizes a way of thinking à la Metzger is: (1) a particular a priori (the systems of organizing concepts as well as certain presuppositions, hypotheses, mental attitudes and non-rational factors that are shared by the members of a society in a certain epoch) (2) the specific way all these factors interact in the process of production of knowledge.

2.5 Bachelard's 'Eras of Scientific Thought'

It can be said that Bachelard, who was professor at the University of Dijon from 1930 to 1940 before obtaining the chair of History and Philosophy of the Sciences at the Sorbonne, has been both a philosopher of modern science and of imagination. Indeed, reflections on the great scientific discoveries of physics and chemistry of the late nineteenth and early twentieth centuries and that on the relationship between imagination and rationality are the two main themes of his philosophical trajectory. There is a deep connection between these two themes: like Metzger and unlike the philosophers of positivism and neo-positivism, Bachelard saw the role of imagination as crucial in scientific discoveries, the genesis of which is not reducible to rationality alone. For him, imagination is also at the origin of 'epistemological obstacles', conceived as internal resistances to the very act of knowing.

In works such as *Le nouvel esprit scientifique* [*The New Scientific Spirit*] (1934) and *La formation de l'esprit scientifique* [*The Formation of the Scientific Mind*] (2002 [1938]), Bachelard argued that the 'scientific mind' must take note of the deep ruptures operated by relativistic physics and quantum mechanics. He proposed what he called a 'psychoanalysis of objective knowledge', a kind of analysis directed towards the identification of epistemological obstacles. For example, according to him, certain prejudices that are present in our mind can prevent the progress of scientific rationality. *La formation de l'esprit scientifique* and *La psychanalyse du feu* (1938) identifies the most important epistemological obstacles and show how they are rooted in imagination and emotions. It is by abstracting and mathematizing the physical phenomena that scientists were able to escape prejudices inherent in human nature and arrive at a formulation of quantum mechanics and the theory of relativity. One of these epistemological obstacles was realism, 'the coefficient of reality that the prescientific mind attributes to all that is natural' (Bachelard, 2002 [1938], p. 110), which Bachelard compared to the innate desire to possess riches.

Later, Bachelard expanded his field of investigation to the study of the activity of imagination, as expressed not only in science but in human spirit in general and, in particular, in poetry. In addition to *La psychanalyse du feu* (1938), his studies on imagination gave rise to works such as *L'Eau et les rêves: Essai sur l'imagination de la matière*, (1941), *La poétique de l'espace* (1957), *La poétique de la reverie* (1960), *La flamme d'une chandelle* (1961). What these works show is that the creative activity of mind is not reducible to the construction of rigorous thoughts, according to the demands of scientific reason. It also includes another activity, not less rigorous, that of imagination, 'the faculty of distorting the images provided by perception' (Bachelard, 2009 [1942], p. 11). These two activities of mind appear as antithetical and irreconcilable extremes, not at all complementary, which must guard themselves from the intrusions of the other.

After taking note of the revolution in physics and its destabilizing shocks for classical philosophical thought, Bachelard proposed a new philosophy adapted to the new scientific mind and capable of replacing Kantianism, the limits of which had already been emphasized, before him, by Brunschvicg:

> We therefore believe in a reason constituted before any effort of rationality. Léon Brunschvicg saw the weakness of this position and he often insisted on the essential relativity of reason and of experience: 'We lose sight of the real course ... of this knowledge when we are concerned to push rationality and objectivity out of ourselves, in order to isolate and oppose one to the other absolute reason and absolute object'. (Bachelard, 2004 [1949], p. 21)

Bachelard's point was that there is no fixed Kant's a priori since reason questions the categories and concepts of science by adjusting them to successive scientific revolutions. And there is no 'absolute object' since, after the rupture between the sensitive knowledge and the scientific knowledge operated by the laboratory sciences, the object of knowledge lies beyond the senses. Ultimately, one of the conditions for the existence of a thought in phase with science is 'the relaxation of usual philosophies' [*l'assouplissement des philosophies usuelles*] (Bachelard, 1934, p. 20) which, like Kantianism, not only propose a rigid conception of reason but also of the object of knowledge.

What Bachelard recognized was that the science of the late nineteenth and early twentieth centuries marked the emergence of a new way of thinking in which the objectivity of a concept did not imply the direct grasp of an object or the presentation of an object to the mind, without any designation whatsoever. The point is that, as Teresa Castelão explained,

> [Bachelard] consider[ed] the instruments as materialized theories. This means that, for him, technology is applied science. Thus, the technique includes mathematics as well as the actual material instruments. Both are used with profit in the experimental verification of theories by the scientist. Bachelard then assert[ed] that the study of physical reality becomes technically organized. This kind of realism, technical and constructed, is then the touchstone of scientific objectivity. Scientific knowledge depends both on the instrumental conditions and on the contingency of scientific culture which determine, by tacit agreement, the modes of interpretation of the real. (Castelão, 2010, p. 98)

In other words, 'a knowledge which is not given with its precise conditions of determination is not scientific knowledge' (Bachelard, 2002 [1938], p. 87). Objectivity comes into being with the establishment of these conditions: it is the consensus that emerges within this community that confers objectivity on a statement. As Bachelard put it, 'objectivity is determined in precision' (Bachelard, 2002 [1938], p. 87). However, the coming into being of objectivity is an open-ended process that begins with a break with common sense and proceeds through a continuous rectification of errors:

> Sometimes we marvel at a chosen object; we accumulate hypotheses and daydreams; we thus form convictions which have the appearance of knowledge. But the initial source is impure: the primary evidence is not a fundamental truth. In fact, scientific objectivity is only possible if one has first broken with the immediate object, if one has refused the seduction of the first choice, if one has stopped and contradicted the thoughts that arise from the first observation. Any objectivity, duly verified, belies the first contact with the object. It must first criticize everything: sensation, common sense, even the most constant practice [...]

Thus,

> We must accept, for epistemology, the following postulate: the object cannot be designated as an immediate 'objective'; in other words, a march towards the object is not initially objective. (Bachelard, 2002 [1938], p. 286)

Ultimately, for Bachelard a concept has become objective 'to the extent that it has become technical or is associated with a technique of application' (Bachelard, 2002 [1938], p. 75). During this process epistemological obstacles rooted in our subjective experience are increasingly overcome. For him, objectivity is built through 'a set of corrected errors' [*un ensemble d'erreurs rectifiées*]: 'a psychology of the attitude objective is a history of our own errors' (Bachelard, 2002 [1938], p. 285). In his view, the fact that the alchemists described their procedures in terms of desires, copulation and marriages shows that they had not overcome certain epistemological obstacles. Unlike Metzger, Bachelard viewed the animist theories not as rational but as 'pre-scientific' and primitive, i.e. rooted in emotions or desires. We can schematize the process of the emergence of objectivity as an arrow that goes from subjective and intuitive experience to objective discovery, or from inaccuracy to precision, or from 'the science of the solitary' to 'the socialized science' (Bachelard, 2002 [1938], p. 289). At some points of this process radical revisions of the conceptual foundations of scientific knowledge have happened and can happen, although the past can be transformed rather than completely rejected.

Bachelard looked at the process of objectification from two different perspectives. From a historical perspective, it is possible to distinguish three stages of development (Bachelard, 2002 [1938], p. 9): the pre-scientific era [*état préscientifique*], from Antiquity to the eighteenth century; the scientific era [*état scientifique*], from the eighteenth to the beginning of the twentieth century; and the era of the new scientific mind [*l'ère du nouvel esprit scientifique*], from 1905 onwards. The pre-scientific era is characterized by an understanding of the scientific object that is in continuity with appearances and a way of thinking rooted in imagination and preconcepts. The scientific era is marked by the divorce with common knowledge and an increased abstraction. However, realism remains an epistemological obstacle not yet overcome and the human mind is still confident in the first and basic truths and dependent on a Cartesian epistemology of intuition and simple natures. The era of the new scientific mind is permeated by growing abstraction, sophistication of technology, abstract formalism and calculation.

From the perspective of the human mind in its individual development (Bachelard, 2002 [1938], p. 11), Bachelard distinguished 'three stages of scientific thought' (p.11): the concrete stage [*état concret*] in which the human mind 'enjoys the first images of the phenomenon and relies on a philosophical literature glorifying Nature, singing curiously both the unity of the world and its rich diversity' (p.11); the concrete-abstract state [*état concret-abstrait*] in which 'the mind is still in a paradoxical situation: it is all the more certain of its abstraction as this abstraction is more clearly represented by a sensitive intuition'; (p.11); and the abstract state [*état abstrait*] in which 'the mind uses information obtained from the intuition of real space, voluntarily detached from immediate experience and even in open polemic with the primary reality, always impure, always formless' (p.11).

The way of thinking of the new scientific mind contrasts with that of the pre-scientific era: the former is quantitative and precise, the latter is qualitative; the former is a collective activity, the latter depends on mood, desires and subjective quirks; the former systematizes by making use of laws and theories and by developing notions which are absolutely not reminiscent of immediate experience, the latter simply seeks to unify all the known facts. On the other hand, the way of thinking of the scientific era [*état scientifique*] is still rooted in our common intuition and makes use of notions that need to be revised. It is only a step forward in the construction of objectivity. The crucial point is that, for Bachelard, the way of thinking adopted from 1905 onwards is the sole way of thinking that makes it possible for the emergence of objectivity. It is a way of thinking that differs radically from that of pre-scientific era since it is produced by an epistemological break with immediate experience.

Although Bachelard did not use this terminology, we may say that the era of the new scientific mind is characterized by a new way of thinking that makes possible the emergence of objectivity. Vice versa we may say that objectivity requires the emergence of a 'special' way of thinking that marks a break with sensitive knowledge. Thinking in this new way involves the establishment of norms and standards of truth, the introduction of quantitative determinations into the statements of physical laws and also a new way of doing, i.e. the use of instruments and the conducting of experiments according to certain standards and methods. And, above all, thinking in the new way of the era of scientific mind means to rely on different categories of thought since the theory of relativity emerged with the consequent rejection of all the basic concepts of classical physics: as Bachelard put it: 'contemporary science, in inviting [the mind] to a new form of thought, conquers for it a new form of representation, and hence a new world' (Bachelard, 1940, p. 122) For him, epistemological breaks like this were the consequence of a 'new scientific mind', a way of thinking that had emerged and superseded the 'old Newtonian mind', another way of thinking that was proved not to be eternal and universal. We can also say that the way of thinking of the scientific era replaced the old Kantian a priori of the pre-scientific era. Ultimately, in the mind of Bachelard a way of thinking of an era plays the role of Kant's a priori, involving certain categories of thought, standards of truth and ways of doing such as the use of certain instruments.

It is important to note the notion of 'era' (with its relative three-stage model) can also be viewed as a framework that helps us to highlight some differences between old and new ways of thinking. These differences are crucial to understand how and why objectivity emerges or changes. The notion of 'pre-scientific era' highlights features of human ways of thinking that allow the emergence of images in individual minds. That of 'scientific era' foregrounds the beginning of the use of instruments and therefore an increasing socialization which allows 'a certain degree of objectivity'. Finally, the notion of an era of the new scientific mind calls attention to the use of instruments as an extension of the mind, which shatters intuitive and personal inadequacies and makes possible the emergence of objectivity. As Bachelard put it 'in the proportion where a science becomes social [...] it conquers its objective bases' (Bachelard, 2002 [1938], p. 291).

2.6 Koyré's 'Types of Thinking'

In the works of Alexandre Koyré the concept of 'ways of thinking' is never the focus of discussion. However, his analyses of the transformations from one conceptual system to another, and his comments on the changes of the patterns of our thinking in the course of the history of science, provide clear indications towards his particular notion of this concept. The first period of his activity was oriented towards research into the history of religious ideas, with contributions on Anselm of Aosta (1033–1109), René Descartes (1596–1650), Paracelsus (1493–1541) and Jacob Boehme (1575–1624). He then extended his research to the history of the scientific ideas. Starting in 1922, he became friends with Meyerson, to whom he dedicated his first book on the history of science, *Études galiléennes* (1939). Like Meyerson, Metzger and others French scholars, he refused the positivist orientation which wanted to limit itself to a description of phenomena and their laws. For him, the history of scientific theories could not be separated from that of social and religious ideas.

His background, experiences and even nationality differed from all other French historical epistemologists, with many of whom he had intense intellectual exchanges. Lévy-Bruhl's research on primitive mentality and Brunschvicg's works aroused his interest strongly. He also directed the academic journal *Recherches philosophiques* to which several French philosophers collaborated, including Bachelard. Metzger, who had attended his lectures for five years, belonged to the circle of French thinkers with whom he was in direct contact (see also Zambelli (2016, p. 248) and Chimisso (2019a, pp. 235–236)). Koyré shared with her and many other historical epistemologists the belief that there are different ways of thinking which can be revealed by studying the texts of the past. In works such as *A Documentary History of the Problem of Fall from Kepler to Newton* (1955), *La révolution astronomique* (1961) and *From the Closed World to the Infinite Universe* (1968 [1957]), he compared sources from different periods in order to bring to light the differences between patterns of thought and the deep intertwining between scientific research and metaphysical speculation. He presented the scientific revolution as a discontinuity in the history of science that wiped out the medieval conception of an ordered cosmos and replaced it with an infinite universe.

The stated goal of his research was to understand how some concepts appear absurd in certain epochs and plausible in others: he observed that 'the principle of inertial motion, which to us appears so clear and even self-evident [...], for the Greeks as well for the thinkers of the Middle Ages [...] appeared as evidently false and even absurd'; and added that 'we have to understand why *modern physics* was able to adopt this principle' (Koyré, 1943, p. 403). In our terminology, we can express this goal by saying that Koyré wanted to understand how certain concepts, which were unthinkable, or were simply considered absurd, became objective in a certain epoch, i.e. how they came to be used, shared and discussed by a community of people. This goal is comparable to that of Metzger, who wanted to detect those concepts of a certain era that allowed people to organize their experience. But while Metzger focused on the history of chemistry and crystallography, Koyré wanted to study the conceptual changes that led to the Scientific Revolution. To this aim, he introduced the concept of 'type of thinking':

> the history of the scientific thought may be, *grosso modo*, divided into three stages, which correspond to three *different types of thinking*: the Aristotelian physics; the physics of *impetus*, inaugurated, like everything else, by the Greeks, and elaborated in the current of the fourteenth century by the Parisian nominalists; and finally modern Archimedean or Galilean physics. (Koyré, 1943, p. 406 my emphasis)

In other passages, when referring to the same concept, instead of 'type of thinking', Koyré used phrases such as 'framework of thinking' (Koyré, 1968 [1957], p. VII) or 'worldview' (Koyré, 1968 [1957], p. VIII) and even 'mentality' (Koyré, 1968 [1957], p. 348). In his work, the concept of type of thinking served some interconnected roles. According to him, what made the principle of inertia possible was the geometrization of space and the replacement of the idea of a hierarchically ordered finite world-structure by that of an infinite universe governed by the same laws (Koyré, 1968 [1957], p. VIII). There could not be a principle of inertia without these two circumstances. Thus, his task became that of constructing an explanatory narrative that showed the conceptual changes that led to the 'infinitization of the universe' and 'the geometrization of space'.

The concept of 'type of thinking' served as a framework, i.e. a guide for constructing the explanatory narrative, contained in *From the Closed World to the Infinite Universe*, of what made the inertial principle possible. Indeed, to begin with, the framework of 'type of thinking' divided the history of scientific thought into three periods, as suggested by the passage above. Furthermore, it helped to select certain historical facts: for example, the disregard of Aristotelian scholastics within Italian Renaissance humanism, the rediscovery of Lucretian cosmology by Giordano Bruno (1548–1600) and Nicholas of Cusa (1401–1464) (Koyré, 1968 [1957] chapter I) or the invention of the telescope (Koyré, 1968 [1957] chapter IV) were facts that led to the emergence of a new Galilean physics. In addition, the framework of type of thinking helped to order different organizing concepts: for instance, divine hierarchy and the structure of the heavens belong to the Aristotelian type of thinking, while geometric space and exact measurement pertain to the modern type of thinking. Finally, the framework of type of thinking also served as a guide for an interpretation of the history of science:

> when we study a thought which is not ours anymore [...] it is necessary not only to forget certain truths which have become part of our thought but also to adopt certain modes of thought, or at least certain metaphysical principles that for men of a past epoch were bases of reasoning. (Koyré, 2004 [1933], p. 10)

For Koyré, the Scientific Revolution meant first a change of categories of thought as well as attitudes, beliefs, implicit assumptions—it was an *intellectual* revolution. In the preface of *From the Closed World to the Infinite Universe* (1968 [1957]) he described 'the changes brought forth by the revolution of the seventeenth century' (p. VIII) as a 'spiritual change' (p. VIII) and added that they were reducible to: (1) 'the substitution for the conception of the world as a finite and well-ordered whole [...], that of an indefinite and even infinite universe [...] unified only by the identity of its ultimate and basic components and laws' (p. VIII) (2) the emergence of a new organizing concept of space, i.e. 'the replacement of the Aristotelian conception of space—a differentiated set of innerwordly places—by that of Euclidean geometry' (p. VIII). In turn, this 'spiritual change' made it possible for the emergence of new questions, theories, methods of measurement and concepts, such as the principle of inertia.

The above considerations suggest that for Koyré, as well as for Metzger, a way of thinking is mainly constituted by a system of organizing concepts peculiar to a certain epoch. Indeed, both these thinkers had been influenced by the ideas of 'collective representations' and 'prelogical thinking' elaborated by the sociological school of Durkheim and by Lévy-Bruhl. As

Paola Zambelli (1995) has documented, collective representations were part of Koyré's method from his earliest studies on religion. However, there is an important difference between Metzger's and Koyré's notions of the concept of 'ways of thinking'. The former believed that prelogical thinking was not the exclusive prerogative of primitive people. Incoherence, contradiction and, in particular, active analogy were characteristic of spontaneous thinking in every epoch and culture. Koyré, on the contrary, believed that logical thinking was not the exclusive prerogative of modern societies. He was absolutely convinced of the invariability and immutability of the basic structures of human thinking, i.e. of the identity of rules and categories of logic. Commenting on an example provided by Lévy-Bruhl, the story of certain African social groups who believed that some human beings were leopards at the same time, and therefore should not be hunted [Lévy-Bruhl 1927, p. 204], Koyré wrote that the arguments used by primitive peoples were perfectly logical and sound, although the identification of man with a leopard is absurd for us (Koyré, 1930, p. 2299). For him, there were no cultures that were illogical, although an individual might commit logical mistakes and be illogical. He added that

> Nothing, really, is more unstable than the 'truths' that are accepted and believed in various times and in different social groups [...] But, inside their mentality and their beliefs, the scholastic and the scientist think in the same way. Material differences let subsist the formal identity of thought [...] The scholastic thought uses the same logical categories as we do, but it applies them to different facts and in different ways [...] (Koyré, 1926, p. 466)

That is, while organizing concepts, beliefs, attitudes and preconceptions ('mentality') may change from one social group to another; the principles of logic are invariant ('formal identity of thought'). In other words, different ways of thinking share the same logical rules although they are characterized by different sets of organizing concepts.

Another important difference between Metzger's and Koyré's notions of the concept of 'ways of thinking' concerns the transcendence of thought with respect to *praxis*. For Metzger, the practical activity of medics and pharmacists has been crucial for the birth of chemistry as thinking and doing played inseparable roles in the process of production of knowledge. Koyré, on the contrary, attributes to theory a superiority over practice both in terms of the value of explanation and the source of creation. In several essays,[7] Koyré argued that the influence of external factors to *theoria* invoked by historians is illusory. It was not the appearance of the cannon that caused the birth of a new dynamic, and it was not the needs of navigation or astrology that overturned our conception of the heavens. Greek technique remained at the level of technique since the precision of the heavens did not 'descend' to the earth. However, by mathematizing nature, the type of thinking of the modern world intertwined a physical theory with a technological theory, thus laying the foundations for a development of machinism. Ultimately, according to Koyré, types of *thinking*, rather than of types of thinking ***and*** doing, play the main role in the production of scientific knowledge. A type of thinking can be viewed as constituted by two layers: an invariable layer of logical categories and rules and a variable layer of organizing concepts and implicit assumptions. Once this latter layer changes, it makes possible the emergence of new concepts, questions and theories.

2.7 Fleck's 'Thought Styles'

Ludwik Fleck was another thinker who posed the question as to how facts and concepts emerge as objective for a community of people. He published his main work *Entstehung und Entwicklung einer wissenschaftlichen Tatsache* [*Genesis and Development of a Scientific Fact*] in 1935, but in 1927 he had already published a work on the concept of a medical way of thinking (Fleck, 1986 [1927]) in which he also mentioned the problem of incommensurability. These works only found a broader reception only 50 years after they were written. Both the title *Genesis and Development of a Scientific Fact* and the opening question 'What is a fact?' (Fleck, 1979 [1935], p. XXVII) make it clear that explaining the possibility of objectivity was also Fleck's goal. For him, the study of the emergence of facts and concepts requires a historical investigation. Facts emerge from a ternary and always changing relation between the knower, the object to be known and a 'missing component' (Fleck, 1979 [1935], p. 38), the 'thought style', which is described by these words:

> If we define a 'thought collective' as a community of persons mutually exchanging ideas or maintaining intellectual interaction, we will find by

[7] See for example: 'Les philosophes et la machine' in *Études d'histoire de la pensée philosophique* (Koyré, 1981 [1961]).

> implication that it also provides the special "carrier" for the historical development of any field of thought, as well as for the given stock of knowledge and level of culture. This we have designed thought style. (Fleck, 1979 [1935], p. 39)

In other words, the framework of thought style, which presupposes the cognate concept of thought collective, was introduced by Fleck in order to take into account an element that is crucial for explaining the emergence of a fact. Indeed, if a fact is described by statements such as 'someone recognizes something', then:

> The statement 'Someone recognizes something' [...] is no more meaningful as it stands than the statements 'This book is larger' [...]. Something is still missing, namely the addition, 'than that book', to the statement. Analogously, the statement 'Someone recognizes something' demands some such supplements as 'on the basis of a certain fund of knowledge' or, better, [...] 'in a particular thought style'. (Fleck, 1979 [1935], p. 38)

For Fleck what makes certain propositions meaningful is a thought style, just like for the thinkers examined above what makes certain concepts objective is the emergence of a 'way of thinking', whatever they mean by this phrase. Once a proposition is meaningful, those who have the same thought style can discuss it to ferret out the sources of their disagreements ('interactive objectivity') or agree on its truth-value ('concordant objectivity'). Kuhn wrote that 'what the thought collective supplies its members is somehow like the Kantian categories, prerequisite to any thought at all' (Kuhn, 1976). Indeed, Fleck referred to Kant by quoting the philosopher Wilhelm Jerusalem (1854–1923): 'Kant's belief in a timeless, completely immutable logical structure of our reason [...] has not failed [...] but proved to be definitely erroneous' (Fleck, 1979 [1935], p. 47).

In the few years that preceded the publication of Fleck's book *Genesis and Development of a Scientific Fact*, the scientific community of microbiologists debated how to interpret photographs obtained by shining light on bacteria positioned on light-sensitive film: is the image of the object an observation or an artificial creation? What should a bacterium look like? As a microbiologist, Fleck was involved in those discussions and even presented his own considerations at conferences (see Zittel, 2012, p. 69). He realized that 'to see [means] to recreate a picture, at a suitable moment created by the mental collective to which it belongs' (Fleck, 1986 [1927],

p. 78) and that a lay person looking through the microscope would not be ready to perceive what the expert considered to be the object of observation.

It is plausible that reflecting on similar experiences led him to the idea that a good explanation of how a fact comes about should necessarily appeal to *sociological* factors; indeed, already in 1929 he wrote that 'any new epistemology must [...] be brought into a social and historical context' (Fleck, 1986 [1927], p. 48). Later, Fleck quoted sociologists such as Durkheim and Lévy-Bruhl who recognized 'the importance of sociological methods' and highlighted 'the force exerted on the individual by social structures' (Fleck, 1979 [1935], p. 46) (see also Jean-Francois Braunstein, 2003, p. 410). Furthermore, Fleck's experiences could have also led him to conceiving of a thought style, among other things, as a way of perceiving the world: microbiologists perceive bacteria and cells in their own way just as astrophysicists have their own way of perceiving stars and planets. Within a thought collective, say that of microbiologists, physicists, theologians or astrologists, all the members understand each other very well but, between members of different thought collectives, there is no understanding. Each way of perceiving the world relies on the use of particular instruments and methods as well as criteria of evidence. For example, microbiologists can perceive the images of bacteria by examining light-sensitive films according to certain shared rules of thumb.

As documented by Claus Zittel, the debate on the concept of thought style that took place in 1939 between Fleck and the historian Tadeusz Bilikiewicz (1901–1980) showed that the former 'sought to emphasise the peculiarities of individual thinking styles and to describe their differences' (Zittel, 2012, pp. 69–74). This means two things: Fleck did not want to appeal to the existence of general rules of discourse that are valid for all areas of society in a given epoch or to pay attention to forms of reasoning. What he invoked was a narrow understanding of thought styles: different domains of science or communities of scientists can have distinct ways of thinking. Furthermore, for him it was crucial to compare different thought styles. From their comparative analysis he expected to find a continuous development of knowledge:

> Thoughts pass from an individual to another, each time a little transformed [...] Words which formerly were simply terms become slogans [...] They no longer influence the mind through their logical meaning—indeed, they often act against it—but rather they acquire a magical power and exert a mental influence simply by being used. (Fleck, 1979 [1935], p. 42)

In conclusion, Fleck put forward a notion of the concept of 'ways of thinking', named 'thought style', which replaced Kant's a priori. It is through some initial knowledge provided by the thought style that certain statements can acquire a truth-value and become objective. A thought style involves the use of certain methods, criteria of evidence and it is also a way of perceiving the world. In order to provide an answer to the question of how facts emerge, Fleck appealed to certain explanatory expectations: an epistemology sociological in character, a minor role played by broad rules of discourse and a continuous process of knowledge. The concept of 'thought style' embodies Fleck's explanatory expectations: a 'thought style […] calls for a sociological method in epistemology' (Fleck, 1979 [1935], p. 64); it must be understood in a narrow sense, and it is conceived as continuously changing. How do these features act as principles of selection that shape Fleck's narrative? *Genesis and Development of a Scientific Fact* can be divided into two parts: the first is a history of the conception of syphilis over 500 years; the second one focuses on the origin of the Wassermann test between 1906 and 1936. The fact that in the Renaissance people believed that a conjunction of planets could cause syphilis is foregrounded; but historical events such as crucial experiments are backgrounded to obey explanatory expectations that are sociological in character: Fleck held that religious thought had 'a persuasive effect upon then-current research' (Fleck, 1979 [1935], p. 2), so discoveries were not primarily the result of experiments. The idea of corrupted blood, which slowly undergoes modifications, answers to a continuist stance, and the very choice of Wassermann's reaction as a central subject is designed to show the influence of thought styles; in turn, it selects certain facts leading to its discovery. Finally, events such as the discovery of a causative agent and the attendant epistemological consequences are compared with previous events and ideas in different thought styles in order to show that their social significance is limited compared to other sociological circumstances.

2.8 Foucault's 'Epistemes'

Among the factors of great significance for the development of Foucault's thought there was the tradition of classical historical epistemology represented by Canguilhem. The latter sponsored his doctoral thesis on the history of madness and supported him throughout his career. More importantly, Canguilhem's historical epistemology (together to that of

Bachelard) provided Foucault with a strong sense of the historicity of concepts and of the discontinuities in the history of science. Structuralist linguistics and psychology developed, respectively, by Ferdinand de Saussure (1857–1913) and Jacques Lacan (1901–1981), were also factors of great significance in works such as *Les mots et les choses* [*The Order of Things*] (1994 [1966]), which is particularly relevant in this context. In the latter work he developed the concept of 'episteme', which I am going to present as Foucault's notion of the concept of 'ways of thinking'. In *L'archéologie du savoir* [*The Archaeology of Knowledge*] (1972 [1969]), he showed that the units that form the object of the discourse—medicine, political economy and biology—constitute autonomous ensembles, of which it is possible to grasp the structures, the rules of transformation and the discontinuities that have made possible the conditions for their appearance and configuration.

In the preface of *Les mots et les choses*, Foucault described the goal of his project—identifying the 'conditions of possibility' for thought in a certain period—in Kantian language:

> It is an inquiry whose aim is to rediscover on what basis knowledge and theory become possible; within what space of order knowledge was constituted; on the basis of what historical *a priori*, and in the element of what positivity, ideas could appear, sciences be established, experience be reflected in philosophies, rationalities be formed. (Foucault, 1994 [1966], p. XXII)

Here, the historical a priori represents the conditions of the possibility of positivity—the possibility that sentences, ideas, concepts and theories emerge as candidates for truth or falsehood and, possibly, become objectively true (or false):

> [W]hat I mean by the term is an a priori is not a condition of validity for judgements, but a condition of reality for statements. It is not a question of rediscovering what might legitimize an assertion, but of freeing the conditions of the emergence for statements, the law of their coexistence with others, the specific form of their mode of being [...]. (Foucault, 1972 [1969], p. 127)

The a priori of a given period is the 'episteme': 'the epistemic field [...] in which knowledge [...] grounds its positivity and thereby manifests a history' (Foucault, 1994 [1966], p. XXII). To summarize, Foucault's aim stated in the first passage is that of studying how objectivity is possible and the concept of episteme is the framework introduced for achieving it. The citation of Ulisse Aldrovandi (1522–1605) that 'the human face, from afar, emulates the sky' (Foucault, 1994 [1966], p. 19) is one of the many examples of propositions that were candidates for being true or false only in the episteme of the Renaissance. Similarly, as we shall see in the next chapter, Hacking argued that Paracelsus's sentence 'mercury salve might be good for syphilis because mercury is signed by the planet Mercury, which signs the marketplace where syphilis is contracted' was a candidate for truth or falsehood only in the Renaissance way of reasoning (Hacking, 2002, p. 171).

Foucault adopted explanatory perspectives in which social influences are less important than certain underlying structures of human thought. What Foucault considered explanatory of the emergence of objectivity was the existence of implicit rules that constrain the way people think at a given period. '[O]ne cannot speak of anything at any time; [...] it is not enough for us to open our eyes, to pay attention, or to be aware, for new objects suddenly to light up and emerge out of the ground' (Foucault, 1972 [1969], p. 44). Instead, Foucault claimed, an 'object of discourse' can become thinkable if it finds 'its place and law of emergence', a law which is established not by logical or linguistic formal structures but by deeper rules that limit, constrain or make possible what can be said or thought. Foucault called the attempt to bring these rules to light 'archaeological inquiry', likening his work to that of archaeologists, whose interest 'is not in the object studied but in the overall configuration of the site in which it was excavated' (Gutting, 2005, p. 34). As I have mentioned, very early in his career, he developed a strong interest in the discontinuities in the history of thought, an interest that was already evident in the works that preceded *The Order of Things*. In the latter's preface, Foucault claimed that 'archaeological inquiry has revealed two great discontinuities in the *episteme* of western culture' (Foucault, 1994 [1966], p. XXII) that occurred at the beginning of the Classical age (the middle of the seventeenth century) and of the Modern age (the beginning of the nineteenth century).

Foucault's explanatory expectations can be summarized by saying that he wanted to offer an account of human thought that was archaeological and discontinuous. These expectations determined an explanatory narrative which was not so much a *causal* narrative as a *synchronic* analysis of each episteme. Thus, the events that Foucault foregrounded are, e.g. those which reveal the semantic web of relations in the sixteenth century, from the uses of distinct forms of similitudes by certain authors to the deciphering of different 'sympathies' between entities of the world (Foucault, 1994 [1966] chapter 2). The continuity of the narrative is provided by changing the focus of the analysis from one episteme to another and, above all, by the set of contrasts and comparisons that pairs thinkers of different epistemes, through which Foucault sought to reveal deep historical discontinuities. An example is the comparative analysis of the thought of Georges Cuvier (1769–1832) and Jean-Baptiste Lamarck (1744–1829), in which Foucault reversed the traditional assessments of these two thinkers (Foucault, 1994 [1966], pp. 263–279).

There is a substantial difference between Foucault's and Fleck's explanatory expectations: the latter believed that how people think in a certain period may essentially determine the objectivity of a fact; the former held that what is objective is made possible by the existence of implicit rules outside people's awareness. Fleck was a constructivist in that he believed that not only scientific knowledge, but also its objects (e.g. syphilis) are the product of a community's way of thinking. By introducing his framework of thought style he wanted to provide an account of how 'objective reality can be resolved into historical sequences of ideas belonging to the collective' (Fleck, 1979 [1935], p. 40). Therefore, Fleck's 'history of objectivity' is radically externalist because for him the thought collective constructs scientific facts, which entirely have a social content.

On the other hand, one might say that the early Foucault of *Les mots et les choses* (1966) provided a 'history of objectivity' ***internal*** to human thought: he did not want to explain how biological, economical or geographical factors influenced the thought of an epoch. Rather he wanted to identify, within that very thought, the set of rules and constraints that made certain concepts thinkable. These considerations regarding Foucault only apply to the explanatory expectations behind his framework of episteme. When one looks at the other works of Foucault, which are outside the scope of this book, one should conclude that Foucault is an externalist: in 1978, he made it clear that 'the development of scientific knowledge' can be understood in relationship to the 'changes in the ways power operates' (Foucault, 2001, p. 533).

Ultimately, what is peculiar about Foucault's notion of the concept of 'ways of thinking' (episteme) is his idea that there are implicit rules that restrict the range of thought of people of a certain epoch. The point stressed by Foucault is not so much that these rules act as a Kantian a priori, but that they make a given concept unthinkable in a certain epoch.

Those who follow these rules cannot even recognize or formulate them, at least not so easily as it happens in the case of the use of different categories of space and time. Indeed, these rules cannot be searched for in the constraints of mathematical formalism, of grammar and of logic, nor even in the constraints of the social and economic forces. The rules of a way of thinking must be uncovered by archaeology: by the study of the discursive traces left by the past, the structures and the internal links of a written corpus of an epoch, the practice that order the different epistemic fields.

2.9 Feyerabend's 'Styles of Thought'

It is a fact that the intellectual milieu in which Paul Feyerabend elaborated his ideas had nothing to do with that of his contemporary French epistemologists such as Foucault. Yet, as I will explain, in the course of his philosophical trajectory he moved from a defense of the critical spirit of science with clear Popperian influences to a ferocious critique of scientific thought. And during this process of radicalization he elaborated tools and proposed ideas that have some points in common with historical epistemology. Several studies support this interpretation. In his *On Historicizing epistemology* (Rheinberger, 2010), Hans-Jörg Rheinberger included Feyerabend among the postwar thinkers who historicized epistemology. For his part, Anastasios Brenner underlined the affinities between Feyerabend's philosophy of science and those of Hacking and Foucault (Brenner, 2006, p. 119) and affirmed that his 'is a way of facing the question of objectivity: can we formulate, in a contingent situation, knowledge that is universally valid?'. To use Feyerabend's words, what we need to understand is whether 'what *has been found* in this idiosyncratic and culture-dependent way (and is therefore formulated and explained in idiosyncratic, ad hoc, and culture-dependent terms) *exists* independently of the circumstances of its discovery' (Feyerabend, 1989, pp. 393–394).

Feyerabend's thesis, entitled 'Zur Theorie der Basissätze' ('On the theory of basic statements'), was conceived under the influence of logical positivism and one of his first articles was a discussion of Wittgenstein's book *Philosophical Investigations* (1997 [1953]). In 'An Attempt at a Realistic Interpretation of Experience' (Feyerabend, 1957a) and 'Complementarity' (Feyerabend, 1958a) Feyerabend for the first time proposed an argument against positivism and in favour of a realistic view of scientific theories not unlike that of Karl Popper (1902–1944).

Furthermore, in these and other articles (Feyerabend, 1957b, 1958a, 1958b, 1962b, 1968b/1969), he argued that Bohr's interpretation of quantum mechanics had not been affected by the objections raised against it but did not rule out alternative theories.

Popper's influence is also noted in Feyerabend's early articles on what I will call the transition from myth to logos, i.e. from the mental universe of the Homeric poems to that of the early Greek philosophers: in the seventh and eighth centuries BC, the worldview of myth underwent a series of changes that led to the birth of Western philosophy and science. In 1961 Feyerabend defended the Popperian view of the emergence of rationalism by describing the Ionian philosophers as 'very bold and optimistic' and the world of Homeric poems as 'closed and dogmatic' (Feyerabend, 1961) (see also Heit (2016, p. 72) and (Preston, 1997)).

An important point raised in those years concerned the existence of some methodological rules implicit in the process of producing scientific knowledge. One of these consisted in excluding inconsistent hypotheses with well-confirmed theories or facts. This rule, Feyerabend noted, had been violated several times with good results by great scientists and was in contrast with a basic principle of empiricism that Popper had also insisted on: questioning a theory by subjecting it to the test of as many facts as possible. Preventing the possibility of alternative theories to well-confirmed ones resulted in hindering the emergence of facts potentially capable of refuting the latter. It was then necessary, according to Feyerabend, to encourage the proliferation of theories and not exclude anything, not even everything that was considered unscientific and that could provide ideas for new theories, e.g. metaphysics, myth or cosmologies of a religious nature (Feyerabend, 1962a, 1965 [1981]).

If in this phase Feyerabend was focusing on abstract and general arguments to understand why scientists used certain methodological rules and why they violated them, over time he became convinced that this was an impractical road. In *Against Method* (1975), one of his fundamental works, Feyerabend stated that 'the idea of a method that contains firm, unchanging and absolutely binding principles for conducting the business of science meets considerable difficulties when confronted with the results of historical research' (2002 [1975], p. 21). His point was that in the evolution of scientific research situations emerge that have no counterpart in the past and therefore require new methodological rules, new categories of thought, new methods and entities. As is very well known, in a semi-serious tone Feyerabend also proposed a methodology of the non-method according to the principle that 'anything goes'. This was the only methodological rule that could ever work. As science progresses, in addition to elaborating new theories, the researcher must invent new methodological rules.

Thus, like the historical epistemologists, Feyerabend came to criticize the idea of fixed and ahistorical rules. Suffice to recall that Bachelard, e.g. argued that the path of science is studded with epistemological ruptures that come into being, among other things, through the overcoming of methodological rules and old ways of thinking. But Feyerabend went further. While Bachelard believed that, in continually correcting itself, science becomes more and more rational and objective, Feyerabend concluded that science is not a rational activity, in the sense that it does not necessarily follow rules in accordance with reason. Furthermore, while for Bachelard science and non-science were rooted in different mental faculties—rationality and imagination—and therefore differed profoundly, for Feyerabend they could not be distinguished either on the basis of a method, or on the basis of rules of rationality, or on the basis of content. Any criterion of distinction between scientific and non-scientific ways of thinking can only be identified in particular contexts and conditions.

At first, Feyerabend had believed that non-scientific worldviews did not have the same status as scientific ones: they still had to acquire empirical content and their scopes had to be circumscribed. In the course of his studies he realized that non-scientific worldviews were anything but devoid of factual and cognitive content. They could therefore be considered real alternatives to scientific theories. For example, myths conveyed knowledge with the aim of promoting social harmony; the dramas of ancient Greece laid bare the contradictions of society and used particular methods to suggest the reasons. Science could therefore benefit from the comparison with myth and metaphysics which thus became an instrument of criticism (Feyerabend, 1967, 1968a, 1975).

Helmut Heit argued that 'the radicalization of the last phase of Feyerabend's thought, that is, the passage from the proliferation of theories as a means of progress in the sciences to more relativistic views on the scientific tradition as such' matured thanks to studies of myth and the emergence of rational thinking in Ancient Greece (Heit, 2016, p. 71). Indeed, in a 1963 letter to the Australian philosopher Jack Smart (1920–2012), Feyerabend revealed that he had always been interested in the nature of myths, which he considered as completely autonomous and independent ways of perceiving the world. So much so that the interest in these topics remained present in many subsequent works, from *Against the Method* (1988 [1975]) to *Conquest of Abundance* (1999) and *Philosophy of Nature* (2016 [2009]). In 1975, Feyerabend argued that the worldviews of myth and that of the early Greek philosophers were 'incommensurable' and based on two different ways of perceiving the world. The term 'incommensurable' appeared for the first time in 1962 to describe two theories that could not be deduced from each other because there were no deductive relationships between their respective basic concepts. Similarly, Feyerabend considered the worldviews of the Homeric poems and those of the early Greek philosophers to be incommensurable because their respective basic concepts 'could not be used simultaneously and no logical or perceptual relationship could be established between them' (Feyerabend, 1988 [1975], p. 169). This led to a relativistic vision that embraced the entire spectrum of ways of thinking about the world: the scientific vision of nature that arose with the Ionian philosophers had not been defeated by the argument but by history.

In short, Feyerabend gradually abandoned the idea of methodological rules. From there, thanks to the study of the transition from myth to logos, he came to question the existence of a general criterion that demarcated science from non-science. In doing so, he implicitly proposed his own notion of the concept of 'ways of thinking', with precise characteristics, as I am going to explain. First of all, in the aforementioned 1963 letter to Jack Smart, Feyerabend complained that 'for Kant and many other thinkers, myths were a matter of fantasy and had nothing to do with forms of thought' (Feyerabend, 2016 [2009], p. 212). Ernst Cassirer himself (1874–1945) had conceived myth, as well as art, language and religion, as a symbolic tool for interpreting the world, but he had not adequately recognized its cognitive content (Feyerabend, 2016 [2009], p. 212). For Feyerabend, a worldview was instead characterized by specific categories of thought which, instead of being universal and immutable, as Kant believed, could vary from one historical, social or cultural context to another.

> If it was indeed true, as Kant thought it was, that certain reaction patterns, or conceptual schemes are inborn then it would quite obviously be impossible to replace them by different points of view. Still I do not believe that this possibility, which is not yet established as a positive fact, can be used as an argument against total replacement, or at least against the initial attempt at total replacement. [...] There seems to be strong *prima facie* evidence

> *against* existence of insuperable conceptual schemes. [...] I think we shall have to admit, following a careful analysis of the structure of the worldviews which have been connected with different cultures, that they are genuine alternatives, genuinely different manifest images. (Feyerabend, 2016 [2009], pp. 211–212)

Feyerabend tried to make this statement plausible by analysing both the differences between the way of thinking of the mythological writers and that of the first Greek philosophers as well as the differences between art and science. From these studies it is possible to deduce what exactly a way of thinking is for Feyerabend or, to use other expressions of his terminology with the same meaning, a 'style of thought' (Feyerabend, 1983b, pp. 163–189) or a 'form of life' (an expression borrowed from Wittgenstein and used, for example in (Feyerabend, 2016 [2009], p. 3)).

In the preliminary note to *Philosophy of Nature*, referring to the transition between myth and logos, Feyerabend wrote:

> It is interesting that in science, myth, and philosophy we have three different ways of comprehending the world, which are *complete* in that they include both *objectives* and *methods* to achieve these objectives as well as *criteria* to determine success and failure. Ideally, the *transitions* from one state to the next, which from now we can reconstruct in detail only for the occidental part of the world, bring with them not just new *ideas* and *methods* but also new *objectives* and new *perceptions*, which unite to form new perceptual worlds. (Feyerabend, 2016 [2009], pp. 3–4)

We can paraphrase this passage by saying that the emergence of a novel style of thought is characterized by some novelties: new ideas, new methods, new objectives, new perceptions and new criteria of objectivity. The case of the transition from myth to logos, on which Feyerabend dwells, is exemplary. In fact, when the rational thinking style of the first philosophers took hold, it brought with it a 'movement tending towards more abstract and schematic concepts' (Feyerabend, 1983b, p. 140). Thus, if the myth explained God with a series of episodes rich in details, the new style of rational thought replaced this richness of description with 'a concept in which one speaks only of power, or of being' (Feyerabend, 1983a) (Feyerabend, 1983b, p. 141). Thus, new ideas or concepts emerged, such as that of 'soul'. In fact, a man of Homer's time did not have a unified concept of what we call 'soul' or 'personality' (Feyerabend, 2016 [2009], p. 71). In Feyerabend's terminology, that was the transition from a 'universe of aggregates' to a 'universe of substances', one in which a concept was replaced by an aggregate of events, actions, descriptions.

Feyerabend espoused an interpretation of myths according to which 'theories and primitive methods have the same *goal*: to collect and causally explain salient appearances' (Feyerabend, 2016 [2009], p. 40). His idea was that 'the strange events described in the myths, as well as the strange creatures with which they populate the world, are really perceived in it' (Feyerabend, 2016 [2009], p. 38). In other words, he believed that a style of thought represents the world precisely in the way it is seen and experienced by those who adopt it. Thus, in the mythological style of thought 'the world really appears to this early thinking as a "You" not as an "It", the sky as a "picture book" rather than a "computation book"' as in the style of the rational thought (Feyerabend, 2016 [2009], p. 38).

Finally, the method used by mythological writers to represent the world differed from that of the rational way of thinking. This point can be explained by a comparison: 'whereas Euclidean science used circles, squares, lines and points, the inventors of myths [...] used a story represented in pictorial images' (Feyerabend, 2016 [2009], pp. 37–38). Just as a physical theory introduces a model without being interested in all aspects of what it describes, so a myth can use social or zoological episodes with the purpose of illustrating only some general cosmological structures (Feyerabend, 2016 [2009], p. 38). In this sense, if the Greek deities were an inseparable part of the mythological way of thinking, the method of representation of the rational thinking style introduced new abstract entities such as circles, squares, lines and points, which were in use in the geometric proofs of Greek mathematicians. The trend towards more abstract concepts resulted in:

> The achievement of a sort of independence [...] the discovery that with their help it was possible to tell new kinds of stories, so to speak new sorts of myths, with surprising traits. The development of these new myths was no longer subject to the external constraint of a tradition, but was regulated from within, 'followed' by the nature of things. (Feyerabend, 1983a, 1983b, p. 141)

Feyerabend alluded to the discovery of the argument, which had its apogee in the demonstration (Netz, 1999). Between 410 and 360 BC the Greek mathematical texts essentially consisted of diagrams, i.e. graphic representations of geometric figures, of letters, e.g. those placed at the vertices of a triangle, and of words. On the logical level these three elements were combined to form logical-deductive chains that provided knowledge of general validity. Thus, a 'sort of independence' was achieved, a 'development regulated from within' that 'follows from the nature of things', as Feyerabend stated: from the starting points, conclusions were drawn that were both necessary and general.

> The discovery of stories that already tend in themselves to a certain end now made a much better means available to the defenders of a parochial limitation: demonstration (or argumentation). What is demonstrated is not imposed on the disciple from the outside, but follows from the nature of the thing itself. It is no longer the pedagogical methods of a tradition, which are always tainted by the accidentality of history, but the things themselves that point the way and in an 'objective' way, independent of the causally present points of view. For the intellectuals of ancient Greece, a new and extremely fruitful possibility arose of finding one and only one truth in the contrast of traditions. (Feyerabend, 1983a, 1983b, p. 143)

The demonstration, and more generally the rational argument, is therefore the criterion of objectivity that is affirmed with the emergence of the Greek philosophers' style of thought, the one on which a community of human beings in a certain historical era finds consensus.

The 'birth of rationalism' is not a transition from a closed to an open worldview or from one that is the fruit of imagination to another that is the only 'correct' one. On the contrary,

> Science is much closer to myth than a scientific philosophy would admit. It is one of the many forms of thought that have been developed by man, and not necessarily the best. It deserves consideration, it is noisy and impudent, but it is inherently superior only to those who have already decided in favor of a certain ideology, or those who have accepted it without having examined its advantages and limits. (Feyerabend, 1983a, 1988 [1975], p. 249)

These forms of thought can coexist: 'philosophy, science and myth sometimes coexist in peace side by side, other times they deprive each other of the right to exist' (Feyerabend, 2016 [2009], p. 3). In summary, Feyerabend's notion of the concept of 'ways of thinking' is characterized by a criterion of objectivity, a way of perceiving the world, an ontology, as well as by certain ideas, methods and objectives. Science itself has its ways of thinking. The example discussed by Feyerabend concerns the transition from the Aristotelian style of thought to the Galilean one. The latter brought with it not only new ideas but also new criteria. The conclusion is then that science is not distinguished from art. Indeed,

> in art there is neither progress nor decadence, but only different styles. Each style is complete in itself and obeys its own laws. Art is the production of styles and the history of art is the history of their succession. This conception was founded and developed with great clarity by Alois Riegl in 1901. (Feyerabend, 1983a, 1983b, p. 115)[8]

But, for Feyerabend, this statement is also true in the case of the sciences:

> They too have developed a number of styles, including verification styles, and the evolution from one style to another is completely analogous, let's say, to the evolution of Art from Classical Antiquity to the Gothic style. (Feyerabend, 1983b, p. 154)

There is consensus on certain criteria of objectivity only for those who adopt a certain style of thought. Since none of these criteria can be evaluated based on external criteria, the very notion of objectivity becomes void of value. So much so that in his autobiography (Feyerabend, 1995, p. 179) Feyerabend revealed that one of the 'motives for writing Against Method was to free people from the tyranny of philosophical obfuscators and abstract concepts such as "truth", "reality", or "objectivity", which narrow people's vision and ways of being in the world'.

2.10 Kuhn's 'Paradigms'

Kuhn's *The Structure of Scientific Revolutions* (Kuhn 1996 [1962]) is widely presented as the book that produced a decisive transformation in the predominant image of science as exhibiting progress and methodological unity. However, as we have seen, at the beginning of the twentieth century the attempt to identify the credentials of scientific knowledge had

[8] Feyerabend is referring to the book *The Late Roman Art Industry* (Riegl, 1985 [1901]) by the art historian Alois Riegl (1858–1905). The latter regarded the transition from a style to another as a change in the creative intention of the artist as a consequence of a mutation of the worldview. As I have explained in the first chapter, this book played an important role in the development of Mannheim's notion of 'style of thought'.

already been set aside and a space for a historicization of epistemology had already been opened up. This considered, there is no doubt that through the influence of his major work, *The Structure of Scientific Revolutions* (1962), the fruitful interconnection between history and the philosophy of science reached its acme in the philosophy of science. He argued that the history of science is characterized by an alternation of phases of 'normal science' and phases of 'revolutionary rupture'. In the phase of 'normal science', the fundamental theoretical assumptions, i.e. the paradigms, are continuously articulated and extended as a function of an ever more exhaustive application. However, in the course of this activity, inevitable 'anomalies' arise, which 'paradigmatic' science will first try to reabsorb, but which will ultimately determine the 'revolutionary rupture' and the elaboration and acceptance of a new paradigm. Then, the consolidation of the new paradigm will open a new phase of 'normal science'. In his later writings, such as *The Essential Tension* (T. Kuhn, 1977), he attenuated the irrationalist and relativistic consequences deriving from the thesis of the 'incommensurability' of theories belonging to different paradigms, even if continuing to emphasize the function of sociological and psychological aspects and, on the other, to oppose an image of scientific activity as a progressive and cumulative acquisition of truth.

Kuhn too invoked the Kantian categories and attributed to his framework the role Fleck assigned to that of thought style: 'My structured lexicon resembles Kant's a priori [...] Both are constitutive of *possible experience* of the world, but neither dictates what that experience must be' (Kuhn, 1962, p. 331). To clarify, it must be remembered that, although Kantian transcendental conditions make it possible for us to have objective knowledge of the world, their role does not consist in making our representations true but in making it possible that we can represent anything at all (accurately or not). The quotation above states that the notion of paradigm plays the same role (the fact that Kuhn distinguishes paradigm from the 'structured lexicon' that guides research in normal science does not make a difference for my argument). A remark of Paul Hoyningen-Huene, who quotes passages of *The Structure of Scientific Revolution*, clarifies better this point:

> Paradigms "are constitutive" of this world or nature [...] the world is "determined jointly by nature and the paradigms" [...]. Whatever paradigms are, since they are tools "through which to view nature", they must

> have some world-constitutive function […]. "World" and "nature" here coincide to some extent with what Kant calls "nature in the material sense" […]. For both Kant and Kuhn, epistemic subjects are (albeit in different ways) coconstitutive of this world […]. (Hoyningen-Huene, 1993, pp. 32–33)

However, Kant did not conceive of a priori conditions as historically variable collective forms of intuition, whereas Kuhn did: 'I go around explaining my own position saying that I am a Kantian with moveable categories' (Kuhn, 2002 [2000], p. 264). Thus, for Kant, 'what is objective is that to which all human beings—in virtue of their rational faculties of sensibility and understanding—must necessarily agree' (Friedman, 2011). For Kuhn, it is a given paradigm that 'yields generally agreed upon (although perhaps only tacit) rules definitive or constitutive of what counts as a "valid" or "correct" solution to a problem' (Friedman, 2001, p. 42). This is a form of concordant objectivity in that, within a paradigm, scientists agree on the solutions to a problem and on the reliability of certain methods. And it is also a form of interactive objectivity because the emergence of a paradigm makes it possible that certain problems, methods or exemplars are shared and considered important. In *The Road since Structure* Kuhn wrote: 'Hacking […] spoke of the way in which styles introduce into science new candidates for true/false. Since that time I have been realizing […] that some of my central points are far better made without speaking of statements as themselves being true or as being false' (Kuhn, 2002 [2000], p. 99). Indeed, when he elaborated on the concept of 'lexicon', a network of related terms associated with a taxonomy, he described it in Hacking's terms as the condition of possibility of new candidates for truth or falsehood (see Politi, 2020).

In summary, like Kant, Kuhn introduced his framework in order to answer the question of what makes it possible for something to become objective. On the other hand, unlike Kant, Kuhn believed that objectivity comes into being through specific historical processes. Therefore, the notion of paradigm will necessarily have to be different from the Kantian a priori—it must be a 'moveable category', suitable for dealing with a historicized concept of objectivity. An aspect of the emergence of objectivity that Kuhn set out to investigate is related to the problem of scientific discovery. What he wanted to show was that the existence of an entity becomes objective only when certain circumstances happen to exist:

> A discovery like that of oxygen or X-rays does not simply add one more item to the population of the scientist's world. Ultimately it has that effect, but not until the professional community has re-valuated traditional experimental procedures, altered its conception of entities with which it has long been familiar, and shifted the network of theory through which it deals with the world. (Kuhn, 1996 [1962], p. 7)

That is, once the scientific community agrees on new experimental procedures and conceptions of what an entity is, the propositions concerning the discovery and existence of oxygen can be assessed as true or false, becoming objectively true (or false). He also used the term 'discovery' with regard to concepts that become objective once a new paradigm emerges (see for example (Kuhn, 1987)). The concepts of force and mass in Newton's second law of motion are a case in point: for Kuhn, these concepts differed from those in use in the previous paradigm. Ultimately, the latter is the framework introduced by Kuhn to explain the emergence of objectivity.

A few brief autobiographical notes in Kuhn's writings provide some hints about the explanatory expectations that shaped his narrative. In the foreword to *Genesis and Development of a Scientific Fact*, he wrote: 'Fleck's text helped me to realize that the problems which concerned me had a fundamentally sociological dimension' (Kuhn, 1976, p. VIII). He focused on the structure of the scientific community rather than on the logical reconstruction of theories and, for this reason, he has been credited as one of the first proponents of a 'social epistemology of science' (Wray, 2011). Like Fleck, Kuhn did not appeal to the existence of general rules of discourse but rather to what unifies specific scientific communities at a certain point in history. Contrary to Fleck, however, he problematized the very concept of 'scientific community': as he noticed in the *Postscript* to *The Structure of Scientific Revolutions*, scientific communities may exist at different levels—e.g. there is the general scientific community of physicists and then, at lower levels, the specialties and sub-specialties communities. This means that changes affecting a community at the lower level may not affect the community at large (see Politi, 2018). Nevertheless, as Mößner (2011) noted, the influence of social factors is more widespread in Fleck: whereas the latter applies the framework of thought style even to non-scientific areas, Kuhn is reluctant to extend that of paradigm out of the hard sciences, not to mention to ordinary life. Consequently, unlike Kuhn's, Fleck's framework is potentially suited to including factors outside the scientific realm in the study of objectivity.

When, in 1995, Kuhn was asked about the ideas he had in mind before writing *The Structure of Scientific Revolutions*, he answered: 'I didn't know quite what it was going to look like, but I knew the noncumulativeness, and I knew something about what I took revolutions to be' (Kuhn, 2002 [2000], p. 292). Already in 1947, under the influence of works such as Koyré's *Etudes galiléennes* (1939, 1996 [1962], p. IX), Kuhn had developed the view that it is impossible to provide a continuous description of scientific development. As he explained (Kuhn, 2002 [2000], pp. 13–32), at that time he realized that Aristotle looked at the world in a manner completely different from Newton: the illusion of continuity was provided by identical words such as 'motion', which mask the stark contrast between the two different ways of thinking. This discontinuist perspective makes objectivity more the product of scientific revolutions than of a slow process of the development of past ideas, as Fleck presented. For the latter, proto-ideas and old conceptions can survive changes in thought styles, serve as heuristic guidelines for new research developments and play a role at a deep level in the emergence of objectivity.

These explanatory expectations are embodied by the concept of paradigm. The latter is the basis for the set of contrasts and comparisons in *The Structure of Scientific Revolutions*, of which some of the most important concern the Copernican, Newtonian, Chemical and Einsteinian revolutions. For each of these central subjects, Kuhn compared and contrasted the concepts, terms and experimental practices of different paradigms. It is by relying on the explanatory expectations that these historical events are selected. For example, 'noncumulativeness' is a guiding principle for foregrounding the 'scientific revolutions' of Copernicus, Newton, Lavoisier and Einstein, and it is considered explanatory of how objectivity is possible: 'in order to make [...] a discovery one must alter the way one thinks about and describes some range of natural phenomena. The discovery (in cases like these "invention" is a better word) of Newton's second law of motion is of this sort' (Kuhn, 2002 [2000], p. 15) To paraphrase, we could say that the objectivity of certain concepts is not possible until a paradigm is in place.

Kuhn saw himself as an internalist: 'the intellectual milieu reacts on the theoretical structure of a science only to the extent that it can be made relevant to the concrete technical problems with which the practitioners of the field engage' (Kuhn, 1971, p. 280). As I have pointed out, for him the 'discovery' of oxygen became objective when the scientific community reforged its conception of entities, its network of theories and its experimental procedures (Kuhn, 1996 [1962], p. 7). The framework of paradigm embodies his explanatory expectations: in periods of normal science the scientific community is insulated from 'those socially important problems that are not reducible to the puzzle form' (Kuhn, 1996 [1962], p. 37) and the objectivity of new facts and concepts is made possible by a change of paradigm triggered by a 'crisis' within science.

Under the influence of Fleck, Kuhn maintained that different paradigms use concepts that are incompatible. However, as he refined his ideas, he developed a concept of incommensurability not only limited to the idea of translation failure. He argued that incommensurability was not only caused by linguistic factors but also by differences in methods for setting up research and evaluating its results. Furthermore, under the influence of the Gestalt psychologists, he also spoke of differences of perceptions between adherents of different paradigms. It is from the collaboration of all these elements that incommensurability follows. For example, he argued that in the seventeenth century different people might have viewed in a different way the same stone swinging from a string. An Aristotelian would have seen the stone as constrained in its downward motion; an adherent of the new Galilean physics would have seen a pendulum that repeats the same motion forever. Kuhn claimed that these two conceptualizations are incommensurable because there is no neutral way to adjudicate between them (Kuhn, 1996 [1962], p. 150). Ultimately, to think in different paradigms means to see the world in different manners, as well as to use certain organizing concepts and criteria of evidence, to invoke particular models that have gained acceptance in the solution of problems within a particular field, to consider important certain questions, methods and techniques.

2.11 Conclusions

In this chapter I have presented the main notions of the concept of 'ways of thinking' that have been conceived in the wake of the tradition of historical epistemology and I have highlighted their common features. My analysis shows that the objectives of these thinkers' notions of the concept of 'ways of thinking' were to replace Kant's a priori. In other words, shedding light on how objectivity is possible is the necessary goal of their common programme of research—that of historicizing Kant's a priori. In this connection, I recall that historicized Kantianism has been crucial to the formation of French historical epistemology. Since these notions capture different channels through which objectivity becomes possible, it is possible to group them in the same genus within historical epistemology or, alternatively, we may say that all these frameworks provide a 'history of objectivity'. Furthermore, these frameworks answer to different explanatory expectations which foreground distinct episodes and periods of the history of science. It is also important to underline that, since the projects of the thinkers examined in this chapter can be viewed as studies of an organizing concept (objectivity), in this sense they fall into a field of research that today is called historical epistemology ('historical meta-epistemology' in Hacking's terminology). In the next chapter I shall present the notion of the concept of 'ways of thinking' of Hacking and that of Daston and Galison. Their analysis will complete my presentation of the different species of the genus 'ways of thinking'. In the following chapter, I shall draw some comparisons between the distinct species of the genus 'ways of thinking' I have presented in order to provide a conceptual taxonomy.

Bibliography

Bachelard, G. (1934). *Le nouvel esprit scientifique*. Presses Universitaires de France.

Bachelard, G. (1938). *La Psychanalyse du feu*. Gallimard.

Bachelard, G. (1940). *La Philosophie du non: essai d'une philosophie du nouvel esprit scientifique*. PUF.

Bachelard, G. (1941). *L'Eau et les rêves: Essai sur l'imagination de la matière*. José Corti.

Bachelard, G. (1957). *La Poétique de l'espace*. PUF.

Bachelard, G. (1960). *La Poétique de la rêverie*. PUF.

Bachelard, G. (1961). *La flamme d'une chandelle*. PUF.

Bachelard, G. (2002 [1938]). *The formation of the scientific mind. A contribution to a psychoanalysis of objective knowledge*. Clinamen Press.

Bachelard, G. (2004 [1949]). *Le rationalisme appliqué*. Puf.

Bachelard, G. (2009 [1942]). *L'air et les songes: essai sur I'imagination du mouvement*. LGF.

Bouthoul, G. (1961). *Les Mentalités*. PUF.

Braunstein, J.-F. (2003). Thomas Kuhn lecteur de Ludwik Fleck. *Archives de Philosophie, 66*(3), 403–422.

Braunstein, J.-F. (2008). Introduction. In J.-F. Braunstein (Ed.), *L'histoire des sciences. Méthodes, styles et controverses* (pp. 7–20). Vrin.

Brenner, A. (2006). Quelle épistémologie historique? Kuhn, Feyerabend, Hacking et l'école bachelardienne. *Revue de Métaphysique et de Morale, 1*(49), 113–125.
Brunschvicg, L. (1912). *Les étapes de la philosophie mathématique*. Alcan.
Brunschvicg, L. (1922). *L'expérience humaine et la causalité physique*. Alcan.
Brunschvicg, L. (1934). *Les âges de l'intelligence*. Alcan.
Brunschvicg, L. (1939). *La Physique au vingtième siècle*. Hermann.
Canguilhem, G. (1994 [1968]). *Etudes d'histoire et de philosophie des sciences concernant les vivants et la vie*. Vrin.
Castelão, T. (2010). *Gaston Bachelard and les études critiques de la science* (1st ed.). L'Harmattan.
Chimisso, C. (2019a). *Hélène Metzger, historian and historiographer of the sciences*. Routledge.
Chimisso, C. (2019b). Voyage vers le matérialisme: François Dagognet et l'évolution de l'épistémologie historique. In *François Dagognet: Philosophe, épistémologue* (pp. 25–35). Éditions Materiologiques.
Currie, A., & Walsh, K. (2019). Frameworks for historians and philosophers. *HOPOS: The Journal of the International Society for the History of Philosophy of Science, 9*, 1–34. https://doi.org/10.1086/699797
Deprez, S. (2009). *Lévy-Bruhl et la rationalisation du monde*. (Doctorat). Université catholique de Louvain, Louvain.
Douglas, H. (2004). The irreducible complexity of objectivity. *Synthese, 138*(3), 453–473. https://doi.org/10.1023/B:SYNT.0000016451.18182.91
Durkheim, É. (2008 [1912]). *The elementary forms of religious life*. Oxford University Press.
Edwards, T. (2002). *Durkheim, Kant and the social construction of the categories* (pp. 73–83). Brill.
Elwick, J. (2012). Layered history: Styles of reasoning as stratified conditions of possibility. *Studies in History and Philoophy of Science, 43*, 619–627. https://doi.org/10.1016/j.shpsa.2012.07.004
Falasca-Zamponi, S. (2014). Society as representation: Durkheim, psychology and the 'Dualism of Human Nature'. *Durkheimian Studies / Études Durkheimiennes, 20*, 43–63.
Feyerabend, P. (1957a). An attempt at a realistic interpretation of experience. *Proceedings of the Aristotelian Society, 78*, 143–170.
Feyerabend, P. (1957b). On the quantum-theory of measurement. In S. Körner & M. Pryce (Eds.), *Observation and interpretation: A symposium of philosophers and physicists, proceedings of the ninth symposium of the Colston Research Society*. London and New York.
Feyerabend, P. (1958a). Complementarity. *Proceedings of the Aristotelian Society, 32*, 75–122.
Feyerabend, P. (1958b). Eine Bemerkung zum Neumannschen Beweis. *Zs. Physik, 145*, 421–423.

Feyerabend, P. (1961). *Knowledge without foundations. Two lectures delivered on Nelli Heldt Lecture Fund*. Oberlin College.
Feyerabend, P. (1962a). Explanation, reduction and empiricism. *Minnesota Studies in the Philosophy of Science, 3*, 28–97.
Feyerabend, P. (1962b). Problems of microphysics. In *Pittsburgh studies in the philosophy of science* (Vol. I). University of Pittsburgh.
Feyerabend, P. (1965 [1981]). *Problems of empiricism: Philosophical papers* (Vol. 2). Cambridge University Press.
Feyerabend, P. (1967). On the improvement of the sciences and the arts, and the possible identity of the two. In R. Cohen & M. Wartofsky (Eds.), *Proceedings of the Boston colloquium for the philosophy of science* (Boston Studies in the Philosophy of Science) (Vol. 3, pp. 387–415). Springer.
Feyerabend, P. (1968a). Outlines of a pluralistic theory of knowledge and action. In S. Anderson (Ed.), *Planning for diversity and choice*. MIT Press.
Feyerabend, P. (1968b/1969). On a recent critique of complementarity. *Philosophy of Science, 35/36*, 309–331, 82–105.
Feyerabend, P. (1975). Lets make more movies. In *The owl of minerva*. Prentice Hall.
Feyerabend, P. (1983a). Science as art: A discussion of Riegl's theory of art and an attempt to apply it to the sciences. *Art and Text, 12–13*, 16–46.
Feyerabend, P. (1983b). *Scienza come arte*. Laterza.
Feyerabend, P. (1988 [1975]). *Against method*. Verso.
Feyerabend, P. (1989). Realism and the historicity of knowledge. *Journal of Philosophy, 86*(8), 393.
Feyerabend, P. (1995). *Killing time*. University of Chicago Press.
Feyerabend, P. (1999). *Conquest of abundance*. The University of Chicago Press.
Feyerabend, P. (2002 [1975]). *Contro il metodo*. Feltrinelli.
Feyerabend, P. (2016 [2009]). *Philosophy of nature*. Polity.
Fleck, L. (1979 [1935]). *Genesis and development of a scientific fact*. The University of Chicago Press.
Fleck, L. (1986 [1927]). Some specific features of the medical way of thinking. In R. S. Cohen & T. Schnelle (Eds.), *Cognition and fact—Materials on Ludwik Fleck*. Springer.
Foucault, M. (1972 [1969]). *The archaeology of knowledge*. Pantheon Books.
Foucault, M. (1994 [1966]). *The order of things*. Vintage Books Edition.
Foucault, M. (2001). *Dits et écrits: 1954–1988* (Vol. II, pp. 1976–1988). Gallimard.
Friedman, M. (2001). *Dynamics of reason*. The 1999 Kant Lectures at Stanford University.
Friedman, M. (2011). Extending the dynamics of reason. *Erkenntnis, 75*(3), 431–444.

Garlitz, D. (2020). Durkheim's French Neo-Kantian social thought: Epistemology, sociology of knowledge, and morality in the elementary forms of religious life. *Kant Yearbook, 12*(1), 33–56. https://doi.org/10.1515/kantyb-2020-0002
Gutting, G. (2005). *Foucault. A very short introduction*. Oxford University Press.
Hacking, I. (2002). *Historical ontology*. Harvard University Press.
Heit, H. (2016). Reasons for relativism: Feyerabend on the 'rise of rationalism' in ancient Greece. *Stud Hist Philos Sci, 57*, 70–78. https://doi.org/10.1016/j.shpsa.2015.11.007
Hoyningen-Huene, P. (1993). *Reconstructing scientific revolutions*. The University of Chicago Press.
Koyré, A. (1926). Review of Louis Rougier, «La scolastique et le thomisme». [La Scolastique et le Thomisme, Louis Rougier]. *Revue Philosophique de la France et de l'Étranger, 101*, 462–469.
Koyré, A. (1930). Review of Lucien Lévy-Bruhl, «Die Seele der Primitiven». *Deutsche Literaturzeitung, V*(XLVIII), 2294–3000.
Koyré, A. (1939). *Études galiléennes*. Hermann.
Koyré, A. (1943). Galileo and plato. *Journal of the History of Ideas, 4*(4), 400–428. https://doi.org/10.2307/2707166
Koyré, A. (1955). *A documentary history of the problem of fall from Kepler to Newton: De motu gravium naturaliter cadentium in hypothesi terrae motae*. American Philosophical Society.
Koyré, A. (1961). *La Révolution astronomique: Copernic, Kepler, Borelli*. Hermann.
Koyré, A. (1968 [1957]). *From the closed world to the infinite universe*. The Johns Hopkins University Press.
Koyré, A. (1981 [1961]). *Études d'histoire de la pensée philosophique*. Gallimard.
Koyré, A. (2004 [1933]). *Paracelse*. Editions Allia.
Kuhn. (1962). Afterwords. In P. Horwich (Ed.), *World changes* (pp. 311–341). MIT Press.
Kuhn. (1971). The relations between history and history of science. *Daedalus, 100*(2), 271–304.
Kuhn. (1976). Foreword. In *Genesis and development of a scientific fact*. The University of Chicago Press.
Kuhn. (1987). What are scientific revolutions? In L. Krüger, L. J. Daston, & M. Heidelberger (Eds.), *The probabilistic revolution*. The MIT Press.
Kuhn. (1996 [1962]). *The structure of scientific revolutions*. The University of Chicago Press.
Kuhn. (2002 [2000]). *The road since structure* (J. Conant & J. Haugeland, Eds.). The University of Chicago Press.
Kuhn, T. (1977). *The essential tension*. The University of Chicago Press.
Lévy-Bruhl, L. (1903). *La morale et la science del moeurs*. Alcan.
Lévy-Bruhl, L. (1927). *L'âme primitive*. Paris: Librarie Felix Alcan.

Lévy-Bruhl, L. (1923). La mentalité primitive. Conference of February 15, 1923 at the Société française de philosophie: https://www.sofrphilo.fr/activites-scientifiques-de-la-sfp/conferences/grandes-conferences-en-telechargement/#5-8211-sciences-sociales

Lévy-Bruhl, L. (2015 [1910]). *How Natives Think*. Martino Publishing.

Loi, M. (1984). Léon Brunschvicg et les mathématiques Séminaire de Philosophie et Mathématiques, 1984, fascicule 2 « Léon Brunschvicg et les mathématiques». *Séminaire de Philosophie et Mathématiques, 2*, 1–15.

Merllié, D. (2012). Durkheim, Lévy-Bruhl, and "Primitive Thinking": What Disagreement? [Durkheim, Lévy-Bruhl et la « pensée primitive »: quel différend ?]. *L'Année sociologique, 62*(2), 429–446.

Metzger, H. (1914–1939). *La méthode philosophique en histoire des sciences*. Fayard.

Metzger, H. (1926). *Les concepts scientifiques*. Alcan.

Metzger, H. (1969 [1918]). *La genèse de la science des cristaux*. Albert Blanchard.

Mousalimas, S. A. (1990). The concept of participation in Lévy-Bruhl's primitive mentality. *Journal of the Anthropological Society of Oxford, 21*(1), 33–46.

Mößner, N. (2011). Thought styles and paradigms: A comparative study of Ludwik Fleck and Thomas S. Kuhn. *Studies in History and Philosophy of Science Part A, 42*(3), 416–425. https://doi.org/10.1016/j.shpsa.2011.02.001

Netz, R. (1999). *The shaping of deduction in Greek mathematics: A study in cognitive history*. Cambridge University Press.

Paoletti, G. (2004). Durkheim and the problem of objectivity: A reading of Les Formes élémentaires de la vie religieuse. *Revue française de sociologie, 45*(5), 3–25. https://doi.org/10.3917/rfs.455.0003

Piaget, J. (2015). *Structuralism (psychology revivals)*. Taylor & Francis.

Politi, V. (2018). Scientific revolutions, specialization and the discovery of the structure of DNA: Toward a new picture of the development of the sciences. *Synthese, 195*, 2267–2293. https://doi.org/10.1007/s11229-017-1339-6

Politi, V. (2020). Taxonomies, networks, and lexicons: A study of Kuhn's post-'Linguistic Turn' philosophy. *International Studies in the Philosophy of Science, 33*(2), 87–103. https://doi.org/10.1080/02698595.2020.1865784

Preston, J. (1997). *Feyerabend. Philosophy, science and society*. Polity Press.

Rheinberger, H.-J. (2010). *On historicizing epistemology*. Stanford University Press.

Riegl, A. (1985 [1901]). *Late Roman art industry* (R. Winkes, Trans.). Bretschneider.

Schmaus, W. (1996). Lévy-Bruhl, Durkheim and the positivist roots of the sociology of knowledge. *Journal of the History of the Behavioral Sciences, 32*(4), 424–440.

Sciortino, L. (2021). The emergence of objectivity: Fleck, Foucault, Kuhn and Hacking. *Studies in History and Philosophy of Science Part A, 88*, 128–137. https://doi.org/10.1016/j.shpsa.2021.06.005

Vernant, D. (1990). Epistémologie. In S. Auroux (Ed.), *Encyclopédie philosophique universelle* (Vol. 2). Presses Universitaires de France.
Wray, B. (2011). *Kuhn's evolutionary social epistemology*. Cambridge: Cambridge University Press.
Winther, R. G. (2012). Interweaving categories: Styles, paradigms, and models. *Studies in History and Philosophy of Science, Part A, 43*(4), 628–639.
Wittgenstein, L. (1997 [1953]). *Philosophical investigations* (G. E. M. Anscombe, Trans. 20th ed.). Blackwell.
Zambelli, P. (1995). Alexandre Koyré versus Lucien Lévy-Bruhl: From collective representations to paradigms of scientific thought. *Science in Context, 8*(3), 531–555. https://doi.org/10.1017/S0269889700002155
Zambelli, P. (2016). *Alexandre Koyré in incognito*. Leo S. Olschki.
Zittel, C. (2012). Ludwik Fleck and the concept of style in the natural sciences. *Studies in East European Thought, 64*(1), 53–79. https://doi.org/10.1007/s11212-012-9160-8

CHAPTER 3

Notions of the Concept of 'Ways of Thinking': From Hacking to Daston and Galison

3.1 Introduction

We have seen that different thinkers who belong to the tradition of historical epistemology shared the common project of historicizing Kant's a priori, despite disagreeing on certain assumptions. Some of them, e.g. Fleck, Foucault and Kuhn, introduced a notion of the concept of 'ways of thinking' explicitly, and they clearly stated that their aim was to study the emergence of objectivity. Hacking too was clear enough about it when he wrote:

> My study is a continuation of Kant's project of explaining why objectivity is possible [...] Kant did not think of scientific reason as a historical and collective product. We do. (Hacking, 2002 [1992], p. 181)

To this he added that '[Styles of reasoning] are part of what we need to understand what we mean by objectivity' (Hacking, 2002 [1992], p. 181). In the former quotation, Hacking takes for granted Kant's idea that objectivity is made possible by certain a priori conditions but makes a fundamental departure from him by suggesting that what is objective is the result of historical circumstances. Since the notion of style of reasoning is a framework for explaining how objectivity is possible given a historical context, we can say that Hacking's explanatory aim is both a continuation of Kant's project and a radical deviation from it, its historicization.

L. Sciortino, *History of Rationalities*,
https://doi.org/10.1007/978-3-031-24004-1_3

What 'possibility of objectivity' means can be clarified by noting that Hacking used statements such as 'the proposition has a truth-value', or 'has a sense', or 'is positive', or 'is meaningful' or 'is objective' interchangeably. In particular, the expression 'becoming objective' has been used in reference to propositions that become candidates for truth or falsehood or in reference to concepts, questions, conjectures, discoveries or solutions that come to be shared, discussed and debated, i.e. become 'possible'. For example, the emergence of a deductive style of reasoning in mathematics made possible the propositions of geometry. In this respect, as we shall see, Hacking used the notion of a deductive style as a tool for understanding how certain sentences became candidates for truth or falsehood.

Furthermore, we understand why Hacking's styles project can be viewed as a specific study in historical epistemology: its objectives and methods are congruent with those of this latter. Indeed, if 'styles of reasoning [...] are part of what we need to understand what we mean by objectivity', the scope of the styles project is identical to one of the objectives of historical epistemology—to understand an organizing concept, objectivity, which undergoes mutations and evolves along with the changes of our practices. We shall see that it is the mutation of a 'way of being objective', such as the emergence of a new criterion of evidence, which can make possible the emergence of a new style of reasoning. Therefore, the reverse of Hacking's sentence is also true: what we mean by objectivity, its study, is necessary to explain what a style of reasoning is. But for Hacking objectivity is to be studied using the methods of historical epistemology, so the latter is necessary for elaborating the theory of styles of reasoning.

Finally, it must be emphasized that the explanatory goals of Crombie and Hacking were different: the former used the notion of style as a grand framework for describing the Western scientific tradition, while Hacking used it as a framework for addressing a philosophical question about the emergence and the stability of objectivity in different historical periods. He himself made this point clear by stating, 'I adapted the idea [of style of reasoning] to metaphysics and epistemology [...]. The two uses, by historians and philosophers, are complementary but to some extent asymmetric' (Hacking, 2002 [1992], p. 178).

In order to lay bare the content of the concept of 'style of reasoning', I will have to deal with a difficulty: Hacking is unwilling to provide necessary and sufficient conditions of a style and maintains that its characterization should be based on examples drawn from the study of past and present scientific practice (Hacking, 1983a, p. 456) (Lakoff, 2012, p. 227). Indeed, the styles project itself reflects a certain style of *philosophical* thinking that does not embody a perspective based on logic, where reasoning is ensured by stable prescribed definitions rather than by reliance on Wittgensteinian 'family resemblances'. There cannot be any argument that the notion of style of reasoning should fit a narrow criterion. This being the case, it seems appropriate to set out the concept of style of reasoning going from usage to abstract characterization: to grasp the general physiognomy of styles of reasoning as well as to assess the style project in the next chapters will be easier by referring to concrete examples. In particular, I shall not start by presenting Hacking's general remarks on the features of the style of reasoning: without an example, his characterization would remain vague. On the contrary, in the next section I shall show the features that make styles of reasoning distinctive by discussing the genesis of a specific style of reasoning, the statistical style of reasoning. Only later, on the basis of this example, will I present the main claims of Hacking's style of reasoning project.

With the purpose of providing insight into the statistical style of reasoning, in the next section I shall mainly base my analysis on Hacking's *The Emergence of Probability* (Hacking, 2006 [1975]). There is no mention of the concept of style of reasoning in that book: it appears in Hacking's later studies on probability (Hacking, 1992b, 2008 [1990]). However, as it will be clear in the next section, although Hacking got the idea of style of reasoning from Crombie in 1978 (Hacking, 2009, p. 6), three years later the publication of *The Emergence of Probability*, his inquiry into the concept of probability contained in that book can be read, with hindsight, as I am going to argue, as an account of the birth of a particular style of reasoning, namely the statistical one. Only in Sect. 3.3, shall I turn to Crombie's notion of style of reasoning, the starting point of Hacking's reflections on the concept of style of reasoning, and discuss further development of the latter's ideas (Sect. 3.4). Finally, in Sect. 3.5 I shall examine Hacking's thoughts concerning the laboratory style of reasoning and in Sect. 3.6 I shall present Galison and Daston's notion of 'epistemic virtue'. In the last section, to examine further the connection between the notion of style of reasoning and the concept of objectivity, I shall compare the styles project with some ideas of Bernard Williams (1929–2003) as well as Daston and Galison's study of objectivity.

The aim of this chapter is: (1) to connect different suggestions of Hacking about styles of reasoning so to present them in a systematic way;

(2) to briefly present Daston and Galison's notion of the concept of 'ways of thinking' as it emerges from their history of the concept of objectivity. I will give more space to Hacking's notion which I will develop in the next chapters. Furthermore, I shall restrict my analysis to the statistical and the laboratory style of reasoning, leaving aside other styles of reasoning: Hacking has mainly offered suggestions about these two styles and has only mentioned or discussed other styles of reasoning *en passant*. Conversely, in Chap. 5, I shall develop his style of reasoning project by outlining a 'theory of styles of reasoning'. My claim will be that there are at least four other styles of reasoning that share a set of features with the statistical and the laboratory style of reasoning. The conclusion will be that it is possible to give substance to Hacking's notion of style of reasoning by exhibiting six styles of reasoning that share a set of common features, i.e. those features of styles of reasoning suggested by Hacking over the years. It can be said that this chapter presents Hacking's styles project while Chaps. 5 and 6 develop it into a theory. This further step is necessary in order to assess, in the following chapters, Hacking's notion of style.

3.2 The Roots of Hacking's 'Styles Project': The Emergence of Probability as a Study on the Statistical Style of Reasoning

In the late Sixties, when he was an associate professor in Uganda interested in logical questions about statistical inference, Hacking read the first abridged edition of Michel Foucault's *Madness and Civilization* (Foucault, 2006 [1961]) and *The Order of Things* (Foucault, 1994 [1966]). It was under Foucault's influence that in the following years he wrote *The Emergence of Probability* (Hacking, 2006 [1975]) and *The Taming of Chance* (Hacking, 2008 [1990]) (see also (Martínez Rodríguez, 2021)), which examined the transformations in scientific understanding of probability.[1] *The Emergence of Probability*, in particular, applied Foucault's archaeology to the study of the emergence of ideas about induction and statistical inference (Hacking, 2006b, p. 2): Hacking dug into the past in order to understand what made it possible for our world to be dominated by probabilities. By offering a historical account, he argued that the concept of probability was absent before the time of Blaise Pascal

[1] Hacking revealed this autobiographical detail in an interview (Lakoff, 2012).

(1623–1662)—what made the concept of probability possible was the coming to the fore of a new form of evidence. This is a point relevant to understanding the concept of style of reasoning. In order to explain it carefully I shall now summarize Hacking's account.

To start with, Hacking reminds us that, in medieval thought, *scientia* represented the knowledge of universal truths that is obtained by demonstration; conversely, *opinio* was associated with beliefs, resulting from argument or disputation, which could not be demonstrated. The word *probabilis* did not suggest the idea that a hypothesis was better supported by evidence than others but meant 'worthy of approbation', i.e. approval by 'intelligent people' (Hacking, 2006 [1975], pp. 22–23). In Renaissance medical textbooks, an *opinio* was supported by *signs* such as planets in conjunction, comets and other natural phenomena. While physicists were dedicated to demonstrative science, physicians had to predict the likely course of a medical condition on the basis of signs. In the case of an ill person, the causes had to do with efficient causes of the illness and signs were anything by which a doctor could make a prognosis (Hacking, 2006 [1975], p. 28). Ultimately, in the Middle Ages an *opinio* was probable when supported by 'intelligent people', authoritative people or ancient books. Later on, in the writings of Renaissance physicians, opinions were probable when supported by signs of Nature, rather than the written word. There was no antithesis between these two epistemological attitudes: since for the Renaissance authors Nature was the writ of God, its signs represented approval by the ultimate authority.

Beside approbation, there was no other concept of evidence related to opinion. What was lacking was the evidence provided by *things*, not to be confused with that provided by the data of senses. To use the example given by J. L. Austin (1911–1960) and quoted by Hacking (2006 [1975], p. 32), pig-like marks and buckets of food outside a sty represent the *evidence of things* for the statement that in the sty there is a pig; the coming into view of the pig represents the *evidence of the senses*—I can *see* what animal it is. Obviously, in the Renaissance people did use the evidence of things, but for Hacking the concept was absent: 'dogs and boars can tell there is a pig, and do not thereby have a concept of evidence' (Hacking, 2006 [1975], p. 34). For us books and testimony represent indirect evidence reported by other people. The Renaissance had the order reversed: testimony and authority were primary, things counted as evidence only insofar as they resembled the authority of testimony and books. The concept of sign underwent several changes—for instance, Paracelsus

(1493–1541) ignored the fact that the names of stars were conventions and thought that the 'true names' of celestial bodies were signs; but later on, the distinction between conventional signs and natural ones was clearly made. The physician Girolamo Fracastoro (1483–1553) wrote that, among the signs in the sky, air, soil or water that are premonitory

> some are almost always, others are often, to be trusted. Therefore one ought not to consider them all as prognostications, but only as *signs of probability*. (Hacking, 2006 [1975], p. 28)

These changes opened the way for the evidence of signs to turn into the evidence of things. In the passage above, we may note that, since not all signs are to be trusted with certainty, the idea of probability is connected with frequency, i.e. with what happens 'almost always'. Here, we already recognize some of the features of our statistical concept of probability. According to Hacking, in a text published in 1650 by Thomas Hobbes, the concept of evidence of things conjoined with that of frequency had already taken full shape and probability had emerged in all but name (Hacking, 2006 [1975], p. 48). In 1662 also the Port Royal logic, published by members of the Jansenist movement, distinguished arbitrary and conventional signs—the concept of sign as evidence had become endemic; stable and law-like regularities became worthy of observation. Later on, the studies of Jacques Bernoulli (1654–1705) culminated in the central limit theorem; in 1756 Thomas Simpson (1710–1761) applied the theory of errors to the discussion of uncertainty; finally, in 1820s the first public statistics were published (Hacking, 2008 [1990], p. 133) and some years later the statistical laws were used to explain phenomena.

The coming to the fore of the evidence of things produced a new way of reasoning whose features I am going to illustrate. Among the sentences formulated by the alchemists and astrologists of the Renaissance, Hacking takes the following sentence (henceforth called *S1*) of Paracelsus as an example:

> (*S1*) Mercury salve is good for syphilis because mercury is signed by the planet Mercury, which signs the marketplace where syphilis is contracted. (Hacking, 1983b, p. 71)

He claims that for us *S1* is not a candidate for truth or falsehood (Hacking, 2002 [1982], p. 171). How can this claim be justified? The point is that, for Hacking, Paracelsus's reasoning was based on a kind of evidence that died out in the seventeenth century and became unknown afterwards: 'the way propositions are proposed and defended' (Hacking, 2002 [1982], p. 170) by Paracelsus is alien to us. In order to make sense of *S1* it is not sufficient to know the meaning and the connotations of all the words. In short, it is Paracelsus's way of reasoning that we need to reconstruct. Only when we have done so can we talk as Paracelsus talked.

To make Hacking's point clear, let us examine Paracelsus's reasoning behind the sentence *S1*. The starting point is that diseases must be cured by similarity. Bearing that in mind, this is the logic chain of Paracelsus's reasoning: syphilis is contracted in the market place; the planet Mercury has signed the marketplace; this is evidence that the metal mercury, which has the same name, can treat syphilis, given that diseases are cured by similarities. Ultimately, for syphilis, the name 'mercury' of the metal mercury is the sign that a good physician must know. It is a sign that provides evidence for the proposition 'Mercury treats syphilis'. This way of reasoning is not ours: propositions are defended on the basis of a system of evidence (signs), rules (e.g. similarities) and practices now fallen in disuse. Because of that, Hacking maintains that they do not have truth-value for us.

Following Hacking, one could say that, just as some propositions are candidates for truth or falsehood only in the Renaissance way of reasoning, so certain propositions acquire a truth-value only as consequence of the emergence of probability (or, better, of the evidence of things). For example, the sentence:

(*S2*) The adult height for one sex in an ethnic group follows a normal distribution

did not have any truth-value in the Renaissance not only because the term 'normal distribution' was not defined but because the truth-value of the sentence *S2* hinges on the evidence of things. This is a kind of evidence that, as I have said, according to Hacking was not yet conceptualized before the mid-seventeenth century.[2] Moreover, to reason about this sentence means to take all the steps of a complex process—selecting a sample,

[2] Hacking's own example is 'the gross national product of Württemberg in 1817 was 76.3 million adjusted to 1820 crowns'. For him, this sentence did not have a truth-value before 1821 'because there was no procedure of reasoning about the relevant idea' (Hacking, 1992b, p. 143).

measuring, classifying, dealing with errors or assessing the difference between the experimental distribution and the theoretical one. Ultimately, to prove the sentence above involves, for all the means and purposes, not only a way of thinking but also a *way of doing*.

Hacking's point is that certain sentences may be meaningful only in a given epoch not because some terms are not defined but because their having a truth-value hinges on different ways of thinking and doing. To appreciate his point, let us compare *S1* and *S2* with this passage from *The Merchant's Tale*, a bawdy story from *The Canterbury Tales* by Geoffrey Chaucer (1343–1400):

> 'Allas', quod he, 'that I ne had heer a knave
> That koude clymbe! Allas, allas', quod he;
> 'For I am blynd!' 'Ye sire, no fors', quod she;
> 'But wolde ye vouche sauf, for Goddes sake,
> The pyrie inwith youre armes for to take,
> For wel I woot that ye mystruste me,
> Thanne sholde I clymbe well ynogh', quod she,
> 'So I my foot myghte sette upon youre bak'. (In Gooden, 2009, p. 66)

In his *The Story of English* (Gooden, 2009), Philip Gooden says that in the passage there are several challenges for the modern reader. One has to know what certain words stand for, e.g. *quod*, past tense of *queten*—'to say'; *pyrie*—'pear tree'; *knave*—'servant'; *no fors*—'no matter' (Gooden, 2009, p. 66). However, although the passage is full of *double-entendre*, once the archaic forms are defined, its literal sense becomes clear—it becomes meaningful. It refers to what happens when the servant Damyan climbs up a pear tree in the garden of January, the main character, and waits for the arrival of the young May and her husband. Not so for Paracelsus's sentence—understanding it is for us an entirely different exercise that requires an effort beyond that of expressing the sentence in modern English. Not so for the sentence *S2* either: for Hacking a Renaissance thinker would not have been able to assert or deny it, even if expressed in her language.

Taking a cue from these comparisons, another important point can be noted. The sense of both *S1* and *S2* rests on two different kinds of evidence, each of them peculiar to the respective ways of thinking. On the other hand, another passage of *The Merchant's Tale*, which precedes the passage just quoted, reads:

(*S3*) She taketh hym by the hand

(i.e.: 'She takes him by the hand'). Hacking would say that the sense is straightforwardly clear because it hinges on the evidence of senses, which of course does not belong exclusively to a given way of thinking. In any epoch, once the terms are defined, it can be asserted or denied—come and look. Indeed Hacking stresses that, beside sentences whose truth-value rests upon a style of reasoning, there are 'observation statements' (Hacking, 1992b, p. 133), or in other words 'factual statements' (Hacking, 1983b, p. 173). Hacking is adamant that these sentences do not require any reasoning to be assessed as true or false: they rely on the evidence of senses, which is foundational, ahistorical.

The idea of 'observational statements', or 'protocol statements' in the original terminology of the Vienna Circle, has been much criticized in the last decades. For example, Norwood Hanson (1924–1967) and Paul Feyerabend (1924–1994) maintained that observation statements are theory-laden (Hanson, 1958, chapter 1) (Feyerabend, 1988 [1975], chapter 3). Hacking agrees with them only to a certain extent:

> The core objection to a correspondence theory is that there is no way in which to identify the facts to which a statement corresponds, independently of the statement itself. That is true in general, but not of a lot of the run-of-the-mill sentences of the sort beloved by logicians. (Hacking, 1992b, p. 134)

For sentences that make sense only against the background of the 'accepted' way of reasoning, Hacking thinks that 'Moritz Schlick's motto, "the meaning of a sentence is its method of verification", points in the right direction' (Hacking, 1992b, p. 135). In other words, he does not think that 'there is one theory of truth, or one semantics, that applies to all contingent empirical sentences investigated in the sciences' (Hacking, 1992b, p. 135). On the one hand correspondence theories of truth apply to *style-independent sentences* (e.g. *S3*), i.e. sentences whose truth-conditions are not determined by the way we reason. On the other hand, the methods of verification, which may change over time, apply to *style-dependent sentences* (e.g. *S2*), making them candidates for truth or falsehood.

The verificationist account of meaning is particularly important for the styles project, and especially for establishing whether or not certain propositions are style-dependent, that is candidates for truth or falsehood. Hacking's concern is chiefly with the fact that Schlick put forward 'a test for marking off the meaningful utterances' (Hacking, 2006 [1975],

p. 97): as we shall see, within the styles project whether or not a community that adopts a certain style of reasoning possesses methods for assessing a proposition as true or false can be crucial for establishing whether or not that proposition is meaningful for that community. Indeed, by saying that certain sentences are not meaningful Hacking often assumes the verificationist principle. As an example one can consider the sentence *s2* above: for him whether or not it is meaningful for a certain community, i.e. a candidate for truth or falsehood, depends on whether or not that community possesses the methods of the statistical style of reasoning or its kind of evidence.

This leads to another important point. The verification methods that make statistical sentences meaningful are refined versions of the procedures for testing errors and making estimations that have emerged since the eighteenth century. Hacking notes that these methods of verification are themselves couched in terms of probability:

> the conditions of assertibility of statistical hypotheses are themselves to be determined [...] in terms of yet a new layer of sentences that themselves are statistical. (Hacking, 1992b, p. 151)

This remark can be explained by considering once again *S2* (i.e. 'The adult height for one sex in an ethnic group follows a normal distribution'). What makes *S2* true or false is the existence of criteria for assessing whether the normal distribution fits the observed statistical distribution of the adult height. Typically, these criteria are measures of goodness of fit such as the chi-squared tests, which evaluate how well the observed frequencies follow a Gaussian distribution. However, the chi-squared tests are themselves statistical—e.g. to 'verify' *S2* one has firstly to use the differences between the theoretical and empirical frequencies for calculating a value called chi-square; then, on the basis of this value and a statistical distribution (chi-square distribution), one works out a probability. If this probability is lower than a certain number, it is considered very unlikely that *S2* is true.[3] In conclusion we use a statistical distribution and other probabilistic methods to assess a statistical sentence.

Hacking maintains that there is some circularity in the process:

[3] An introduction to the use of the chi-squared test can be found in (Taylor, 1982, chapter 12).

> There is an odd way in which a style of reasoning and truth-conditions of some sentences are mutually self-authenticating. The truth is what we find out in such and such way. We recognize it as truth because of how we find it out. And how do we know that the method is good? Because it gets at the truth. (Hacking, 1992b, p. 135)

In terms of my example, Hacking's claim is that the truth conditions of *S2* are determined by a process of reasoning that includes the use of statistical procedures such as the chi-square test; on the other hand, this way of reasoning has become 'the correct standard' of assessing *S2* because it has, to use a phrase cited of Charles Sanders Peirce (1839–1914) by Hacking, the 'truth producing virtue' (Hacking, 1992b, p. 135) i.e. it is able to identify whether or not a proposition is true. It is to this circularity that Hacking alludes to when he says that a way of thinking is *self-authenticating*.

So far I have illustrated by examples the following points of Hacking. In the Renaissance the evidence of things was lacking and the world testified by its signs. Certain sentences of the Renaissance thinkers (e.g. *S1*) are not among our candidates for truth or falsehood because their sense hinges on a kind of evidence and a way of reasoning different from ours. Later on different scholars maintained that signs can be imperfect: for example, in 1650, in Hobbes's *Human Nature*, signs are considered imperfect and only 'very often' right. Credibility and frequency were thus linked. Then conventional signs were distinguished from natural signs and the concept of the evidence of things emerged. In the seventeenth century the new evidence of things opened a space for a new way of reasoning and, in particular, for new candidates for truth or falsehood. Hacking thinks of this transformation as a historical *discontinuity* in the history of human thought:

> there was hardly any history of probability to record before Pascal, while after Laplace [...] a page-by-page account of published work on the subject became almost impossible. (Hacking, 2006 [1975], p. 1)

He even gives a date for the birth of probability: the decade around 1660 (Hacking, 1992b, p. 141).

Incidentally, the coming to the fore of probability was not only followed by the emergence of new style-dependent sentences (e.g. *S2*) but also by new criteria (e.g. best fit), types of explanation (e.g. statistical distributions), laws (e.g. the law of large numbers), classifications (e.g.

society is divided into genres of people (Hacking, 2008 [1990], p. 134)) and new objects (e.g. the population characterized by a mean and standard deviation).

All these points of Hacking can be summarized by saying that there exists a way of thinking and doing, which he called the *statistical style of reasoning*, with the following *characterizing properties*:

> The statistical style of thinking and doing relies on a new kind of evidence for finding out; it is self-authenticating and introduces new candidates for truth-or-falsehood, criteria, types of explanation, laws, classifications and objects. The emergence of the statistical style of thinking and doing represents a sharp break in the history of Western thought.

Hacking has used the label 'style of reasoning' ever since his first paper on the topic 'Language, Truth and Reason' (Hacking, 1982). However, in one of his most recent papers (Hacking, 2012) he wrote that 'style of scientific thinking & doing' is a better label. I shall stick to his original use, 'style of reasoning' (or 'style' when there will not be any possibility of confusing Hacking's notion with Fleck's). Since Hacking did not want to provide definitions, as I have explained, I shall consider the features listed just above as the characterizing properties of the statistical style. As I shall explain in Sect. 3.4, we can say that a style of reasoning in the sense of Hacking is a way of thinking and doing that possesses most of the features listed just above (which I shall call 'the characterizing properties of style of reasoning').

3.3 From Crombie's Styles of Thinking to Hacking's 'Styles Project'

For his concept of style Hacking acknowledged his debt to the Australian historian of science Alistair Cameron Crombie (1915–1996). In 1978, three years after the publication of *The Emergence of Probability*, Hacking was invited to a conference in Pisa in which Crombie (1915–1996) lectured on what he called 'styles of scientific thinking in the European tradition' (Crombie, 1978). Crombie spoke of exactly six styles of scientific thinking still practised in the modern sciences, each of which emerged at a specific time in history and evolved in its own time frame. That lecture suggested to Hacking the 'idea of a small manifold of style of reasoning' (Hacking, 2009, p. 6), as he revealed later (Hacking, 2012, p. 1). In order to explain how he turned Crombie's ideas into his own philosophical project I shall first summarize the latter's account of styles.

The point that six styles can be distinguished in the history of Western science is a recurrent view in many of Crombie's writings (Crombie, 1978, 1988, 1992, 1994). Since the mid-seventies, his aim had been to write an encyclopaedic history of their growth, but it was only in 1994 that his ideas finally appeared in an ambitious three-volume book, entitled *Styles of Scientific Thinking in the European Tradition* (Crombie, 1994). In that book, Crombie maintained that a 'postulational style of thinking' was adopted by the ancient Greeks in mathematics, Aristotle's works on natural sciences, Hippocratic medicine, ethics, law and metaphysics (Crombie, 1994, p. 74). He conceived this style, the first he described, as a method that consisted of a two-step process: search for the simplest and fewest premises and deduction of their implications by 'proof', a Greek invention (Crombie, 1994, p. 20). For example, the demonstrative power of geometry was used by Claudius Ptolemy (90, 168) to calculate celestial motions and to build a theory of the process of vision in optics. In their scientific works, the Greeks set up a rational scientific system in which 'the sequence of theoretical reasoning from premise to conclusion matched the sequence in nature from cause to effect' (Crombie, 1994, p. 20). In other words, the phenomena were reproduced by an abstract deductive formal system, which embodied postulated principles and offered causes of regularities. The world came to be thought as rational, entirely deducible by first principles.

Crombie also spoke of an experimental style that came about in ancient Greece because observation and measurement were necessary either to control the explanations obtained by the method of postulation or to explore the possible causes of phenomena. For Crombie, the use of experiments became more and more sophisticated from thirteenth to the seventeenth century through the strong relation between the postulational style in the natural sciences and engineering (Crombie, 1994, pp. 311–423). Then, the habit of measurement and observation fully developed in the seventeenth century besides the growth of mathematics.

By 'the hypothetical style of modelling' Crombie meant 'the method of elucidating the unknown properties of a natural phenomenon by simulating the phenomenon with the known properties of a theoretical physical artefact' (1994, p. 74). For him, the style of hypothetical modelling had its origin in ancient Greece, in particular in ancient theology and astronomy. He illustrated in detail this thesis by examining thinkers who lived from 400 AD onwards (i.e. from Augustine to Kant) (Crombie, 1994, pp. 1081–1231). For example, the armillary sphere (astrolabe) was a model of celestial sphere invented by Eratosthenes of Cyrene (276–195 BC) in order to describe the movements of stars and to map the constellations. Hypothetical modelling fully developed in application to early modern perspective, painting and engineering; then, from these fields of study it was transferred to modern science.

Furthermore, Crombie maintained that the fourth style, the taxonomic one, is 'the method of ordering variety in any subject matter by comparison and difference' (Crombie, 1988, p. 11). In the first volume of his 1994 work he pointed out that this style emerged as an explicit logic of classification in Plato and Aristotle and 'in many ways it is the foundation of all natural science, establishing fundamental similarities and difference' (Crombie, 1994, p. 75). Then, in the fifth part of the second volume, Crombie illustrated the search of a natural system throughout history from Andrea Cesalpino (1524–1603), who proposed a classification based on the reproductive parts, to Linnaeus, who funded his natural system on the fixity of genera.

The fifth style described by Crombie is the statistical one. He proposed for this style an origin alternative to that suggested by Hacking. Indeed, Crombie's story of probable reasoning begins in ancient Greece and leads to Pierre Simon Laplace (1749–1827) without any sharp transition in systems of thought such as that emphasized by Hacking (Crombie, 1994, p. 75). On his part, Hacking stressed the fact that his account of the emergence of probability is alternative to Crombie's *continuist* one (Hacking, 2006 [1975], p. X).

Finally, for Crombie the style of historical derivation, originated in ancient Greece, aimed at explaining the present as a development of the past regulated by certain laws and at deducing the past from present regularities (Crombie, 1994, p. 75). Again, Crombie conceived the development of the style of historical derivation as *continuous* from Greeks to Charles Darwin (1809–1882); however, these methods were developed in application first to the history of languages, then to the geological history of the Earth and the evolution of living organisms (Crombie, 1994, p. vol. III chapter seven).

In his seminars at the Collège de France, Hacking referred to Crombie's styles above by using the following *aides-mémoire*:

(a) The *postulational style* (in Hacking's writings the terms 'geometrical style' and 'style of mathematics' are also found (Hacking, 2005b, p. 543) (Hacking, 2012))

(b) The *experimental style*
(c) The *style of hypothetical modelling*
(d) The *taxonomic style*
(e) The *statistical style*
(f) The *historico-genetic style*

Henceforth, I shall refer to this list of styles of thinking as 'Crombie's list'.

Crombie introduced his concept of style to organize his history of Western sciences. Sometimes, he used interchangeably the terms 'method' and 'style' (e.g. Crombie, 1988, p. 10); in other occasions he added the specification that 'styles of scientific thinking are distinguished by their *objects* and their methods of reasoning' (Crombie, 1994, quoted in Hacking 2009, p. 90). He explained what a style is with these words:

> We can define what I call a 'scientific style' in terms of three features: 1) its form of argumentation (its methods of discovery and demonstration); 2) its conception of nature (beliefs about what exists and is to be discovered); 3) the mental habits (especially the expectations and the answers concerning innovation and change; the dispositions of a society and individuals within it). (Crombie, 1992, p. 103 my translation)

Feature one concerns the model of discovery that scientists in distinct periods use (Crombie, 1988, p. 2): for example, in certain circumstances, the geometrical model in which the laws of phenomena are derived from first principles might be preferred to an analogical model, in which phenomena are understood and explained by building their mechanical representations.

Feature two concerns the *conception of nature* associated to every style. For example, the statistical style opened the way to quantum theory (see Hacking, 2006 [1975], p. 174), whose formalism cannot be interpreted as if probabilities reflected our ignorance.[4] As a consequence, statistical laws have become an irreducible fact of the world and classical physics' deterministic conception of nature has given way to an indeterministic one. In this sense, the advocates of statistical style are committed to an indeterministic conception of nature.

[4] The works of the physicists John Von Neumann (von Neumann, 1955), John Bell (Bell, 1966), Simon Kochen and Ernst Specker (Kochen & Specker, 1967) show that it is impossible to provide a 'classical' reformulation of the formalism of quantum theory.

Feature three concerns the fact that distinct societies have been more or less ready to accept change because of their beliefs: for example, as we shall see, according to Geoffrey E.R. Lloyd it was precisely because Greek society was so polemical and open to debate that an axiomatic-deductive mathematics came about in ancient Greece and was put into general practice (Lloyd, 2004).

A point to note is that all of Crombie's styles emerge in ancient Greece and develop continuously. Both Kerry Magruder and Robert Iliffe pointed out that Crombie's continuity thesis about the development of styles is to be linked to his continuism concerning the Scientific Revolution (Iliffe, 1998, p. 20; Magruder, 1995, p. 408). In Crombie's view, the experimental methodology of Galileo and Newton was first developed and practised by medieval natural philosophers. Crombie accepted that the crucial steps in the emergence of modern science took place during the Scientific Revolution, but he thought that the essence of history of science was a continuous development of scientific thought from Greek thinkers onwards. As Kusch noted, 'the tool-box of Crombie-inspired styles contains no ingredients to understand historical change' (Kusch, 2010, p. 166). Indeed, Crombie's notion has been conceived from a historical perspective different from that of 'style of reasoning':

> Crombie's vision of the history of the European sciences favours continuity. My instinct is exactly the opposite. I like to tell the history of each style as having at least one sharp moment of crystallization, a fixing of how to go in the future [...]. I acquired this habit early, in *The Emergence of Probability*, some years before I had ever heard of Crombie. (Hacking, 2009, p. 14)

In this passage Hacking reveals that it was his study on probability that made him sceptical about continuism. So much so that he supposes that other ways of reasoning possess the same feature. Consequently, what he wants is a notion that describes ways of reasoning whose beginnings mark a discontinuity point in the trajectory of Western thought.

In 1982, four years after his encounter with Crombie's ideas, Hacking published his first paper ever on style of reasoning, 'Language, Truth and Reason' (Hacking, 1982), which was included by the English philosopher Martin Hollis (1938–1998) in a collection of essays entitled *Rationality and Relativism* (Hollis & Lukes [Eds.], 1982). A clear point emerged from that paper: despite Hacking's admission that Crombie had been the starting point of his style of reasoning project, their enterprises must be regarded as entirely different. Indeed, Hacking launched his own *philosophical* project by making a *distinctive* use of the latter's notion of style. I am going to illustrate this project in the next section.

3.4 Hacking's 'Styles of Reasoning'

3.4.1 *The 'Styles Project'*

'The style project' is the label used by Hacking in his '"Language, Truth and Reason" thirty years later' (Hacking, 2012) to describe the programme sketched in his earlier papers on styles of reasoning, namely 'Language, Truth and Reason' (Hacking, 1982) and '"Style" for historians and philosophers' (Hacking, 1992c). However, later thoughts expressed in works such as *Scientific Reason* (Hacking, 2009) might be included in this project. In this section I shall present those points of the styles project made by Hacking that mainly concern his characterization of style of reasoning and I shall leave for later discussion other philosophical questions touched on by him. My point will be that the style of reasoning project contains ideas that need to be developed or better clarified.

Ever since his 1982 paper Hacking's reflection on styles of reasoning has centred around the point that 'whether or not a proposition is as it were up for grabs, as a candidate for being true-or-false, depends on whether we have a way to reason about it' (Hacking, 2002 [1982], p. 160). For him styles of reasoning fix the sense of *certain* sentences by making them candidates for truth or falsehood. The sentence *S2* ('The adult height for one sex in an ethnic group follows a normal distribution') in the previous section illustrates well this point: its sense hinges on the style of reasoning appropriate to it, i.e. the statistical style of reasoning. Indeed—Hacking would argue—what determines its truth-value is the statistical style of reasoning: its methods, its way of thinking and doing and its kind of evidence.

Hacking distinguishes *positive* propositions, i.e. propositions that are true or false, from *bivalent* propositions, i.e. propositions that have a definite truth-value, true or false (Hacking, 2002 [1982], p. 166). 'Positive' is a less strong feature than 'bivalent' because a positive proposition is a proposition potentially susceptible of being true or false, given the existence of a style of reasoning; on the other hand a bivalent proposition is a proposition that is actually true or false. Hacking gives this example: once James Clerk Maxwell (1831–1879) said that some propositions about the relative velocity of light were intrinsically incapable of determination; but few years later Albert A. Michelson (1852–1931) invented a technology to give a precise answer to Maxwell's questions (Hacking, 2002 [1982], p. 167). Maxwell's sentences were *only* positive; but after Michelson's experiments they became bivalent. The point is that for Hacking a proposition is positive if it makes sense, even if the correct way of assessing it by the standards of truth and methods of its style of reasoning has not yet been found. Styles of reasoning have to do with positive sentences rather than bivalent sentences.

The fact that Hacking essentially thinks of styles of reasoning as structured classes of propositions that are capable of being true or false marks a profound difference between Crombie's notion of style and Hacking's notion of style of reasoning. What for the historian Crombie is a device for describing the Western scientific tradition, in the hands of Hacking becomes the tool 'we need to understand what we mean by objectivity' (Hacking, 2002 [1992], p. 181). What Hacking means can be grasped by looking at his study on probability: he thinks that the idea of a statistical style of reasoning is a useful tool for understanding how certain sentences became candidates for truth or falsehood and how the concept of probability itself became objective. More generally, Hacking wants to use the notion of a style of reasoning to capture the way in which the past has determined our present scientific conceptions. To adopt Foucault's phrase, Hacking wants to do a 'history of the present' (Hacking, 2009, p. 4), i.e. to dig into the past to reveal the conditions of possibility of our present knowledge. This point will be relevant to tracing the philosophical lineage that has shaped the styles project.

3.4.2 A List of Styles of Reasoning

Over the years Hacking has mentioned different styles of reasoning. Although in 1992 he declared that Crombie's three-volume work provides a historical account useful for characterizing his style of reasoning (Hacking, 2002 [1992], p. 186), he did not accept as style of reasoning all the styles in Crombie's list. For example, he observed that Crombie's styles b) and c) (the experimental style and the style of modelling) on their own cannot account for the considerable sophistication undergone by the experimental method over the centuries. Hacking did not deny that both experimenting and modelling can be in play in particular areas of research (p. 186), but he claimed that the concept captures the genesis and development of experimental thinking is, as he called it, *the laboratory style of reasoning*. I shall characterize it better in the next section. For the time being, I want to outline the evolution of Hacking's thought as regards this style of reasoning.

For a start he has always been convinced that the salient feature of this style of reasoning is that, ever since the time of the Scientific Revolution, humans have built experimental apparatuses to elicit new phenomena, i.e. phenomena that had never been observed before (Hacking, 2009, p. 43). The emblem of this new way of doing is Robert Boyle (1627–1691), who made the air pump in order to investigate the capacity of the air to exert pressure; this is a thesis that Hacking drew from Steven Shapin and Simon Schaffer's book *Leviathan and the Air Pump* (Shapin & Schaffer, 1989 [1985]), as I shall explain in Sect. 3.5. Ultimately, one way to think of this style of reasoning is to regard it as a particular deployment of Crombie's styles b) and c) plus the creation of phenomena by man-made machines (Hacking, 2007, p. 3).

Furthermore, in the paper *The Self-Vindication of the Laboratory Sciences* (Hacking, 1992a), Hacking argued that laboratory sciences tend to produce a sort of self-vindicating structure in that background theories, hypotheses, modelling of apparatus, data analyses and other items of knowledge are mutually adjusted to each other. He referred to this particular internal dialectic between different items of knowledge as a *self-stabilizing technique (or self-vindicating technique)*, the process peculiar to the experimental thinking that, as I shall explain, makes the laboratory style of reasoning self-authenticating.

Concerning Crombie's postulational style, Hacking has always insisted that it can be considered a style of reasoning with this proviso: the real novelty brought into being by Greek thought is 'proof' rather than, as Crombie believed, postulation. In 1992 he had made it clear that a comprehensive account of the postulational style of reasoning, which he never offered, should explain why 'mathematics has the astonishing power to establish truths about the world independently of experience' (Hacking, 2002 [1992], p. 183). Some years later, in *Scientific Reason* (Hacking, 2009), he declared that the *Shaping of Deduction in Greek Mathematics* (Netz, 1999) by the historian of ancient mathematics Reviel Netz is the study he wanted (Hacking, 2009, p. 67): an investigation on the cognitive and social factors which shaped the emergence of proof.

With regard to the remaining styles in Crombie's list, the taxonomic style and the historico-genetic, the styles project says almost nothing, as

Hacking explicitly admits (Hacking, 2012, p. 10). On the other hand, in his 1992 paper he mentioned new style of reasoning that could be added to Crombie's list. For instance, he suggested that 'the Indo-Arabic style of applied mathematics, little interested in postulation but dedicated to finding algorithms' (Hacking, 2002 [1992], p. 185) might represent a distinct style, which he called 'algorismic' style of reasoning (Hacking, 2002 [1992], p. 185). Later, in his lectures at the Collège de France (Hacking, 2006c, p. 7) and in *Scientific Reason* (Hacking, 2009) he brought up this idea again. Finally, he made no mention of it in his 2012 paper, in which he rejected some proposals of other styles of reasoning in literature.

It should not be forgotten that in 'Language Truth and Reason' (Hacking, 1982) as well as in *Representing and Intervening* (Hacking, 1983b), Hacking mentioned another way of reasoning, that of Paracelsus (Sect. 2.2), and in his 2012 paper he also explicitly referred to it as a style of reasoning (Hacking, 2012, p. 9). However, Robert Kowalenko wrote that Hacking, in personal communication, told him that Paracelsus's style cannot be considered as 'scientific' (Kowalenko, 2011).[5] Hacking has never clarified this point. His last word about the number of styles of reasoning was that 'it is better not to multiply six [the number of styles in Crombie's list] beyond necessity' (Hacking, 2012, p. 6). Even so, he never explained why certain styles he mentioned or others that have been proposed by other scholars should be excluded. Is there a deeper reason than the simple observance of *lex parsimoniae*? This is a question I shall consider in Sect. 5.6.

3.4.3 The Characterizing Properties of Styles of Reasoning

In Sect. 3.2 I have introduced the 'characterizing properties of style of reasoning', exemplified by the features of the statistical style. Indeed, over the years Hacking has mentioned each of those properties in his *general* discussions about styles of reasoning. For example, I have already pointed out that for Hacking the statistical and the laboratory style of reasoning had clear beginnings. Might there exist other ways of reasoning with this property? The style of reasoning project contains no answer to this question. On the other hand, I have explained that the statistical style of reasoning introduces new objects, new candidates for truth or falsehood,

[5] Kowalenko's paper was presented at the third Bi-annual Conference of the European Philosophy of Science Association in Cape Town.

criteria and laws, relies on the evidence of things and is self-authenticating. This is exactly what Hacking says *in general* about style of reasoning:

> Every style of reasoning introduces a great many novelties including new types of:
> Objects
> Evidence
> Sentences, new ways of being a candidate for truth or falsehood
> Laws, or at any rate modalities
> Possibilities
> One will also notice, on occasion, new types of classifications and explanations (Hacking, 2002 [1992], p. 189) (see also Hacking, 1995, p. 75).

Further on in the text he added that:

> Styles become standards of objectivity because they get at the truth. But a sentence of that kind is a candidate for truth or falsehood only in the context of the style. Thus styles are in a certain sense 'self-authenticating'. (Hacking, 2002 [1992], p. 191)

Although the sense of these quotations appears clear if one refers to the statistical style of reasoning, some points are still ambiguous. It is unclear whether all the potential candidates for being identified as style of reasoning mentioned by Hacking are self-authenticating and why. This is not a trivial matter. In 1996, in respect of the question 'how do the sciences differ from the humanities?' Hacking suggested that only the sciences exhibit ways of reasoning that have developed self-stabilizing techniques (Hacking, 1996, p. 74). If so, establishing which ways of reasoning are self-authenticating is relevant to characterizing science itself.

Incidentally, I object to Hacking's use of the verb 'to introduce' in relation to evidence. Considering as an example the statistical style of reasoning, it is not that it 'introduced' new evidence; if anything, the order is reversed: only when the new evidence of things was conceptualized did a 'space' for the birth of the statistical style of reasoning open. So I think this verb does not really do the trick, at least when we want to describe the internal mechanisms of the statistical style of reasoning. One should simply say that to adopt a new style of reasoning involves the use of a new type of evidence.

In 1992 Hacking wrote that there is 'a necessary condition for being a style of reasoning: each style should introduce novelties of most or all of the listed types [above]' (Hacking, 2002 [1992], p. 189). In other words, according to Hacking, a style of reasoning is a way of thinking that has most of the properties above. In 2012, interviewed by Andrew Lakoff, Hacking added:

> It is not surprising that one can't give a definition, a set of necessary and sufficient conditions, for being in the Crombie's list. You can't define science: a fortiori you cannot define styles of scientific reasoning. […] There are general things that one can say, but they tend to be too general. (Lakoff, 2012, p. 227)

3.4.4 *Doing*

In his later writings (e.g. Hacking, 2009) about style of reasoning Hacking insisted on another point that I have already touched on: styles of reasoning also are ways of ***doing***. Crombie focused on 'knowing by doing' in the style of hypothetical modelling and the experimental style: for example, perspective was a new way of seeing, depicting and manipulating (Crombie, 1994, pp. 444–460) and imitation of nature was a new way of knowing. In Hacking this inextricable combination of knowing and doing is much more developed: he is often keen to remark that humans are embodied creatures that use both their minds and bodies to think and act in the world (Hacking, 2012, p. 2). For example, he stressed that there is a strong link between thinking and doing in mathematical proofs (Hacking, 2011). In order to underline this, he favoured the locution 'style of reasoning' over 'style of thinking' because 'thinking is too much in the head' (Hacking, 2002 [1992], p. 189) and he wanted to stress that assessing the truth of a style-dependent sentence has to do with 'doing' as well as 'thinking'.

Some of Hacking's critics have passed over this remark. I am with James Elwick, who said:

> [There is] a tendency of many scholars to transform 'styles of reasoning' into 'styles of thinking' despite Hacking's insistence that hand-work [sic] is just as important as head-work [sic]. (Elwick, 2012, p. 1)

This tendency might be ascribed to the fact that Hacking has not clarified why all the styles of reasoning he mentioned can also be viewed as ways of doing. This is another suggestion that it will be necessary to develop.

Finally in 2012 Hacking made it clear that styles of reasoning should be understood as *scientific* ways of finding out

> Another available moniker [for style of reasoning] is: [...] ways of finding out in the sciences. Not only finding out 'that' so and so, but also, finding 'how to'. Finding out what's true, and finding out how to change things. (Hacking, 2012, p. 3)

In this sense, to develop the styles project means to explain how we have found out about our world and altered it. This leads us to viewing the style of reasoning project from another perspective.

3.4.5 The 'Styles Project' as a Study in Cognitive History

Ever since his lectures at the Collège de France (Hacking, 2006a), Hacking flirted with the idea that every style of reasoning is grounded in one or more modules, sorts of separate structures of the mind, which are the product of evolution by natural selection and can be studied from the point of view of cognitive science and neuroscience (Hacking, 2009, p. 48). For example, he conjectured that there might be a module for geometrical reasoning and one for numerical and combinatorial reasoning (Hacking, 2011) (see also Spelke, 1994).

In addition to that, Hacking also regarded the style of reasoning project as an 'ecological' study (Hacking, 2012) (see also Crombie, 1994, p. 6). The term 'ecology' comes from the Greek 'oikos'—'house', and indicates the branch of biology that studies the relations of organisms to their surroundings, their 'house', so to speak. So Hacking means that the style of reasoning project is not only about human beings and their capacities but also about the relations to their intellectual, moral and physical environment (Hacking, 2012, p. 9). Finally, Hacking also used the term 'cognitive history' to describe the style of reasoning project, borrowing the term from Netz (Netz, 1999, p. 6). Hacking thinks of his theory of styles of reasoning as a further exploration of how certain cognitive capacities, within certain historical contexts, led to certain ways of acquiring knowledge (see also Netz, 1999, Introduction).

In Chap. 5 I shall argue that there is a set of styles of thinking and doing that, although different from one another in their histories and in their historical trajectories, share all the features suggested by Hacking over the years and presented above. For clarity's sake I list all these features (the characterizing features (or properties) of styles of reasoning) below:

(1) Being ways of thinking and ***doing***. In particular, a way of ***intervening*** in the world and 'finding out "that" so and so, but also, finding "how to". Finding out what's true, and finding out how to change things' (Hacking, 2012, p. 3).
(2) Relying on a new kind of evidence for 'finding out in the sciences' (Hacking, 2012, p. 3)
(3) Introducing new candidates for truth or falsehood, new types of explanations and/or new criteria, laws, classifications
(4) Being self-authenticating
(5) Representing a sharp break in the history of Western thought

In Sect. 3.2 I had pointed out that the statistical style of reasoning possesses the characterizing features *1)... 5)*. In this section I have explained that Hacking has never listed together all the characterizing features of style of reasoning *1)...5)*: first he listed features *2)*, *3)* and *4)*; in his more recent writings he added and stressed that styles are not only ways of thinking but also ways of doing. We can invoke Wittgenstein's idea of family resemblance and think of styles of reasoning as a variety of ways of thinking belonging to a single species and connected by many overlapping similarities (to possess most of the characterizing properties above), where no one property is common to all of them. Ultimately, a 'style of reasoning' in the sense of Hacking is a way of thinking that possesses most of features *1)...5)*, ***which I will call the characterizing features of styles of reasoning or the characterizing properties of styles of reasoning.***

It is important to note that the set of styles of reasoning is not empty: not only does the statistical style of reasoning belong to it, as I have illustrated, but also does the laboratory style of reasoning, as I am going to show. In Chap. 5, I shall argue that other ways of thinking have most of the characterizing features of styles of reasoning (*1)...5)*) and can be called styles of reasoning. It is my contention that by this move it is possible to provide a better characterization of scientific thinking: there will be a set of ways of thinking that will be called styles of reasoning while other ways of thinking will satisfy only few of the characterizing features of styles of reasoning.

3.5 The Laboratory Style of Reasoning

In this section I shall show how some of Hacking's ideas can be connected with one another and used to argue that there is a way of thinking that satisfies the characterizing properties of styles of reasoning: the laboratory style of reasoning. Crombie placed the emergence of the experimental style in ancient Greece and was disinclined to see any sharp break, any discovery of a new human potential in the long development of experimental thinking. On the contrary, Hacking asserted that the programme of experimentation developed in the seventeenth century introduced a novelty in the way humans used experimenting and modelling to know: the invention of a place, the laboratory, in which instruments are built in order to elicit phenomena that do not exist in isolation. This point deserves an explanation.

In his *Representing and Intervening* (Hacking, 1983b), Hacking had highlighted certain features of laboratory research. For instance he had distinguished between *effects* and phenomena. Substantially, Hacking separated phenomena that can be recognized and observed without a direct intervention in the world from phenomena that do not exist in isolation but require special devices to be elicited (i.e. effects). He gave the example of the Hall effect, i.e. the voltage difference induced by a magnetic field perpendicular to an electrical current. The Hall effect occurs anytime and anywhere in nature once there is a particular arrangement of conductors and magnetic fields: for Hacking, it is a kind of regularity different from, say, the rise and fall in sea level produced by the attraction of the moon and the sun. The Hall effect requires a piece of equipment to be 'noted', whereas the phenomena connected with the tidal force are 'given in nature' and only need to be observed. Hacking often uses the verb 'to create': by saying that certain phenomena are 'created' in laboratory or are 'artificial' he just wants to say that nowhere in nature is there an arrangement such as the apparatus that produces the Hall effect. More importantly, '[an effect] can be seen as a regularity only against the further background of theory' (Hacking, 1983b, p. 225), i.e. as I interpret him, only against a further theory of the working of the apparatuses that produce it.

The concept of the laboratory style of reasoning is meant to account for this feature of scientific research: the production of effects. When did it start? As I have said, Hacking mooted that the production of effects settled into routine in scientific research in the seventeenth century. He drew his answer from Shapin and Schaffer's *Leviathan and the Air Pump*, which he considers a book about the origin of the laboratory style of reasoning, as I am going to explain. This book presents the thesis that in the late 1650s Boyle and the English experimental community, still in its infancy, established new rules of discourse by which matters of fact had to be both generated and defended. They maintained a certain form of discourse by agreeing upon what should be considered a fact or a hypothesis, how disputes had to be settled, how matters of facts had to be explained.

Beside the air pump, the other protagonist of *Leviathan and the Air Pump* is the philosopher Thomas Hobbes (1588–1679), who takes the role of Boyle's opponent: he did not accept the conventional basis of the latter's way of producing knowledge. Hobbes was an opponent of the social setting and the practice of acquiring knowledge, with its conventions and standards of truth, which was about to be widely adopted in scientific research.

Shapin and Schaffer tell that Boyle instructed Robert Hooke (1635–1703) to build for him an air pump, a closed vessel (receiver) from which air could be extracted by using a pump-mechanism. Boyle produced an approximate vacuum in the receiver so to be able to perform a series of experiments to investigate the composition and compressibility of air (e.g. he repeated Torricelli's barometric experiment enclosing a Torricellian apparatus in the receiver). These experiments were carried out in what Shapin and Schaffer call the new *place*, the laboratory, namely a site for *finding out* by *making*: for example, finding out that the air is necessary for a candle flame by 'creating' the phenomenon of vacuum in a vessel. The nascent laboratory was a 'public' space: to be public meant that the validation and the constitution of a piece of knowledge was obtained when all the individuals who attended the performance of an experiment agreed about what they saw. An aspect that Shapin and Schaffer contrast with 'the private work of the alchemists [...], who produced their knowledge claims in a private and undisciplined space' and with 'radical individualism—the state in which each individual set himself up as the ultimate judge of knowledge' (Shapin & Schaffer, 1989 [1985], p. 78). However, the new place was also 'private': only the opinion of those who were able to understand how an instrument works and followed certain conventions could contribute to the validation of knowledge.

Shapin and Schaffer's book is considered one of the most influential approaches to social constructivism. For instance, Jan Golinski wrote that '[for Shapin and Schaffer both the laboratory] and the wider realm beyond its walls can be viewed as an arena in which knowledge is constructed' (Golinski, 2005 [1998], p. 37). Indeed Shapin and Schaffer argued that 'Boyle's technologies [of persuasion] contributed to a common strategy for the constitution of the matter of fact [...] That is to say, each technology functioned as an *objectifying resource*' (Shapin & Schaffer, 1989 [1985], p. 77). In this sense, Hobbes against Boyle, as conceived by Shapin and Schaffer, is a quarrel about what should count as evidence: is evidence what we find among us, as Hobbes thought, or is it what apparatuses produce, as Boyle suggested? Both Hobbes and Boyle, argue Shapin and Schaffer, had justified and rational arguments, but those of Hobbes had no force: had the social context been different, Hobbes's objections would have found a different reception. For Shapin the social world of seventeenth-century England with its values of civility, honour and integrity formed the basis in which knowledge was established (Shapin, 1994a). These ideas fit the constructivist standpoint according to which truth is the product of contingent historical or social forces.

3.5.1 New Evidence and Discontinuity

Hacking wrote that '[Hobbes] saw that [Boyle's] was a new and threatening style of reasoning' (Hacking, 1991, p. 240). 'Hobbes saw' because he was *aware* of the conventions by which knowledge was acquired and validated in the new place that developed into the laboratory. Put differently, Hobbes was able 'to see from outside' the features of the new emerging style and did not accept them. The first of these features is represented by a new type of *evidence*: for Boyle 'evidence' was not only what can be found as 'given' in nature but also what can be *made* by man. Hobbes, for his part, refused to accept that the result of an experiment could represent the evidence on which to base our claims of knowledge. His point was that all the experiments conceal theoretical assumptions, e.g. about the functioning of the apparatus and the setup of the experiment, whose validity could always be disputed. For him, whatever explanation was made to account for an artificial phenomenon, there is always a superior explanation that proceeds from different assumptions about the experiment.

The emergence of a community of people that used as a new type of evidence phenomena produced by devices such as the air pump, which did not previously exist in isolation, is a punctuated event in the history of science. To illustrate this point, let us look at the social context in which Galileo worked. Although just a handful of years separates Galileo and

Boyle (Boyle was 16 years old when Galileo died) their milieus could not have been more different. Mario Biagioli has argued that 'Galileo and his readers did not belong to a professional community that could provide the kind of credit and rewards he sought' and that in his field there was a 'lack of consensus about style of argumentation and standards of evidence as well as [a] scant interdependence among the members' (Biagioli, 2006, p. 97). Biagioli shows that Galileo's discoveries were discussed in an intellectual milieu radically different from that of Boyle. The former's social setting included scholars dispersed over different universities and courts linked through correspondence networks; the latter included a tiny elite of people in London, who explored the 'secrets of nature' *in the same public place*. Galileo's milieu could do little to favour the establishment of new standards of truth; the mutual trust of the gentlemen of London's scientific societies was crucial for the coming into being of a new form of evidence. It must also be added that for Galileo 'experience' did not mean anymore to draw generalizations by induction and intuition from the observation of 'ordinary' things, as Aristotle advocated, rather it meant to 'construct' observation by mathematics. With Galileo a new criterion for selecting which observations were meaningful emerged: what could not be quantified and mathematically analysed had to be considered less important, especially with a view to proving or disproving a theory.

How radical was the metamorphosis brought about by Boyle's community appears even more evident when one considers that until the first decades of the seventeenth century it was impossible to trace a demarcation line between 'magicians' and 'scientists', i.e. between the hermetic tradition and the emergent laboratory style. Many scholars, included Boyle, accepted various hermetic doctrines (see Trenchard More, 1941) *and* were committed to the laboratory style. But in the span of Boyle's life, 'secrecy for science became a *disvalue*' (Rossi, 2000, p. 34)—making one's knowledge *public* became a value whereas keeping one's opinions secret came to be considered akin to treason. Did the creation of phenomena mark out the beginning of a new style of thinking and doing? I shall answer this question in the next subsection.

3.5.2 A New Way of Thinking and Doing

Once a question about the world is asked, e.g. whether or not the flame of a candle needs air to burn, finding out the answer in the style of laboratory means, one can imagine, to perform three types of mental and physical actions. The first type comprises actions such as to design and calibrate a machine that, as Shapin and Schaffer say, 'stands between the perceptual competences of a human being and natural reality itself'(Shapin & Schaffer, 1989 [1985], p. 77). These actions involve not only the brain but also the body:

> Doing is more important for the laboratory than for mathematics [...] In the hypothetical-deductive reasoning of laboratory science I include the making and the calibrating as well as the designing and the deducing. (Hacking, 1991)

The second kind of mental actions is illustrated by this passage:

> The final 'working' of air pump technology is a matter of using a lot of different plastic resources—the material apparatus, the background theory, ideas about what the machine does, and how things in it behave—each of which is adapted to the other. (Hacking, 1991, p. 237)

In brief, the actions of the second type are represented by choices that make different epistemological elements of the experiments cohere: what is considered a fact, theories about nature, the setup of the apparatus and the pieces that constitute it, the interpretation of data. For example, the actions of the second type include: to distinguish matters of fact from speculative hypotheses and to establish how theory and interpretation of the results of the experiment are connected.

Finally, the previous two types of mental and manual actions are made visible to everybody so that the knowledge produced becomes the result of a collective enterprise. This is the aim of the third type of actions. Particular rules of discourse and conventions of social relations are followed, e.g. what is to be considered as 'the experimental fact' is isolated and described in detail so as to facilitate its replication by as many people as possible. In Shapin's account, the third type of actions evolved out of the early programme of Boyle and the Royal Society, who wanted to 'build a solid factual foundation for a reformed natural philosophy by soliciting more and more testimony and extending networks of justified trust further and further' (Shapin, 1994a, p. 205)—by the third type of actions the fact becomes objective.

3.5.3 *Techniques of Self-vindication*

The actions of the second type are relevant for grasping Hacking's thesis that the laboratory style of reasoning is self-authenticating.[6] Shapin and Schaffer explained that Boyle, in his *Two Essays Concerning the Unsuccessfulness of Experiments*, offered a repertoire of reasons why an experiment might not produce the expected outcome: from the construction defects of the materials to their composition, from the context of the experiment to the background theories involved (Shapin & Schaffer, 1989 [1985], p. 185). However, they noted, he never considered any of these reasons to reject the idea of air's pressure: for example, when two cohered thin discs placed in the bulb of the air pump did not separate as expected for the absence of air, Boyle blamed the leakage and continued to refuse his opponents' idea of *horror vacui*. Shapin and Schaffer concluded:

> Any physical hypothesis could be saved from an admitted experimental failure either by pointing to a range of subsidiary hypotheses or by modifying the key hypothesis so that this could be seen as unaltered for all practical purposes. (Shapin & Schaffer, 1989 [1985], p. 186)

In 1992, in the paper 'The Self-Vindication of the Laboratory Sciences' (Hacking, 1992a), Hacking made a point that can be put in relation to Shapin and Schaffer's: laboratory sciences tend to produce a sort of self-vindicating structure in that background theories, hypotheses, modelling of apparatus and data analyses are mutually adjusted to each other. To start with, the concept Shapin and Schaffer express in the passage above could be reformulated in terms of the Duhem thesis: in the *modus tollens* if p entails the evidence e then—e implies—p; however, Pierre Duhem (1861–1916) observed that it is always p plus a, where a is the whole of auxiliary hypotheses and background knowledge, rather than only p, which implies e. Therefore, if—e, one can *only* conclude that either p and/or one of the auxiliary hypotheses included in a are/is false. Going by the *logical* point made by Duhem, Hacking pointed out 15 different 'plastic resources', included in a, which can be modified in order 'to keep p alive'. Here is his full list of the plastic resources:

[6] I should make it clear that Hacking has never distinguished different types of actions within a style of reasoning.

(1) *questions* about theories or the subject matter of the experiment;
(2) *background knowledge*, that is those beliefs that are taken for granted;
(3) *systematic theory*, namely general theories of high level about the subject matter of the experiment (e.g. quantum theory can be taken as the systematic theory for the Stern-Gerlach experiment);
(4) *topical hypotheses*, that is rules that connect theoretical concepts or symbols with observational terms (e.g. how the concept of speed vector is connected with the observed values of the motion of a body);
(5) *modelling of the apparatus*, theories or background lore about how the apparatus works;
(6) *target*, which indicates a part of the experimental material, such as a population of bacteria to be studied;
(7) *source of modification*, it is that part of the experimental apparatus that interferes with the target, e.g. a beam of light;
(8) *detectors* to measure the modification of the target;
(9) tools, which are the objects on which an experimenter must count, such as a chemical substance or a ruler;
(10) *data generators*, which are particular tools such as a camera or anything that counts events;
(11) *data* such as graphs, tables, photographs, measurements of physical magnitudes and so on;
(12) *data assessment* such as statistical or not statistical estimations of errors;
(13) *data reduction* such as theory neutral mathematical methods to transform types of data into other kind of data, e.g. Fourier transforms;
(14) *data analysis*, a kind of data reduction that is not theory neutral;
(15) *interpretation* of the data, which is always made by using a theory (Hacking, 1992a) (Hacking, 1988).

Hacking thinks that in laboratory research there is an interplay not only between theory and observation, as certain scholars had suggested before him (Pickering, 1984, p. 8), but also among *all* the 15 elements listed above:

> We can 1) change questions; more commonly we modify them in mid-experiment. Data 11) can be abandoned or selected without fraud; we con-

> sider data secure when we can interpret them in the light of, among other things, systematic theory 3) [...] Data processing is embarrassingly plastic [...] in the case of data assessment and reduction, 12) and 13) [...]. Data analysis is plastic in itself; in addiction, any change in topical hypotheses 4) or modelling of the apparatus 5) will lead to the introduction of new programs of data analysis. (Hacking, 1992a, p. 54)

Hacking's conclusion is that 'our preserved theories and the world fit together so snugly less because we have found out how the world is than because we have tailored each to the other' (Hacking, 1992a, p. 31). For him theories are not refuted or confirmed by a *direct* comparison with the world but by making all the other 14 items in the list above cohere. If anything, he thinks, we should limit ourselves to saying that the sciences of laboratory are true to the phenomena elicited by man-made machines. This feature makes the laboratory science a sort of self-vindicating and stable structure. Turning to the terminology of styles of reasoning, I express his point by saying that there is a circularity in the way of thinking and doing of the laboratory style: in order to confirm theories, experimental apparatuses that provide data are made; and how do we know that the experimental apparatus has been designed properly and the data are correct? By checking whether or not the data fit our theories. That is to say, the laboratory style possesses a technique of stabilization that consists of the iterative back and forth, whereby the 15 elements above are compared and discussed. In a nutshell, the symmetrical exchanges between theories and data that go on in laboratory research make its style self-authenticating.[7]

The practice of climate science is a case in point. It would be wrong to say simply that empirical data validate climate models, as it is shown by a study of Hélène Guillemot (Guillemot, 2010). She explained that the simulations in climate science are confronted with data provided by three types of sources: networks of meteorological stations; instruments on satellites; data collected during field experiments. These sources do not provide 'raw data': at least three models that combine the initial data with other factors are necessary to acquire the surface temperature from a signal

[7] It is important to note a point that will be clearer later: Hacking does not draw constructionist conclusions from the plasticity of the 15 items. He only notes that this plasticity makes the style of reasoning self-authenticating and therefore science stable (see Hacking, 1988). For him only a little number of the possible combinations of the 15 items above can endure. As he says, 'Not all the descriptions [of a phenomenon] can be accepted but there is no reason for thinking that there is only one possible description' (Hacking, 2005a, p. 5). His metaphysics is anti-constructivist *and* pluralist.

captured by a satellite; and computer models are employed to harmonize all the data coming from the surface stations (Guillemot, 2010, p. 245). Among the different kinds of validation of models, Guillemot outlines the features of the so-called bottom-up validation used for example for the parametrization of clouds in the global climate model. In that case, the model equipped with the parametrization is tested by comparing it with another model validated by observation (Guillemot, 2010, p. 247). In other cases models are used to complete the data or to test the methods of their interpretation (Guillemot, 2010, p. 249). Therefore, the practice of climate studies reveals that there are symmetrical exchanges between many elements mentioned by Hacking.

3.5.4 *New Candidates for Truth or Falsehood*

Up to now, I have described some features of the laboratory style. It is a new way of thinking and doing that involves three types of mental and physical actions originally practised within a small community of thinkers in the seventeenth century. These thinkers brought to the fore a new type of evidence—the evidence produced by man-made phenomena—by which the knowledge produced was validated. Hobbes refused to embrace this new criterion of validation of knowledge and did not accept the rationale of the three types of mental and physical actions. He maintained that the philosopher's model for knowing consequences and causes was provided by geometry.[8] Since the postulational style of reasoning adopted in geometry is a style of reasoning, as I shall show, it is possible to say that Hobbes adopted a style of reasoning, the postulational one, different from that adopted by Boyle, the laboratory one. Notice that Boyle opened a new space of discourse in which new questions were admissible, a feature that, as I shall explain, belongs to all the style of reasoning. In Shapin and Schaffer's words:

> Was the Torricellian space a vacuum? Did the exhausted receiver constitute a vacuum? The platform from which Boyle elected to address these questions was *experimental*: the way of talking appropriate to experimental philosophy was different in kind to existing natural philosophical discourse. (Shapin & Schaffer, 1989 [1985], p. 45)

[8] Shapin and Schaffer cite Hobbes's *Leviathan*: 'As when we know, that, if the figure shown be a circle, then any straight line through the centre shall divide into two equal parts [...] And this is the knowledge required in a philosopher' (Shapin & Schaffer, 1989 [1985]).

The last feature of the laboratory style I shall deal with concerns the point that the birth of the laboratory style brought into being new candidates for truth or falsehood. Hacking provided examples of both observational statements whose truth-value is independent of the laboratory style and theoretical statements whose truth-value is determined by that style itself (style-dependent statements). He noted that the sentence 'my skin is warmed', which is used by William Herschel (1738–1822) in his theory of heat to describe the effect of filters of some colours on his skin, is of the former class and referred to it as a 'sense-datum sentence' (Hacking, 2002 [1982b], p. 173). Conversely, Hacking cited another sentence formulated by Herschel:

The heat which has the refrangibility of the red rays is occasioned by the light of those rays.

According to Hacking this sentence is style-dependent:

> If another culture has acquired the styles of reasoning enumerated by Crombie, it can perfectly well learn Herschel's physics from the ground up—that is just what I do in making sense of Herschel's text. The problem is that the sufficiently foreign person will not have Herschel's kind of sentence as the sort of thing that can be true-or-false, because the ways of reasoning that bear on it are unknown [...] [Archimedes] would not be able to effect a translation until he had caught up on some scientific method. (Hacking, 2002 [1982a], p. 173)

In 1800 Herschel proved that the different colours of the spectrum are of different temperatures and discovered the infrared radiation of sunlight (Herschel, 1800b). He used prisms or filters (i.e. 'man-made devices') in his telescopes to create a 'new effect': the separation of the rays of light; then he measured the temperature of each of them by a thermometer. In some cases he even tried to feel the rays from different filters as heat on the skin.

Hacking argues that Archimedes did not possess the style of reasoning of Herschel and therefore he could not make sense of Herschel's sentence. To interpret Hacking one can focus on the fact that the important novelty introduced by the laboratory style—that phenomena elicited by apparatuses constitute the evidence for certain propositions—is also an essential element of what, since the seventeenth century, is considered the scientific method. Bearing this in mind, Hacking's point would amount to say that Archimedes could not make sense of a sentence such as 'the heat which has the refrangibility of the red rays is occasioned by the light of those rays' because its sense hinges on a new kind of evidence which emerged only at Boyle's time.

Although it is clear enough that Archimedes did not possess the methods that bear on Herschel's sentence, I do not think that Hacking's is the best example that can be given. Herschel's effect was not really a phenomenon that 'did not exist in isolation': it is well known that a spectrum of light is *naturally* produced in the sky by reflection and refraction of sunlight by rain droplets in the atmosphere. A better example might concern the early experiments on the scattering of alfa-particles in Rutherford's laboratory. These experiments elicited a phenomenon never seen before: we need a particular arrangement of experimental apparatuses to see the effects of the scattering of a beam of charged particles. So, an example of style-dependent (theoretical) sentence in the laboratory style of reasoning is:

The scattering has cylindrical symmetry about the beam axis and therefore the cross section is independent of the azimuthal angle.

Among the style-*independent* sentences, there is a particular class that I want to bring to the fore for my purposes in Chap. 7. In his book *Representing and Intervening* (Herschel, 1800a, 1800c), Hacking defended a particular form of realism according to which the word 'real' refers to the possibility for a class of entities of affecting macroscopic objects. He is a realist about entities such as electrons because we can use them to intervene in the world affecting other objects. For example, he says that what convinced him of realism was the fact that in some experiments scientists succeed in spraying positrons to increase the charge of a niobium ball:

> If you can spray them then they are real. (Hacking, 1983b, p. 22)

The point I want to mention here and explain in detail later is that for Hacking the causal effects produced by particles are described by observational propositions, i.e. propositions that are not 'theory-loaded', a catchword used by Hacking to say that they do not carry a 'load of theory'. For him propositions such as:

> The gauge indicates that the niobium ball is altered.

must be considered observational and consequently style-independent because their sense, Hacking claims, hinges on a kind of evidence—the 'evidence of senses'—which is ahistorical. To sum up, in this section I have connected some ideas of Hacking scattered in different writings in order to make this point: around the time of Boyle a new style of thinking and doing, which satisfies all the characterizing properties satisfied by the statistical style of reasoning, emerged. From now on, I shall refer to it as the laboratory style of reasoning. It is a corollary of my argument that Shapin and Schaffer's account in *Leviathan and the Air Pump* can well be reread as a quarrel between two men that adopted different styles of reasoning.

3.6 Daston and Galison's 'Epistemic Virtues'

Daston and Galison's book *Objectivity* (Daston & Galison, 2007) charts the history of nature of new conceptions of objectivity that emerged in the last three centuries. They based their study on the analysis of scientific atlases from different fields of research. Atlases are 'those select collections of images that identify a discipline's most significant objects of inquiry' (Daston & Galison, 2007, p. 17). Daston and Galison think that their centrality comes from the fact that they 'set standards for how phenomena are to be seen and depicted' (Daston & Galison, 2007, p. 19). They analysed atlases in order to gain insights into three conceptions of objectivity that came about—they argue—in the last three centuries: *truth-to-nature*, *mechanical objectivity* and *trained judgement*. Truth-to-nature emerged in the eighteenth century when scientists started to be afraid of being overwhelmed by the chaos of sensations and the intricacy of natural phenomena: too many non-essential details could hinder the knower from describing properly a phenomenon. Lisa Jardine described some technological factors that became catalysts for scientific advance in the seventeenth century: enhanced microscopic observation, chemical substance analysis, autopsy, *camera obscura* (Jardine, 2000 chapter three). What her account reveals is how these new techniques produced a blinding intricacy of details that made it easy to fail to see the whole picture of a phenomenon. As a reaction to all that, Daston and Galison argue, scientists appealed to reason in order to find the hidden order behind the monstrosity of nature; they trained themselves to prune, shape, select and represent things in such a way that only what was *typical* of a certain class of objects could be recognizable. For example, eighteenth-century atlases show images of *ideal types* that mean to illustrate the fully developed features of objects and be truer to nature than a specimen (Daston & Galison, 2007, p. 56).

In the first decades of the nineteenth century, scientists started to criticize the artistic portrayals and praise the techniques of photography that were emerging. They ceased being concerned about being overwhelmed by nature and started to fear to project their theories into nature, deforming it. As a consequence, they tried to repress their predisposition to idealizing, pruning and perfecting. For example, Daston and Galison quoted the Berlin bacteriologist Carl Fraenkel (1861–1915) as saying:

> A drawing can only be the expression of a subjective perception and therefore must, from the beginning, renounce the possibility of an objection-free reliability. [...] The photographic plate, by contrast, reflects things with an inflexible objectivity as they really are. (Daston & Galison, 2007, p. 177)

Daston and Galison considered objectivity as a 'passionate commitment to suppress the will, a drive to let the visible world emerge on the page without intervention' (Daston & Galison, 2007, p. 143) as the manifestation of another scientific ideal that they called *mechanical objectivity*.

In the twentieth century mechanical images became complex and needed to be interpreted and compared. According to Daston and Galison many scientists began to question the current way of representing (and knowing) nature: they regarded certain images as unreliable because of the abundance of incidental details and artefacts. As an example of the inadequacy of mechanical objectivity, as perceived by scientists, consider that the radiologist Rudolf Grashey (1876–1950) wanted to classify x-ray images of human organs and bones in order to distinguish the normal, with all its variations, from the pathological. The mechanical registration of images could not be the solution: he was forced to decide which of them represented deviations within normality and choose some of them as a boundary post of the normal (Daston & Galison, 2007, p. 344). In short, the insistent drive to repress the wilful intervention of the scientist did not prove adequate to answering the need of interpreting, highlighting, comparing and handling images.

In doing research, the images that any machine could produce were not sufficient to understand phenomena; in communicating and teaching science, some representations uncontaminated by human intervention were of no help to explain what was to be considered as pathological in a body or to distinguish a new particle among the tracks of myriads of decay particles. For example, in the *Atlas of Electroencephalography* (1941), the neurologists Frederic A. Gibbs (1903–1992) and Erna L. Gibbs (1906–1987) declared that their book had been written 'to train the [reader's] eye so that he can arrive at diagnoses from *subjective* criteria' (Daston & Galison, 2007, p. 321). In many cases a *subjective* decision was a necessary supplement to mechanical images or the mathematical techniques. Daston and Galison called this supplement *training judgement*. The word judgement stands for the ability of the researcher to make decisions without following standard procedures; it is a *training* judgement because it is a kind of perception one can be trained for. Lastly, in the twentieth century, fidelity to nature gave way to an engineering-inspired way of knowing in which making, manipulating or simulating, more than depicting, are considered the best method to gain insight into the phenomena. In *Objectivity* Daston and Galison also mentioned another form of objectivity, *structural objectivity*, which counters the hermetic privacy of the self: the visual was to be distrusted and only relationships and structures were to be relied upon (Daston & Galison, 2007, pp. 253–307). All these different conceptions of objectivity can be viewed as other transient substitutes of the timeless Kantian conditions of possible experience: by imposing them on the raw sense data they make a set of data objective experience.

At first it would seem that Daston and Galison prise the concept of objectivity away from any other notion. According to this misreading, they would *limit* themselves to identifying different ontological, epistemological, methodological or moral senses of objectivity and to tracing their genesis and development in scientific practices. Actually, as I am going to explain, each conception of objectivity is rooted in an *epistemic virtue*, as Daston and Galison call the scientific ideal to which scientists are committed in a particular period or circumstance. Indeed, consider first that for Daston and Galison

> Scientific objectivity always counters some aspect of the self, but not always the same one. This is why the genus objectivity embraces the species of both mechanical and structural objectivity—and no doubt potentially others as well. Because the subjectivity is multifarious, objectivity must be too. (Dear et al., 2012, p. 31)

For example, mechanical objectivity is the expression of the fear of one's subjectivity (Daston & Galison, 1992, p. 83): it opposes the post-Kantian unified self, organized around the will (Daston & Galison, 2007, p. 199). On the other hand, the enemies of truth-to-nature and trained judgement are external, Daston and Galison argue: the overwhelming wealth of irrelevant variations of nature for the former and the clutter of incidental details and artefacts introduced by mechanical objectivity for the latter. The battle is fought by what Daston and Galison depict as different prototypical knowers of nature, which in turn correspond to the different epistemic virtues: the 'insightful sage' (truth-to-nature), the 'diligent worker' (mechanical objectivity), the 'trained expert' (trained judgement) (Daston, 1999, pp. 98–100) (Daston & Galison, 2007, pp. 216–233). These scientific *personas* cultivate particular traits of the self and pursue different values at the expense of others. Epistemic virtue is therefore the term with ethical overtones that Daston and Galison use to describe the values pursued by scientists in the battle for knowledge against powerful enemies such as those just mentioned. In this sense truth-to-nature, mechanical objectivity, trained judgement can be seen either as epistemic virtues *or* as different forms of objectivity. Although the epistemic virtues oppose different sorts of subjectivities, they all serve a common ideal, that of being faithful to nature. So, one can say that Daston and Galison present a history of the ways of being faithful to nature, a point that I will develop in the next chapter from a different perspective.

I wish to insist on the fact that in Daston and Galison the concept of epistemic virtue is inseparable from the concept of objectivity: there is an epistemic virtue for each form of objectivity. Rather than saying that the concept of epistemic virtue is prior to the concept of objectivity, it would be better to say that it is simply the other side of this latter or, *rectius*, that each epistemic virtue represents the other side of *a* conception of objectivity. As Daston and Galison put it, each conception of objectivity is 'rooted' in an epistemic virtue or, equally, in an epistemic fear (Daston & Galison, 2008, p. 671). In conclusion, there is a plurality of epistemic virtues or, alternatively, different forms of objectivity, which oppose different sorts of subjectivities and serve the common ideal of being faithful to nature.

3.7 Reinterpreting Styles of Reasoning

3.7.1 Truth and Truthfulness

Hacking's writings that date from after 2000 present an interesting rereading of the concept of style of reasoning inspired by the philosopher Bernard Williams. In 2002, the latter published *Truth and Truthfulness* (Williams, 2002) an essay that explores the value of truth and the meaning of truthfulness. For the benefit of his argument, Williams reflected on different commitments to truthfulness that originated in certain historical periods, took over one from another or coexisted. Hacking found this set of reflections on truth-telling very interesting (Hacking, 2004) and redescribed his notion of style of reasoning from Williams's perspective. In this and the next section, I shall compare Williams's account of truth and truthfulness with Hacking's views in order to explore the philosophical import of the notion of style of reasoning. This analysis will not be conceptually separated from the content of the previous section: once I shall have explained Hacking's point that styles of reasoning are different 'ways of being truthful', the notion of epistemic virtue will look much more as tight related to that of style of reasoning.

In his book Williams points out that some philosophers, called 'deniers' (Williams, 2002, p. 5) are suspicious of the fact that any historical account can aim at truth: the stories that are told about the past, e.g. that of European anti-Semitism or the French Revolution, are always biased or ideological. Therefore, they say, we should admit that the concept of 'pursuit of truth' is empty and truth or truthfulness cannot be considered as values. Williams defends the opposite thesis that *any* human society needs to possess truth and truthfulness as values. His move is to look at truthfulness from a historical perspective in order to provide a narrative that reveals the role of truthfulness in different societies, the way it came about and developed. Indeed, according to Williams, truthfulness has a history and telling it will show that all the societies need dispositions of truthfulness and the sense of the value of truth, otherwise they 'lose everything' (Williams, 2002, p. 7). I shall not examine the whole of Williams's reply to the deniers; instead, I shall focus on those parts of his argument that are interesting for my purposes. In particular, I shall make it clear that both Williams and Hacking start from a common premise; then, discussing an example of Williams about how truthfulness changes, I shall make some more points about the concept of style of reasoning.

The premise of Williams's account of truthfulness and truth is that they are radically different categories—truthfulness has a history, truth does not. For him, truthfulness essentially means to tell stories respecting truth, i.e. to tell the truth about something. In turn this means to be accurate and sincere: *accuracy* implies doing one's best to acquire true beliefs and *sincerity* implies telling exactly these beliefs to other people. These two dispositions have a history because the way of being sincere or accurate has changed throughout different cultural situations and ways of thinking: for example, Williams argues that '[...] a set of ideas, which associate sincerity with personal authenticity, [...] came into distinct existence in the eighteenth century' (Williams, 2002, p. 172) and, as we shall see in the next subsection, Denis Diderot (1713–1784) and Jean-Jacques Rousseau (1712–1778) had two distinct ideas of sincerity and authenticity.

While sincerity and accuracy, and thus truthfulness, are historically changeable, the concept of truth has never changed. By saying this, Williams means that the idea of what truth is, the role that this idea plays in talking, thinking, arguing and so on has always been the same in history: the concept of truth for an Egyptian or a Greek is absolutely the same of that for a European in the Middle Ages or for us; even if, telling their stories, they all have been answering to different standards or values. This view is entirely shared by Hacking. Although in his review of *Truth and Truthfulness* he points out that Williams's claim that truth is a human universal is 'more of an assertion than an argued statement' (Hacking, 2004, p. 140) it remains the key premise of the styles project: 'I accept as a convention that truth is a formal concept, without content and history' (Hacking, 2006d, p. 3). I invite the reader to bear this important claim in mind as my argument proceeds in the next chapters. For the time being I just want to make it more precise: by saying that styles of reasoning settle what to be objective means, Hacking does not want to say that logic and the concept of truth change through time. Indeed, one's exercise of rationality involves logical argument and evidence (in different forms); it is only what is considered as amenable to public evidence, i.e. what is objective, which changes switching from a style into another, not the types of inferences (deduction, induction and abduction) used in arguments and the concept of what truth is.

Shapin's view in *A Social History of Truth* (Shapin, 1994b) is not at odds with Williams and Hacking's claim that the idea of what truth is, the role that this idea plays in talking, thinking, arguing, does not change through time. Indeed, Shapin argued that in the seventeenth century the decision of whether or not to believe in the interpretation of an experiment was largely a question of what kind of person to trust. For him, in the community of the Royal Society it was the gentleman who was counted trustworthy because of his moral qualities:

> Gentlemen were truth-tellers because nothing could work upon them that would induce them to be otherwise. (Shapin, 1994a, p. 84)

therefore:

> If one wished to make experimental *knowledge*, then *here* [in the laboratory] were the technical, social and discursive means with which it might be made. (Shapin, 1994a, p. 127 my emphasis)

Conversely, for Hobbes:

> The geometer [...] could do better than the experimentalist. (Shapin & Schaffer, 1989 [1985], p. 151)

in that:

> when its methods [are] rightly followed, geometry yield[s] irrefutable and incontestable *knowledge*. (Shapin & Schaffer, 1989 [1985], p. 100)

These passages show that Hobbes and Boyle possessed the same concept of knower: it was someone who possessed a true belief and a property *X*, no matter which one, which correlated well with it. They only disagreed on *which* property *X* had to be attributed to the knower: for Boyle it was the property of possessing the methods of the laboratory style; for Hobbes it was the property of possessing the methods of the postulational style. In other words, Boyle and Hobbes disagreed on the answer to the following question: who were the trustworthy informants? However, they did not have a different concept of what a true belief is. For both of them 'knowledge' had the same meaning: it was a 'tag', so to speak, that could be attributed to the experimenter (for Boyle) or to the geometer (for Hobbes). Whether or not Hacking's styles project implies that scientific knowledge *is* social knowledge, i.e. no more and no less that knowledge about which people agree, as Shapin thinks, will be discussed in Chap. 7.

3.7.2 Styles of Reasoning and Epistemic Virtues as Ways of Telling the Truth

Having discussed Williams's premise, I shall now examine more closely what he means by saying that truthfulness can change. In *Truth and Truthfulness* he illustrates the different conceptions of the self associated with Jean-Jacques Rousseau (1712–1778) and Denis Diderot (1713–1778). Rousseau's project in his *Confessions* was to reveal himself to his readers. One of his main questions was to understand how to bring about a concord between what one is for oneself and what one is for others (Williams, 2002, p. 175). A passage of Williams explains what Rousseau's answer was

> Sincere, spontaneous, non-deceitful declaration, the product of his presence to himself, will guarantee a true understanding of [one's] motives. Moreover, what is revealed and understood in this way will represent a character, a whole person, and this implies that it will be coherent, or, as one might say, steady. (Williams, 2002, p. 178)

A precise conception of sincerity underlies this answer: sincerity means to declare what is immediately evident to oneself. In turn this conception implies an idea of the self as a set of constant unchanging features ('steadiness') that can be revealed once and for all.

Williams contrasts Rousseau with Diderot. The latter wrote *Rameau's Nephew* (published in 1805), an imaginary conversation between 'moi', the narrator, and 'lui'. Diderot wanted to understand how the interaction with other people could stabilize the whimsical inconstant mental constitution of people. Indeed, Rameau is described as a person who does not know who he is, '[who] expresses different things at different times' (Williams, 2002, p. 190). As Williams explains:

> Diderot was always attracted to a picture of the self as something constantly shifting and reacting and altering [...] for him the declaration at a given instant of self can be only a declaration of self at that instant. (Williams, 2002, p. 190)

To sum up, while for Diderot sincerity meant uninhibited expression, for Rousseau it meant to report the finding of self-examination.

While Williams, given his aims, limited his analysis to truthfulness about domains such as the past or the self, Hacking tried to generalize it to other domains of knowledge: 'I interpret [Williams's account] as a genealogy of the possibility of telling the truth in a domain or another' (Hacking, 2006d, p. 5). He proposed this schema:

> the emergence of a new style of reasoning is represented by a change in conceptions of what it is to tell the truth about X. (Hacking, 2004, p. 142)

In the case of Williams's case studies, X was 'the self' or 'the past', but for Hacking X extends to many other domains of knowledge: e.g. the geometrical relations in the case of the geometrical style. For example, following Hacking suggestion we can say that the style of geometry represents a shift of the conception of what means to tell the truth about the geometrical relations: whereas the Egyptians used algorithmic methods, the Greeks introduced geometry, their new method for telling the truth. Ultimately, the schema above, rather than being a proper definition of styles of reasoning, represents a different perspective suggested by Williams from which styles of reasoning are viewed as ways of telling the truth, distinct modes of being truthful.

Let us now consider Diderot's and Rousseau's ideas of the self: Diderot was committed to the uninhibited expression of himself, whereas Rousseau was committed to the non-deceitful declaration of feelings. By paraphrasing Daston and Galison it is possible to say that just as in telling the truth about the self. Diderot and Rousseau were committed to two different 'non epistemic virtues', two different conceptions of both authenticity and sincerity, so in telling the truth about nature scientists are committed to different *epistemic* virtues. Williams's account in ethics parallels Daston and Galison's in epistemology.

Moreover, according to Williams's schema, just as Diderot's and Rousseau's conceptions of the self represent two different ways of being truthful to it so styles of reasoning represent 'different ways of being truthful to nature'. Since there is no significant change of sense resulting from the substitution of the adjective 'truthful' with the adjective 'faithful' it is possible to say that according to Williams's schema styles of reasoning are 'ways of being faithful to nature'—and this is exactly one of the ways in which Daston and Galison have characterized the commitment to epistemic virtues. In short, styles of reasoning as well as the commitments to epistemic virtues can be considered as different ways of telling the truth about nature or, likewise, as different ways of being objective.

The comparison between styles of reasoning, epistemic virtues and ways of being truthful can be pushed further. Let us reconsider Diderot's and Rousseau's pictures of the self instead than their conceptions of sincerity. For Diderot the self was 'disintegrated' (Williams, 2002, p. 189), the result of spontaneity, its manifestation, it consisted of all that can be 'seen' and is 'external' in a given moment: e.g. jokes, lies and expressions. For Rousseau, it was an unchanging unity, all that is 'internal' and in need of being expressed: e.g. feeling, emotions and states of mind. In other words, both Diderot and Rousseau *selected* a particular aspect of the self and *excluded* others: the former directed his attention to the exterior product of spontaneity; the latter directed his attention to interior states of the self. In the previous section I had pointed out that also styles of reasoning and epistemic virtues *select* and *exclude*: each styles of reasoning has its own questions and each epistemic virtue teaches how to make sense of a certain sliver-world. Ultimately, styles of reasoning and epistemic virtues can be viewed not only as ways of being truthful but also as particular ways of looking at the portion of the world under study, which shed light on particular aspects of it and put others in the shade. This may well be the consequence of human inability of encompassing all the aspects of things.

3.8 Conclusions

In this chapter I have presented two different notions of the concept of 'ways of thinking' named 'style of reasoning' and 'epistemic virtue', which have been proposed, respectively, by Ian Hacking and by Peter Galison and Lorraine Daston. To explain how the former characterizes the notion of style of reasoning, I have reread his book *The Emergence of Probability* in terms of the notion of 'statistical style of reasoning' and Shapin and Schaffer's *Leviathan and the Air Pump* in terms of the notion of 'laboratory style of reasoning'. My analysis has shown that many points of Hacking's styles project need to be developed in order to make his suggestions into a full-fledged theory. Such a theory could help to characterize scientific thinking by exhibiting the styles of reasoning that compose it. A salient point of my analysis is that Hacking's is another of the projects of historical epistemology that historicize Kant. I have also shown that there are substantial convergences between the styles project and Daston and Galison's study on objectivity. Different conceptions of objectivity can be viewed as transient substitutes of the timeless Kantian conditions of possible experience and each conception of objectivity is rooted in an epistemic virtue. Finally, Williams's account of truth and truthfulness helps to show that styles of reasoning are ways of telling the truth and ways of looking at the world.

Bibliography

Bell, J. S. (1966). On the problem of hidden variables in quantum mechanics. *Reviews of Modern Physics, 38*, 447–452.

Biagioli, M. (2006). *Galileo's instruments of credit*. The University of Chicago Press.

Crombie, A. C. (1978, 1981). *Philosophical perspectives and shifting interpretations of Galileo.* Paper presented at the Theory Change, Ancient Axiomatics, and Galileo's Methodology: Proceedings of the 1978 Pisa Conference on the History and Philosophy of Science, Pisa.

Crombie, A. C. (1988). Designed in the mind: Western visions of science, nature and humankind. *History of Science, 26*(1), 1–12.

Crombie, A. C. (1992). *Stili di pensiero scientifico agli inizi dell'Europa moderna.* Bibliopolis.

Crombie, A. C. (1994). *Styles of scientific thinking in the European tradition* (Vol. 1,2,3). Duckworth.

Daston, L. (1999). The moralized objectivities of science. In C. Wolfgang & D. Lorraine (Eds.), *Wahrheit und Geschichte* (pp. 78–100). Vandenhoeck & Ruprecht.

Daston, L., & Galison, P. (1992). The image of objectivity. *Representations, 40*, 81–128.

Daston, L., & Galison, P. (2007). *Objectivity.* Zone Books.

Daston, L., & Galison, P. (2008). Objectivity and its critics. *Victorian Studies, 50*(4), 666.677.

Dear, P., Hacking, I., Jones, M., Daston, L., & Galison, P. (2012). Objectivity in historical perspective. *Metascience, 21*(1), 11–39.

Elwick, J. (2012). Layered history: Styles of reasoning as stratified conditions of possibility. *Studies in History and Philosophy of Science.* https://doi.org/10.1016/j.shpsa.2012.07.004

Feyerabend, P. (1988 [1975]). *Against method.* Verso.

Foucault, M. (1994 [1966]). *The order of things.* Vintage Books Edition.

Foucault, M. (2006 [1961]). *History of madness.* Routledge.

Golinski, J. (2005 [1998]). *Making Natural Knowledge.* The University of Chicago Press.

Gooden, P. (2009). *The story of English.* Quercus Publishing Plc.

Guillemot, H. (2010). Connections between simulations and observation in climate computer modeling. Scientist's practices and "bottom-up epistemology" lessons. *Studies in History and Philosophy of Science Part B: Studies in History and Philosophy of Modern Physics, 41*(3), 242–252.

Hacking, I. (1982). Language, truth and reason. In M. Hollis & L. S. (Eds.), *Rationality and relativism* (pp. 48–66 [Reprinted in Hacking 2002]). Blackwell.

Hacking, I. (2002 [1982a]). Language, truth and reason. In *Historical ontology* (pp. 159–177). Harvard University Press.

Hacking, I. (1983a). The accumulation of styles of reasoning. In H. v. D. Heinrich (Ed.), *Kant oder Hegel? Über Formen der Begründung in der Philosophie* (pp. 453–465). Klen-Cotta.

Hacking, I. (1983b). *Representing and intervening*. Cambridge University Press.
Hacking, I. (1988). On the stability of the laboratory sciences. *The Journal of Philosophy, 85*(10), 507–514.
Hacking, I. (1991). Artificial phenomena. *The British Journal for the History of Science, 24*(02), 235–241.
Hacking, I. (1992a). The self-vindication of the laboratory sciences. In A. Pickering (Ed.), *Science as practice and culture* (pp. 29–64). Harvard University Press.
Hacking, I. (1992b). Statistical language, statistical truth and statistical reason: The self-authentication of a style of scientific reason. In E. McMullin (Ed.), *The social dimension of science*. University of Notre Dame Press.
Hacking, I. (1992c). 'Style' for historians and philosophers. *Studies in History and Philosophy of Science, 23*(1), 1–20.
Hacking, I. (1995). Immagini radicalmente costruzionaliste del progresso matematico. In A. Pagnini (Ed.), *Realismo/Antirealismo* (pp. 59–92). la Nuova Italia.
Hacking, I. (1996). The disunity of science. In P. Galison & D. J. Stump (Eds.), *The disunity of science: Boundaries, contexts, and power* (pp. 37–74). Stanford University Press.
Hacking, I. (2002 [1992]). 'Style' for historians and philosophers. In *Historical ontology* (pp. 178–199). Harvard University Press.
Hacking, I. (2004). Review of 'Truth and Truthfulness: An Essay in Genealogy'. *Canadian Journal of Philosophy, 34*(1), 137–148.
Hacking, I. (2005a). Les philosophes de l'expérience. *Tracés. Revue de Sciences humaines*. Retrieved February 11, 2008, from http://traces.revues.org/201
Hacking, I. (2005b). Philosophie et histoire des concepts scientifiques. http://www.college-de-france.fr/media/historique/UPL5880_Hacking2005_2006.pdf
Hacking, I. (2006 [1975]). *The emergence of probability*. Cambridge University Press.
Hacking, I. (2006a). Cognition. Retrieved January 7, 2013, from http://www.college-de-france.fr/media/historique/UPL32429_cognition.pdf
Hacking, I. (2006b). Introduction 2006: The archaeology of probable reasoning. In *The emergence of probability*. Cambridge University Press.
Hacking, I. (2006c). La Démonstration. http://www.college-de-france.fr/site/historique/ian_hacking.htm-|p=../historique/ian_hacking.htm|
Hacking, I. (2006d). Véracité. http://www.college-de-france.fr/site/historique/ian_hacking.htm-p=../historique/ian_hacking.htm
Hacking, I. (2007). The laboratory style of thinking and doing. http://stm.ym.edu.tw/files/u1/HW_Ian_Hacking.pdf
Hacking, I. (2008 [1990]). *The taming of chance*. Cambridge University Press.
Hacking, I. (2009). *Scientific reason*. NTU Press.

Hacking, I. (2011). How did mathematics become possible? http://www.youtube.com/watch?v=PbKUsAR_8DY
Hacking, I. (2012). 'Language, Truth and Reason' 30 years later. *Studies in History and Philosophy of Science Part A, 43*(4), 599–609.
Hanson, N. R. (1958). *Patterns of discovery*. Cambridge University Press.
Herschel, W. (1800a). Experiments on the refrangibility of the invisible rays of the sun. *Philosophical Transactions, 90*, 284–292.
Herschel, W. (1800b). Experiments on the refrangibility of the invisible rays of the sun. [Abstract]. *Abstracts of the Papers Printed in the Philosophical Transactions of the Royal Society of London, 1*(1800–1814), 22–23.
Herschel, W. (1800c). Investigation of the powers of the prismatic colours to heat and illuminate objects. *Philosophical Transactions, 90*, 255–283.
Hollis, M., & Lukes, S. (Eds.). (1982). *Rationality and relativism*. Blackwell.
Iliffe, R. (1998). Rational artistry. *History of Science, 36*, 329–357.
Jardine, L. (2000). *Ingenious pursuits*. Abacus.
Kochen, S., & Specker, E. (1967). The problem of hidden variables in quantum mechanics. *Journal of Mathematics and Mechanics, 17*, 59–87.
Kowalenko, R. (2011). Styles of scientific thinking can kill. https://wits.academia.edu/RobertKowalenko
Kusch, M. (2010). Hacking's historical epistemology: A critique of styles of reasoning. *Studies in History and Philosophy of Science, 41*(2), 158–173. https://doi.org/10.1016/j.shpsa.2010.03.007
Lakoff, A. (2012). Ian Hacking interviewed by Andrew Lakoff. *Public Culture, 24*(1 66), 217–232.
Lloyd, G. E. R. (2004). *Ancient worlds, modern reflections: Philosophical perspectives on Greek and Chinese science and culture*. Clarendon Press.
Magruder, K. (1995). Review of Crombie, styles of scientific thinking in the European tradition. *Sixteenth Century Journal, 26*(2), 406–410.
Martínez Rodríguez, M. L. (2021). *Texture in the work of Ian Hacking. Michel Foucault as the Guiding Thread of Hacking's Thinking*. Springer.
Netz, R. (1999). *The Shaping of deduction in Greek mathematics: A study in cognitive history*. Cambridge University Press.
Pickering, A. (1984). *Constructing quarks: A sociological history of particle physics*. Edinburgh University Press.
Rossi, P. (2000). *La nascita della scienza moderna in Europa*. Feltrinelli.
Shapin, S. (1994a). *A social history of truth*. Chicago University of Chicago Press.
Shapin, S. (1994b). *A social history of truth: Civility and science in seventeenth-century England*. University of Chicago Press.
Shapin, S., & Schaffer, S. (1989 [1985]). *Leviathan and the Air Pump*. Princeton University Press.
Spelke, E. (1994). Initial knowledge: Six suggestions. *Cognition, 50*(1–3), 431–445.

Taylor, J. R. (1982). *An introduction to error analysis.* Oxford University Press.

Trenchard More, L. (1941). Boyle the alchemist. *Journal of the History of Ideas, 2*(1), 61–76.

von Neumann, J. (1955). *Mathematical foundations of quantum mechanics.* Princeton University Press.

Williams, B. (2002). *Truth and truthfulness. An essay in genealogy.* Princeton University Press.

CHAPTER 4

Taxonomy of the Notions of 'Ways of Thinking'

4.1 Introduction

In the previous chapters, I have explained how different thinkers in the wake of historical epistemology have given shape to the concept of Kant's a priori. With the purpose of understanding how objectivity emerges and changes in the course of history, they have put forward different characterizations of the concept of 'ways of thinking', which can be regarded as distinct 'historicized Kant's a priori'. In the first part of this chapter, I shall draw some comparisons between these different species of the genus 'ways of thinking' in order to provide a conceptual taxonomy. Continuing with the biological metaphor, over time the evolution of historical epistemology has provided more and more sophisticated species of the concept of 'ways of thinking', such as Kuhn's and Hacking's. These species have inherited some 'characters' from their 'ancestors', e.g. from certain notions of the concept of 'ways of thinking' put forward by the classical historical epistemologists. I shall start by tracing these common characters.[1]

[1] In the biological metaphor, among the ancestors of the recent species of the concept of 'ways of thinking' one should mention Mannheim's notion of 'style of thought' [*Denkstil*] (see Chap. 1). Since, as far as I know, neither Fleck nor other thinkers have referred to Mannheim's prior usage of the concept, one could say that, just as flight emerged, disappeared and reappeared with different characteristics in new contexts several times in the course of the evolution of living beings, so the concept of 'ways of thinking' made sometimes its appearance in some fields of study only to be almost forgotten and then reconceived.

L. Sciortino, *History of Rationalities*,
https://doi.org/10.1007/978-3-031-24004-1_4

Following that, I shall compare and contrast the more recent and sophisticated versions of the concept of 'ways of thinking'.

In the second part of the chapter, I shall make another important point: the analytical notions introduced by Fleck, Foucault, Kuhn and Hacking, if used together, provide a more profound understanding of the reasons why sentences, concepts or scientific discoveries become objective. Furthermore, although I shall advocate pluralism when discussing frameworks, I shall stress that, as compared to the notions of Fleck and Kuhn, Foucault's and Hacking's can play a crucial role in revealing fundamental aspects of the process by which objectivity becomes possible. Finally, I shall point out that the historicization of epistemological concepts carried out by the historical epistemologists contains in embryonic form important philosophical issues concerning relativism, contingency and convergence to truth. In the following chapters, I shall use Hacking's notion of 'style of reasoning' as a case study for dealing with these problems and investigating whether given features attributed to the concept of 'ways of thinking' yield certain consequences.

4.2 'Characters' of the Notions of the Concept of 'Ways of Thinking'

4.2.1 *Techne*

As I have pointed out in the first chapter, Bachelard complicated the subject–object relationship by claiming that the production of knowledge is fundamentally instrumentally mediated. He focused his attention on the dialectic between science, technology and the social structure of the scientific community from which objectivity emerges. Hacking developed in his own way this aspect of Bachelard's philosophy: he put the accent on the instrumentally mediated production of knowledge through his idea that styles of reasoning must be understood as ways of thinking *and* doing. In other words, the notion of style of reasoning inherited from Bachelard's views one of its fundamental traits, the emphasis on *techne*: experimentation, intervention and the technical transformation of the world.

Daston and Galison's epistemic virtues possess this identical trait in that they relate to specific practices of capturing the real which can be documented through an analysis of scientific atlases. However, in the case of Hacking, the new standards of evidence which can make possible the emergence of styles of reasoning, such as the statistical and the laboratory style, were not necessarily only the result of the development of technology: in the case of the emergence of the evidence of things (see Chap. 3) many other factors played a crucial role. This trait of Hacking's notion seems more closely linked to Canguilhem's view that concepts unfold according to their own internal logic of change.

4.2.2 *Sociological and Archaeological Perspectives*

For Fleck,

> [a thought style] is characterized by common features in the problem of interest to a thought collective, by the judgement which the thought collective considers evident, and by the methods which it applies as a means of cognition. (Fleck, 1979 [1935], p. 98)

In addition, thought styles are maintained by certain communities whose individuals share the same common knowledge and have the same communicative behaviour and literary style. The styles project proposes no equivalent of the notion of 'thought collective' and the idea of a community is rarely explicit or spelled out. By contrast, epistemic virtues embody the ways of seeing of certain communities of researchers who share sets of practices and normative rules. Indeed, for Daston and Galison, the atlas is always—and fundamentally—an exemplary form of collective empiricism: [...] (Daston & Galison, 2007, p. 27).

Still, communicative behaviour and literary styles are peculiar aspects of the notion of thought collective which have been inherited by the notion of paradigm: Kuhn cites as a source of authority in normal science the popularizations and the textbooks of science, which are committed to an articulated body of problems and theories (Kuhn, 1996 [1962], p. 136). Thought styles do not imply the existence of general rules of discourse that are valid for all areas of society in a given epoch. On the contrary, '[styles of reasoning] are built on fundamental cognitive capacities' (Hacking, 2009, p. 27), although these capacities might have been tapped in different ways had the historical circumstances been different.

It is worth noting that Fleck, before Hacking, had discussed Paracelsus's way of thinking, which he considered as an example of thought style. Commenting on a passage of the latter he characterized Paracelsus's way of thinking in terms identical to Hacking (see Fleck, 1979 [1935], p. 126). Compare:

> [Paracelsus's] style of reasoning is alien. (Hacking, 1983, p. 71)

> Comprehending objects and phenomena in a way similar to our own was completely alien to [Paracelsus's] way of thinking. (Fleck, 1979 [1935], p. 127)

The fact that Fleck considered Paracelsus's way of thinking a thought style invites a question: can we consider it a style of reasoning too? If the answer were positive the two notions would overlap in the characterization of a particular way of thinking. I shall address this question in Chap. 5. For the time being let me point out that both the notions have similar implications concerning the communications between people who adopt different styles of reasoning. Fleck asserted that direct communication between the adherents of different thought styles is impossible (Fleck, 1979 [1935], p. 36) and justified this point both in terms of impossibility of translating certain terms from a thought style into another and in these terms:

> The statement 'Someone recognizes something' [...] is no more meaningful as it stands than the statements 'This book is larger' [...]. Something is still missing, namely the addition, 'than that book', to the statement. Analogously, the statement 'Someone recognizes something' demands some such supplements as 'on the basis of a certain fund of knowledge' or, better, [...] 'in a particular thought style'. (Fleck, 1979 [1935], p. 38)

In the case of Hacking, in order to ***understand*** a sentence what we need is to recognize the presuppositions of the style of reasoning, e.g. those questions, methods, standards of truth, upon which the truth-value of that sentence rests;[2] in the case of Fleck, what we need is the 'fund of knowledge' of the thought style, the shared knowledge of the thought collective. However, Fleck made other strong claims that are not implied by the notion of style of reasoning. He thought that the perception of an individual changes when it is embedded in a certain thought style (or 'paradigm' in Kuhn's version of the same claim). Even facts are created by thought styles; truth is a function of the particular thought style that has been accepted—there is no objective truth. Not only does Hacking, as we shall see, reject these claims but also insists that sense-datum statements are represented by style-independent propositions.

[2] I will say more about this point in Chap. 6.

One of the reasons for the differences between Fleck's and Hacking's notions is ascribable to the fact that Hacking's way of tackling the problem of identifying the a priori conditions of objectivity has a more direct link with the methods of analysis of Foucault. Hacking considered the notion of style of reasoning as a particular element of the episteme, the *decisive* element that is necessary to understand the emergence of new true-or-false sentences. As he put it:

> I do not say that 'style of reasoning' means Foucaultian 'episteme' or whatever. On the contrary, if one could adequately define an episteme, one would surely have to include, as one of its elements, the styles of reason that bear on the positive propositions of that field of knowledge. (Hacking, 2002, p. 181)

In other words, for Hacking, of all the elements that characterize an episteme, e.g. assumptions, unspoken truths, institutions, knowledges, social activities, the material traces left behind by a particular epoch, what really counts in order to understand the emergence of new true-or-false statements is what he calls the 'style of reasoning'. Therefore, Hacking works in Foucault's perspective in that he aims to discover the conditions of validity of statements by using what might be considered an element of the episteme.

Secondly, for Foucault.

> Archaeology does not seek to rediscover the continuous, insensible transition that relates discourses, on a gentle slope, to what precedes them, surrounds them, or follows them [...] On the contrary, its problem is to define discourses in their specificity; to show in what way the set of rules that they put into operation is irreducible to any another. (Foucault, 1969 [1972], p. 139)

In a similar way, the project of styles of reasoning is 'archaeological' in nature. It is not interested in those transitional zones between the phase that precedes the emergence of a style of reasoning and its crystallization, for example in describing the continuous stages of refinements undergone by certain methods, instruments and practices. Rather, what counts is to discover the standards of truth, the questions, the criteria of evidence, the way of thinking and doing, i.e. the presuppositions that characterize the new style of reasoning; and to do so it might be necessary to set up the presuppositions of a new style of reasoning against those of others. In

Sect. 2.1 I have suggested that the notions of Fleck, Foucault, Kuhn and Hacking can be regarded as frameworks that enable us to emphasize certain aspects of the past and construct explanatory narratives; and in Sects. 2.7 and 2.10 I have argued that Fleck and Kuhn appealed to explanatory expectations sociological in character. Thus, we can summarize the main point of this section by saying that Kuhn's framework belongs to a *sociological* explanatory perspective and Hacking's to an *archaeological* one, and that these explanatory perspectives have been inherited, respectively from Fleck and Foucault.

4.2.3 Discontinuity and Accumulation

Bachelard thought that scientific knowledge and common sense differ radically because they are produced by two distinct ways of thinking. As we have seen, in order for scientific knowledge to emerge, an 'epistemological break' with immediate experience that is rooted in imagination and emotions is necessary. With respect to discontinuity, thinkers such as Koyré, Foucault, Kuhn and Hacking took Bachelard's tack by arguing for sharp breaks in the history of science. Hacking paid his obeisance not only to Bachelard but also to Kuhn and Foucault by saying:

> I have a revolutionary temper, perhaps under the excessive influence of Bachelard, Kuhn and Foucault [...] I am interested in ruptures that radically transform our methods of reasoning. (Hacking, 2006a, p. 9)

However, the reasons that underlie the 'revolutionary' attitudes of these thinkers are different, as it can be inferred from the dissimilarities between their notions of the concept of 'ways of thinking'. Take the case of Hacking and Bachelard. While styles of reasoning accumulate, for Bachelard there is no monotonic increase of knowledge: at certain points of history radical revisions of the conceptual foundations of scientific knowledge can happen, although the past can be transformed rather than completely rejected. For example, in the beginning of the twentieth century, the theory of relativity emerged with the consequent rejection of all the basic concepts of classical physics. For him, epistemological breaks like this were the consequence of a new 'scientific' way of thinking that had emerged superseding the 'old Newtonian' way of thinking, which therefore was not eternal and universal. In Bachelard's thought ancient ways of thinking now gone extinct such as the alchemic way of thinking are non-scientific. His philosophical attitude was normative: he used modern science as a standard in order to judge certain ways of thinking as 'non-scientific'. Conversely, as I have pointed out in the third chapter, Hacking's historical epistemology is not normative—he did not put forward any criterion to discard Paracelsus's way of thinking as non-scientific.

Like Bachelard's stages of historical development, paradigms and epistemes replace one another. In the preface of *The Order of Things*, Foucault claimed that 'archaeological inquiry has revealed two great discontinuities in the *episteme* of western culture' (Foucault, 1994 [1966], p. XXII): at the beginning of the Classical age (and the Modern age) a way of thinking became extinct and another emerged. On the contrary, as styles of reasoning have made their appearance they have often been adopted together in order to solve single scientific problems: for example, the statistical style and the laboratory style are not mutually exclusive.[3] Here, the resemblances between the notion of epistemic virtue and that of styles of reasoning are striking: epistemic virtues are made possible by contingent reasons; they can conflict each other or intertwine, and, in principle, vanish; when a new epistemic virtue comes into being, the old ones do not necessarily pass away.

As to discontinuity, Fleck's notion differs from the notions just mentioned. The formation of a thought style is ascribable to the change in meaning that emerges from a continuous communication between the members of a thought collective.

> Thoughts pass from an individual to another, each time a little transformed [...] Words which formerly were simply terms become slogans [...] They no longer influence the mind through their logical meaning—indeed, they often act against it—but rather they acquire a magical power and exert a mental influence simply by being used. (Fleck, 1979 [1935], p. 42)

Ultimately, the process of development is *continuous*:[4] thought styles emerge from communicative interactions, mute continuously and die out in a brief span of time. On the other hand, rather than the slow and continuous transformations of thoughts that pass from an individual to another, what might cause the emergence of a style of reasoning is, e.g. a

[3] In Sect. 6.2.1 I shall discuss the accumulation of styles of reasoning in more depth by giving an example.

[4] Nicola Möβner supports my point (Möβner, 2011, p. 7).

rapid change in the kind of evidence that people use. In comparison with other notions of the concept of 'ways of thinking', Hacking's styles persist for a long period—they are a matter of *longue-durée.*

The claim that discontinuity results in the extinction of a way of thinking calls for an explanation. Hacking wrote that complaints have been addressed to Foucault because he never explained why epistemes die out (Hacking, 2002, p. 195). Unlike the notion of 'episteme', his notion is described, among other things, in terms of the concept of self-vindication. The existence of techniques of self-vindication supplies part of the answer to the question of why styles of reasoning do not vanish. Why the Renaissance episteme of resemblance expired is a question that has no ready answer within Foucault's intellectual framework. On the contrary, why the statistical style endures is a question that admits an answer in terms of the absence of techniques of self-vindication.

4.2.4 Selection and Exclusion

Daston and Galison wrote that 'atlas users become the people of a book, which teaches them how to make sense of their sliver-world and how to communicate with one another about it' (Daston & Galison, 2007, p. 27). It is a sliver-world because, as matter of fact, atlases *select* what is important to ask, see, depict, teach and learn so that they exclude portions of the world, questions and perspectives. This may mean that for the members of a community who pursue certain epistemic virtues some aspects of phenomena (those that have been excluded) cannot play any role in the evidential considerations that bear on their knowledge. For example, the physicist Arthur Worthington (1852–1916), by relying only on his eyes, made symmetrical drawings of the impact of drops of liquids falling vertically on a surface of water (Daston & Galison, 2007, pp. 11–16). In other words, Worthington discarded some features of the phenomenon in the name of the ideal of symmetry. Later on, after having succeeded in stopping a droplet's splash with a photograph, which showed that the phenomenon was anything but symmetrical, he converted to the 'objective view', i.e. to the epistemic virtue of mechanical objectivity. Arguably, under the new possibilities of a 'mechanical objective vision', those variations that had been discarded under the 'truth-to-nature vision', could have become the *evidence* for more accurate theories that describe the splash of a drop. Ultimately, epistemic virtues select end exclude portions of the world and shape what can and what cannot be considered as 'evidence'. Similarly, both epistemes and styles of reasoning select and exclude, opening a specific space of possibilities and relegating to the oblivion questions, concepts and kinds of evidence. A case in point is the historico-genetic style of reasoning: the community of scientists that adopted it turned their eyes to a *previously unnoticed* portion of the world that had been excluded by previous styles of reasoning, fossils, minerals, rocks and geological layers. Likewise, a new episteme introduces constraints on how people are able to think.

In brief, there are different notions of the concept of 'ways of thinking' which have been put forward by different thinkers in the wake of a tradition that historicizes epistemological concepts. Each notion plays the role of a historicized Kantian a priori: it is a collective and historical product that constitutes the basis of objectivity. In this sense, the expression 'way of thinking', does not express so much the idea of the singularity of the way in which a single person thinks in order to find out about the world, but rather that of an entire community of people.[5] Here is where the first differences between these notions appear. The specific social entities to which the different notions draw our attention are often not the same: primitive societies, cultures and communities of scientists to mention just a few. Other differences between the notions of the concept of 'ways of thinking' concern, e.g.: the emphasis placed on *techne* as an important element in the process of production of knowledge; the components of human mind that are considered variable throughout history, e.g. reasoning, logic, imagination, perception, thought categories and values; and whether or not ways of thinking have sharp beginnings. Finally, viewed as frameworks central to historical and philosophical enquiry, all these notions have the same explanatory goal, i.e. to understand how objectivity emerges or changes, but different explanatory expectations, central subjects and sets of contrasts and comparisons (see Sect. 2.1). In Table 4.1, I have summarized the main features of the notions of the concept of 'ways of thinking' discussed in Chaps. 2, 3 and 4. What is shown in Table 4.1 is the 'genus ways of thinking', whose main trait is historicized Kantianism, or, to use Wittgenstein expression, the family resemblances between different notions.

[5] According to Jean Gayon, in the hands of Hacking the meaning of the term 'style' acquired the idea of universality. In his view, the previous uses expressed the idea of the singularity and the locality of the ways of knowing (Gayon, 1996).

Table 4.1 Main features of different notions of the concept of 'ways of thinking' discussed in Chaps. 2, 3 and 4

	Historicized Kantianism	*Deep knowledge*	*Sociological dimension*	*Cumula tiveness*	*Discontinuity*	*Techne*
Mentality (Lévy-Bruhl)	x		x	x	x	
Age of intelligence (Brunschvicg)	x				?	
A priori (Metzger)	x			?	?	x
Era of scientific thought (Bachelard)	x		x		x	x
Type of thinking (Koyré)	x				x	
Thought style (Fleck)	x		x	x		
Paradigm (Kuhn)	x		x		x	
Style of thought (Feyerabend)	x		x	x	x	
Episteme (Foucault)	x	x			x	
Style of reasoning (Hacking)	x	x		x	x	x
Epistemic virtue (Daston and Galison)	x		x	x	x	x

I have used the question mark in all those cases where there are not enough elements to assign a given feature to a certain notion

4.3 A Combined Framework for Studying Objectivity

In this section, I shall argue that the notions I have discussed so far can be used as a toolkit for studying the emergence of objectivity. For simplicity's sake I shall restrict my discussion to four notions: those of Fleck, Foucault, Kuhn and Hacking. The argument could be extended further to cover the cases of other notions. The main upshot of my discussion will be that the frameworks of style of reasoning and of episteme illuminate conditions of possibility that are crucial for the emergence of objectivity and lie at a deeper level than those considered by the other two notions. I shall start by giving an example to demonstrate how a paradigm has been made possible by the coming into being of a style of reasoning. My example will also show that, in turn, that very style has been made possible by the emergence of a new kind of evidence. I shall then draw a general conclusion about the relationships between the four different notions. I recall that, as explained in Chap. 3, in *The Emergence of Probability*, Hacking argued that what was lacking in the Renaissance was the evidence provided by things, not to be confused with that provided by the data of the senses. In our understanding, books and testimony are indirect evidence that is secondary to the experience, but in the Renaissance testimony and authority were primary, and things counted as evidence only insofar as they resembled the authority of testimony and books. It was a conceptual change of what a sign is and represents that opened the way for the evidence of signs to turn into the evidence of things. According to Hacking, the concept of probability fully emerged in 1650 when the concept of evidence of things conjoined with that of frequency took full shape in a text published by Thomas Hobbes (Hacking, 2006b [1975], p. 48).

Hacking noted that 'the advent of the quantum theory has made it possible to conceive of statistical regularity as a brute and irreducible fact of nature' (Hacking, 2006b [1975], p. 174). Indeed, the formalism of quantum mechanics cannot be interpreted as if probabilities reflected ignorance.[6] Statistical laws have become an irreducible fact of the world, and the deterministic conception of nature has given way to an indeterministic one. This transition has been presented by Kuhn as a scientific revolution in which there has been a shift from the classical to the quantum paradigm.[7] For him, it is the change of paradigm that explains how a new

[6] John von Neumann's (von Neumann, 1955), John Bell's (Bell, 1966), Simon Kochen's and Ernst Specker's (Kochen & Specker, 1967) works show that it is impossible to provide a 'classical' reformulation of the quantum formalism.

[7] Kuhn (2002 [2000], p. 25) gave as an example of scientific revolution the origins of the quantum theory: the black-body problem 'caused the reconstruction of a good deal of physics'. Neither this essay nor his book *Black-Body Theory and the Quantum Discontinuity* (1987a [1978]) *explicitly* mentions the term 'paradigm'. However, Kuhn later added a foreword to his book to clarify that he had *not* repudiated this concept: 'when I look back, I have generally been satisfied by the extent to which my narrative fits the developmental scheme that *Structure* provides. *Black-body theory* is no exception' (1987a [1978], p. 363).

ontology of the world, new puzzles, methods and concepts have become possible. In the quantum paradigm, the question about whether or not a particle behaves like a wave is meaningful and objective, whereas the concept of a trajectory of a particle has no meaning.

Ultimately, the emergence of the evidence of things made possible the coming to the fore of the statistical style, which in turn made possible the transition from the classical to the quantum paradigm. If probability had not emerged, there could not have been a quantum paradigm, although the coming to the fore of the latter would not have been a *necessary* consequence. The quantum paradigm inherited from the statistical style standards of evidence, methods, practices, concepts and propositions. This example shows that it is possible to use the notions of style of reasoning and paradigm together to track objectivity. It could also be argued that the analysis of such a case study would benefit from the use of the notion of episteme to reveal, for instance, how in the seventeenth century language could be conceived of as a transparent structure, articulated into an arbitrary binary relation between signified and signifier. This change has been crucial for the emergence of mathematical formalism.

It is possible to give many other examples in which these notions can be used together. Like the notion of paradigm, that of thought style is parasitic upon that of style of reasoning. In the previous chapter, I have shown that it was the emergence of a new kind of evidence (the 'creation of effects') that made possible the emergence of the laboratory style of reasoning. In turn, it was the emergence of the latter that made possible, e.g. the thought style of the microbiologists and therefore the discovery of Wassermann's reaction. In this case, one could use both the notion of style of reasoning and that of thought style together to study how Wassermann's discovery became objective. As another example, the notions of thought style and paradigm could be preliminary for the study of the 'discovery' of oxygen by Priestley and Lavoisier. And one could employ both the notion of style of reasoning and episteme to account for the emergence of Darwin's evolutionary ideas: the former would provide insights into his historico-genetic thinking, the latter would place the emphasis on the role of time in his view of nature.

Let us tie the argument together. The process by which facts, concepts or sentences come into being as objective can be depicted by a diagram of layers, the 'layers of scientific objectivity' (Fig. 4.1). Styles of reasoning and epistemes sit at the lowest layer of the diagram, paradigms and thought styles at the second layer and all the epistemological items which become objective sit at the highest level. Each layer represents conditions of possibility for the existence of the successive layer. The higher layer instantiates the lower layer. There will also be other layers of various contingent factors, of which the emergence of a new kind of evidence is only one, that make possible the level of styles of reasoning or epistemes. Ultimately, we could use the metaphor of the 'machine of objectivity': there is a bottom layer of contingent factors, which trigger the emergence of styles of reasoning or epistemes, which in turn trigger the emergence of paradigms or thought styles and lead to objectivity. Within the metaphor, 'triggering' stands for 'making possible' and not 'causing'.

Models in which lower layers make possible the higher ones are known as 'stratigraphical metaphors'. James Elwick has shown how they have been used by several scholars in history and science studies and how they

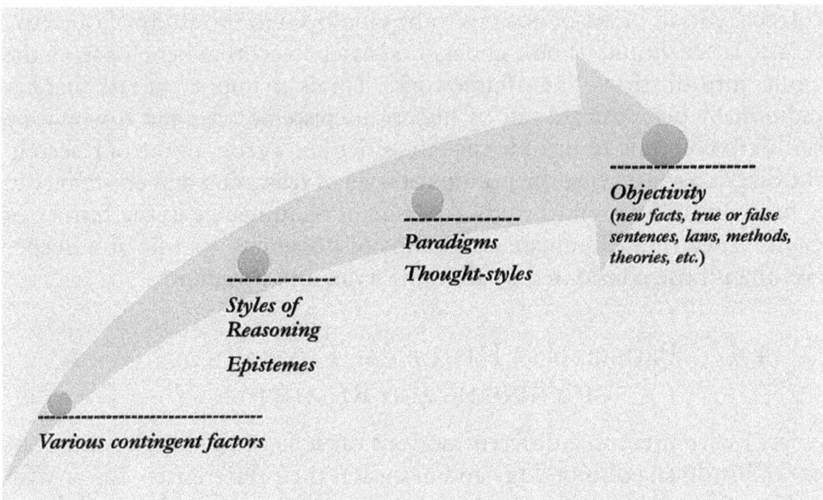

Fig. 4.1 The layers of scientific objectivity: each layer represents conditions of possibility for the existence of the successive layer. There is a bottom layer of contingent factors, which trigger the emergence of styles of reasoning or epistemes, which in turn trigger the emergence of paradigms or thought styles and lead to objectivity

are useful for speculating on degrees of contingency. In particular, he has depicted 'styles of reasoning as conditions of possibility—circumstances that are necessary for other phenomena to occur' (Elwick, 2012, p. 619). On my part, I have used the stratigraphical model to depict the relations between four different notions and their respective roles in making possible the emergence of objectivity. The fact that in Fig. 4.1 frameworks lie in the same layer does not mean that they both will necessarily be applicable to any case study. For example, consider the layer of paradigms and thought styles. Kuhn's discontinuist explanatory expectations make it impossible to apply his notion to those cases in which objectivity is the product of a slow process of development of past ideas. On the other hand, any extended discussion of how objectivity emerges would necessarily have to include notions in lower layers, i.e. notions such as styles of reasoning and epistemes.

Finally, the four notions I have discussed so far do not exhaust all the different possible explanatory perspectives from which to consider a given episode in which scientific objectivity emerges—there are as many frameworks as all the possible histories we can construct. There is no single 'correct' notion or set of notions to be employed in the study of objectivity. The latter should be best understood as a project that benefits from the regular introduction of new frameworks. This is an important task that has traditionally been carried out by historical epistemology, and it would be fruitful to continue to include such inquiries in its programme of research. The diagram shows that the notions of style of reasoning and episteme can be more fruitfully used to study the issue of contingency in the history of science since they illuminate conditions of possibility that lie at a deeper level than those considered by thought styles and paradigms.

4.4 Problems in Historical Epistemology: Ways of Thinking and Relativism

So far, I have presented different notions of 'style of thinking' in the context of historical epistemology and discussed their differences. I now wish to draw attention to some hotly debated problems in the philosophy of science that are related to the concept of 'ways of thinking'. One of the thorniest is whether relativism is a philosophical consequence of the historicization of epistemology. Martin Kusch quite rightly reminded us that the relationship between historicism and relativism already represented a problem for discussion a century ago (Kusch, 2010, p. 168). Cristina

Chimisso pointed out that the historicization of epistemological concepts carried out by French thinkers such as Lévy-Bruhl opened the door to relativism (Chimisso, 2008, p. 70). After including Feyerabend under the label 'historical epistemology', Anastasios Brenner remarked that he posed the 'problem of objectivity' ['*la question de l'objectivité*'] in a particular way (Brenner, 2006, p. 118): 'how can information that is the result of idiosyncratic historical changes be about history-independent facts and laws?' (Feyerabend, 1989, p. 393). As a matter of fact, Feyerabend highlighted a problem which hangs over the whole historicization of epistemology and that can be described as a 'relativism issue': whether our scientific findings transcend their place of origin and become independent of it. In our case, this place of origin is a given way of thinking, whatever notion of this concept one has in mind.

I shall define the 'relativism issue' more precisely in Chap. 6. For the time being, I only wish to point out that three kinds of relativism might be relevant in relation to the different notions of the concept of 'ways of thinking': ***alethic relativism***, according to which what is true in a way of thinking is not true in another; ***epistemic relativism***, which denies the existence of epistemic norms over and above those operative in the single ways of thinking; and ***perceptual relativism***, which affirms that there are no universally objective criteria for judging different ways of perceiving the world.

By dividing human thought in two with his notion of mentality, Lévy-Bruhl was one of the first thinkers to pose the question of relativism in historical epistemology. According to him, 'we have to bear in mind that which their collective representations [of the primitives] instil into all their perceptions' (Lévy-Bruhl, 2015 [1910], p. 43). For him, collective representations act as categories which shape our perceptions by infusing them with emotions. Indeed, as we have seen, in the primitive mentality certain objects are identified with those that mystically symbolize them. By virtue of this violation of the principle of non-contradiction, the primitives do not justify their beliefs as the moderns do. When something new presents itself to the mind of certain primitive peoples, they can see the manifestation of an invisible force in it (Lévy-Bruhl, 2015 [1910], p. 60). This mystical perception could lead to identifying what is presented to their eyes with another object. It is an experience that involves the sensory perception of the new object and the mystical perception of a supernatural power. These two perceptions intertwine and give shape to explanations of what the new object is which are different from those provided by the modern peoples.

Thus, in principle the notion of mentality poses three relativism issues: 'alethic', in that primitive and modern peoples disagree about what the new object is; 'epistemic', in that they provide different justifications; 'perceptual', in that they perceive the new object in different ways. However, as a matter of fact, Lévy-Bruhl's philosophical attitude was normative and therefore anti-relativist: he used the modern mentality as a standard to judge the thought of primitive peoples as based on false premises. He wrote that 'for people like ourselves, speaking the language familiar to us, there is insurmountable difficulty in entering into their way of thinking. The longer we live among them, the more we approximate to their mental attitude' (Lévy-Bruhl, 2015 [1910], p. 14). What he meant was that the primitive mentality must be understood on its own terms: once we know the collective representations of primitive peoples and the way they are connected to each other by the *law of participation*, we can make sense of their beliefs, explanations or practices. But the fact remains that, for Lévy-Bruhl, the law of participation was a wrong assumption in that it was indifferent to the principle of non-contradiction and blurred distinctions such as natural and supernatural, human and non-human and singular and plural. His reflections on the transition to the 'higher mental types' (Lévy-Bruhl, 2015 [1910], pp. 361–387) are indicative of these views: for him, 'progress' in human thought has been achieved by purifying the cognitive process from emotions and mystical perceptions.

Like Lévy-Bruhl, Brunschvicg believed in the existence of a framework-independent vantage point from which different ages of intelligence could be judged. In the history of human thought, he saw ages of progress and ages of regress: for example, the age of the early humans was prelogical in the sense of Lévy-Bruhl, but the following age, when mathematics, geometry and astronomy became object of study in different civilizations, represented an age of progress. On the other hand, the age of intelligence between Aristotle and Descartes, when philosophers replaced the primacy of mathematical thought with that of logical expression, marked another age of regress (Brunschvicg, 1934, p. 43). Ultimately, for Brunschvicg the use of mathematics in a given age of intelligence was a norm of progress: there was no possibility of contributing to the advancement of knowledge without looking at the world as a 'book written in mathematical language'.

In the case of Metzger's notion, the relevant question is whether there are universal and atemporal criteria for assessing the different and incompatible a priori of different epochs. To my knowledge, Metzger never discussed it, but surely her historical epistemology was not normative—she was reluctant to saying that a given a priori was 'incorrect', although she was perfectly aware that its use could lead to wrong conclusions. However, one thing she pointed out was that certain organizing concepts of a given a priori can be part of another a priori. For example, according to her, certain fundamental tendencies crucial in the alchemic thinking have been present in the thought of Newton and other protagonists of the Scientific Revolution. Thus, there was no point in saying that certain concepts on which the thought of a certain epoch was based were in principle 'wrong' assumptions. What Metzger wanted to study was the simultaneous use of sets of concepts and the way they interact with each other and with certain presuppositions or emotional considerations. According to her, the alchemists adopted an a priori that was different from modern science, which included organizing concepts such as analogy and evolution. For her their system of thought had an internal coherence and was not absurd. On the other hand, Bachelard's philosophical attitude, like Lévy-Bruhl's and Brunschvicg's, was normative: he used modern science as a standard to judge certain eras as 'non-scientific'. He saw certain intellectual tendencies and mental attitudes of the alchemists as an epistemological obstacle rooted in imagination. For example, he defined 'animist obstacle' as the tendency to provide explanations of phenomena by reference to processes of the human body.

Like Metzger, Koyré was not interested in assessing the claims established by adopting a certain type of thinking. His concern was to analyse the coherence between these claims and the web of beliefs, concepts and presuppositions which characterize a given type of thinking. For example, only in the light of a particular set of organizing concepts, those of space, time and causality, could the law of universal gravitation be conceived and comprehended. In a letter to the philosopher Herbert Spiegelberg dated 1953, Koyré declared his anti-relativist attitude with these words: 'I inherited from him [Husserl] the Platonic realism that he discarded; the anti-psychologism and the antirelativism' (quoted in Stump, 2001). However, although one could argue that the truths discovered within a certain type of thinking have an objective validity that transcends it, epistemic relativism would remain an open question, i.e. is the justification of these truths universal and atemporal? A negative answer is a possibility because it could happen that the justification of a certain proposition rests on presuppositions of a certain type of thinking and therefore is not intelligible for those who adopt other types of thinking. If this were the case, the justification of that proposition would be relative to a given type of thinking.

These latter considerations can also be applied to Foucault's notion of episteme. In this regard, Gary Gutting pointed out that 'we need to distinguish the relativity of the means whereby truths can be attained from the relativity of their validity' (Gutting, 1991 [1989]). In other words, the criteria of evidence and the methods for assessing a positive sentence might depend on the episteme but, one could argue, the truth thus found out could be universally valid. It might also be the case that the same truth could be obtained by using methods and criteria of evidence of two or more episteme. Therefore, the point is not that the notion of episteme implies alethic relativism but that it may imply *epistemic* relativism: there might be objective truths justified within one episteme but unjustified in another episteme. Of course, this remains only a possibility until one provides a concrete example.

In order to extend our discussion of relativism to the notions of Fleck, Kuhn, Feyerabend and Hacking, it is worth introducing the concept of 'incommensurability'. The term originated in the sixteenth century from Medieval Latin *incommensurabilis* [not-measurable-together]. It was and it is still used in mathematics and physics to indicate the concept of two entities not measurable by the same standard. Typically, two lengths are *commensurable* if one can express their ratio as that of two integers *m* and *n*. So far as I am aware, Fleck has been the first to use the word 'incommensurable' in philosophy of science. In his papers in Polish *he borrowed from mathematics* the term *niewspółmierność* (Sady, 2012), *which was translated into the German term inkommensurabel in his Genesis and Development of a Scientific Fact*. He wanted to express the idea that certain concepts, which are not replaced with adequate substitutes in an emerging thought style, cannot be brought into the usual logical relations of inclusion, exclusion, overlap with new concepts (see for example Fleck, 1979 [1935], p. 62).

It is a fact that only with Kuhn and Feyerabend did the term 'incommensurable' catch on in philosophy of science. Kuhn matured the notion of incommensurability under the influence of philosophers such as Ludwik Fleck[8] and Michael Polanyi (1891–1976) and Gestalt psychologists such as Wolfgang Köhler (1887–1967). According to Eric Oberheim, Feyerabend introduced the term 'incommensurable' more than a decade earlier than Kuhn, 'as part of a criticism of the conceptual conservativism

[8] Kuhn read the *Genesis and Development of a Scientific Fact* in 1949–1950, as he said in his foreword of 1976 to the English edition of the book.

that he found in Nagel's theory of reduction and the Hempel–Oppenheim theory of explanation' (Oberheim, 2005, p. 372). He wanted to describe the relation between two *non instantial-theories*, i.e. theories that have ontological implications about the nature of the world. According to him, two such theories are incommensurable because the concepts of one theory can 'neither be defined on the basis of the primitive descriptive terms of the other [theory], nor related to them via a correct empirical statement' (Feyerabend, 1962, p. 59). Concepts do not 'keep intact' from a theory into another. Feyerabend illustrated this idea by comparing geometrical and wave optics or physics and relativity. For example, he maintained that the term 'mass', as used in classical physics, denotes a constant property of a body, while in theory of relativity it denotes the total quantity of energy divided by the speed of light squared.

To grasp the cause of meaning change, it is necessary to consider that for Feyerabend the framework of a theory gives meaning to its theoretical and empirical terms. In turn, these meanings affect our ontology, our conception of what exists. For example our ontological concept of what light is made of has changed from the corpuscular theory to the wave theory; each of the theories proposed has given meaning to its theoretical terms and this, in turn, has determined a different ontological concept of what light is. Since the framework of a theory gives meaning to its empirical terms (as well as its theoretical ones) it is not possible to count on empirical facts in order to find a universal criterion to judge between different theories. As Feyerabend put it, 'we no longer assume an objective world that remains unaffected by our epistemic activities' (Feyerabend, 1988 [1975], p. 19). Therefore theories are incommensurable: there is no common criterion to decide which ontological consequences are correct.[9]

In a similar way, what Feyerabend considered explanatory of the incommensurability between styles of thought is the fact that, just like incommensurable theories, they cannot be adopted simultaneously and 'neither logical nor perceptual connections can be established between them' (Feyerabend, 1988 [1975], p. 169). Indeed, the emergence of a style of thought, I remind the reader, is characterized by novelties such as new ideas, new methods, new objectives, new criteria of objectivity and new ways of perceiving the world. The strange creatures that populate the Homeric world were truly perceived, according to Feyerabend: 'they are

[9] Beside conceptual inconsistence, Feyerabend also argued for the *logical* inconsistence between theories. I will ignore this point.

elements of a *world* in which the Homeric person as such lived and which was for them as *real* as our own world is *real* for us today' (Feyerabend, 2016 [2009], p. 77). In another, passage Feyerabend wrote that 'atoms exist *given* the conceptual framework that project them' (Feyerabend, 1999). There is no method, no criterion of objectivity, no way of perceiving which is preferable to any other. Thus, there is nothing, not even common ontology or empirical content, on which to base a universal and timeless criterion to judge different styles of thought.

Like Feyerabend, Kuhn believed in the impossibility of defining the terms of one theory on the basis of the terms of the other. According to him, the idea of incommensurability sprung to his mind in 1940 when, studying Aristotle's physics he found some passages unsound and senseless (T. Kuhn, 1987b). Kuhn realized that by changing his way of reading the text, by changing some concepts and the meaning of some words, he was able to understand what Aristotle said. Moreover, under the influence of Fleck, he maintained that different paradigms use concepts that are incompatible. As he refined his ideas, Kuhn developed a concept of incommensurability not only limited to the idea of translation failure. He argued that incommensurability was not only caused by linguistic factors but also by differences in methods for setting up research and evaluating its results. Finally, under the influence of the Gestalt psychologists, he also spoke of differences of perceptions between adherents of different paradigms. In *The Structure of Scientific Revolutions* he argued that in the seventeenth century different people might have viewed in a different way the same stone swinging from a string. An Aristotelian would have seen the stone as constrained in its downward motion; an adherent of the new Galilean physics would have seen a pendulum that repeats the same motion forever. Kuhn claimed that these two conceptualizations are incommensurable because there is no neutral way to adjudicate between them:

> [This is the] most fundamental aspect of the incommensurability [...] The proponents of different paradigms practice their trades in different worlds. One contains constrained bodies that fall slowly, the other pendulums that repeat their motions again and again. [...] Before they can hope to communicate fully, one group or the other must experience the conversion that we have been calling a paradigm shift. (T. Kuhn, 1996 [1962], p. 150)

As a consequence, paradigm change is like a Gestalt switch in perception.

To sum up, in both Kuhn's and Feyerabend's versions of incommensurability I have sketched out it can be recognized a *semantic aspect*: two paradigms (or theories) are incommensurable when there is no neutral language into which their content can be translated. In this sense by a lack of a 'common measure' it is meant as a lack of 'common language'. However, Kuhn's version of the concept of incommensurability also contains an emphasis on *epistemological and methodological aspects* concerning how scientific claims have to be justified: there is no universal standard that serves to justify a paradigm against its rivals. Finally, it is important to bear in mind that Kuhn and Feyerabend believed that observational propositions are theory laden. This point is crucial for their claim of incommensurability: for example, in Kuhn the replacement process of paradigms cannot be a falsification process *à la Popper* in which certain observational data falsify a paradigm.

The two kinds of incommensurability I have distinguished in the works of Kuhn and Feyerabend have been progressively recognized in the contemporary debates as different in principle. In 2004 Léna Soler distinguished two aspects of incommensurability (Soler, 2011):

(1) The *semantic incommensurability*, which concerns semantical or descriptive non-translatability between rival theories, paradigms, thought-styles.
(2) The *methodological incommensurability* which concerns changes in the conceptions of *how a scientific claim can be justified*.

Epistemic relativism as defined earlier is an issue of methodological incommensurability (case *2*)): there is no common criterion of justification above those operative in the single ways of thinking.

For his part, Martin Kusch explained the positions of Feyerabend and Kuhn by arguing that they supported 'non-neutrality' and 'locality', but not 'equal validity'. They claim that when we rank different ways of thinking, we must always rely on, or take our starting point from, some other way of thinking. However, they do not go further by declaring all ways of thinking to be equally correct or valid (Kusch, 2020). In my terminology, we may say that the notions of Feyerabend and Kuhn do not imply alethic relativism but only epistemic relativism. Can we reach the same conclusions in the case of Hacking's notion? When we compare Hacking's claims with Feyerabend's and Kuhn's we can note that nothing in the styles of reasoning project leads to semantic incommensurability. It is true that by saying that certain style-dependent sentences have no meaning within another style of reasoning Hacking often assumes verificationism, a thesis about *meaning*. But to assume that only statements about the world that are verifiable by a certain method are meaningful does not imply that the meaning of a sentence changes from a style of reasoning to another. That is: within the styles project verificationism is a semantic thesis that does not imply semantic incommensurability as in the case of Kuhn and Feyerabend. Hacking's point is that there are sentences that are meaningful in certain styles of reasoning (e.g. because there are methods to assess their truth) and meaningless in other styles of reasoning—the meaning of sentences does not change from a style of reasoning to another. Like in other cases discussed above, the suspect is that a claim could be justified in a style of reasoning and unjustified in another.

Hacking has been trying to allay any suspicion of relativism ever since he wrote his first papers on styles of reasoning. For example, he insisted on the idea already mentioned that a style is characterized by a set of propositions that are true or false, not a set of true propositions, i.e. the actual truth-value of style-dependent sentences is external to the style. He distinguished *form* from *content* of science and remarked that, although the question of science is historically variable, the answers are not (Hacking, 1999, 2000). That is: in a given historical context, once a question is fixed, there is only one correct answer to that question; no relativism about answers.

By putting forward these ideas Hacking also distanced himself from thinkers such as Fleck who, by historicizing a great deal of elements involved in the knowledge of the world, provided arguments for the thesis of incommensurability between thought styles. From a thought style to another everything changes: the way we perceive reality, the language used to communicate, the research problems, the standards of truth and the reality created by the thought style itself (Fleck, 1979 [1935], p. 100). After Fleck, social constructionists such as Shapin and Schaeffer spoke of the social construction of the actual answers to well asked questions (Hacking, 2000). As I shall show in Chaps. 7 and 8, Hacking's defence is not sufficient to escape the accusation of *epistemic* relativism, but the styles project does not imply *alethic* relativism.

Relativism issues continue to be relevant today in relation to historical epistemology. In *Objectivity* Daston and Galison took care of remarking that relativism has nothing to do with their account of objectivity. I interpret this *excusatio non petita* as the symptom of the general worry, among historical epistemologists, about the relativistic implications of their conclusions. Just like Williams and Hacking (see Sect. 3.7), Daston and Galison raised the question of the permanence of truth over time. They remarked that whether or not truth has a history cannot be concluded from the fact that the means devised to attain it vary over time (Daston & Galison, 2007, p. 377). Maybe so, but this does not provide an answer to the tricky question of establishing how far we can go with our theses about what varies without ending up stating that there is no universal and atemporal justification for our epistemic claims.

4.5 Problems in Historical Epistemology: Ways of Thinking and Contingency

Besides the incommensurability problem there is another aspect of historical epistemology that has important implications. Since most of the thinkers I have discussed postulate the existence of different ways of thinking whose emergence is often contingent, and since these ways of thinking play a role in the production of knowledge, the achievement of certain scientific results might not be an inevitable fact. In literature this issue is framed in many other different forms: for example, often it is asked whether a particular field of science could have taken a different route from the actual one or whether the achievement of certain scientific results has been contingent. Apart from the problem of framing it, the contingency issue includes many sub-problems since different items can be contingent: questions, methods, theories, laws of physics, form of equations, constants of physics, experiments. Moreover, each of these items requires a specific analysis and its being 'contingent' is always relative to something to be specified. The immense importance of the contingency issue can be grasped by recognizing its bearing on the question of convergence: if contingency is a relevant feature of the evolution of the sciences, we might never achieve a complete description of a part of the world.

As we have seen, most of the historical epistemologists did not believe in any historical laws and considered the history of thought as an open-ended process characterized by periodic conceptual changes or reorganizations. Even when, as in the case of Brunschvicg, they judged certain ways of thinking as progressive or regressive, they did not regard the history of thought as a process governed by laws of development. This absence of constraints in the development of thought left little room for inevitability. The contingency issue can be found in embryonic form in Fleck's notion of the concept of 'ways of thinking' (Fleck, 1979 [1935]): his doctrine that scientific facts exist only within thought styles raised the doubt that they are not inevitable at all, rather determined by *contingent* social forces. Although the idea of contingency was central to the work of many French philosophers such as Foucault, it did not play any fundamental role in the debates on the growth of knowledge that raged soon after the Second World War in the Anglophone philosophy of science. With the advent of social constructionism the idea of contingency gained momentum. Shapin's conclusion that reality is capable of sustaining different accounts given of it suggested that our theories might have been different from the actual ones. Pickering's book *Constructing Quarks* (Pickering, 1984) encouraged the thought that a quarky physics was not *inevitable* at all: the theories about quarks and the experiments performed to test them, which convinced the scientists of their existence, represent the product of a historical process. The theme of contingence also played a role in some discussions on the interpretation of Quantum Theory. James T. Cushing made a case for thinking that a plausible temporal reordering of certain scientific achievements could have engendered a deterministic worldview instead of an indeterministic one (Cushing, 1992) (Cushing, 1994). His point raised the important question as to whether or not science is bound to converge on a single answer to a given well-posed question or can take alternative routes at certain critical junctures.

Hacking, reflecting on these thinkers, brought the contingency issue to the fore in its full-fledged form. He framed the issue in (Hacking, 1999) and developed it with some variations in (Hacking, 2000, 2014) on the basis of ideas expressed also in (Hacking, 1988, 1992, 1996) and later in (Hacking, 2005a, 2005b). In particular, he prompted the following formulation of the contingency issue:

> If the results R of a scientific investigation are correct, would any investigation of roughly the same subject matter, if successful, at least implicitly contain or imply the same results? (Hacking, 2000, p. S70)

This formulation generalizes the question posed by Pickering's point as to whether a non-quarky physics, as triumphant as the high-energy physics that evolved after the Second World War, could have developed and become our current physics. In March 2006 the philosopher Léna Soler organized a conference on the contingency issue (Soler, 2008a, 2008b)

(Trizio, 2008) (Sankey, 2008) (Franklin, 2008b), and a second one in 2009 with the participation of Hacking and Pickering. One of the upshots of those discussions was that Hacking's formulation of the contingency issue is only one of the many possible ways of framing the issue (Soler, 2008a, pp. 222–227). Those conferences made evident how vast the range of philosophical sub-problems of the contingency issue is (Soler, 2006 note 3 p. 364). For example, Soler showed how Hacking's formulation of the contingency issue above is ambiguous (Soler, 2008b). Hacking substantially defined contingentism as the claim that a science *radically different* from the actual one, but as *successful and progressive*, would be possible. He also asked whether such a science, although different, should imply or reach, at least in the long run, the same results as the current science. The problem with this formulation is that, terms such as 'results', 'successful science' and 'correct' need clarification.

Soler reformulated contingentism in terms of two successful physics that have developed in complete isolation from each other (Soler, 2008b). Then, by means of a thought experiment, she focused on what 'different' should mean, that is what kind of differences these two physics should present for giving rise to a form of contingentism. However she left out the problem of clarifying what a 'successful' physics really is. Emiliano Trizio investigated the connection between inevitability and success by examining three conceptions of scientific success: truth, empirical adequacy and robust fit (Trizio, 2008). One of the implicit results of his analysis has been to show how Hacking's formulation of the contingency issue above is not neutral. Indeed, of course, an attribute such as 'successful' depends on philosophical commitments—e.g. a conception of success as truth implies inevitabilism—and Trizio clarified how. Very recently Hacking showed more awareness of the fact that the inevitability question is inextricably connected with the question of success although he did not provide any in-depth analysis (Hacking, 2014, p. 117).

Sankey understood Hacking's formulation in terms of convergence on a single unified theory of the word: for him a contingentist would hold that science, properly conducted, might well have led to a completely different theory of the world from that of contemporary science (Sankey, 2008). So he ended up discussing contingency in relation to realism. His conclusion was that, although realism does not entail inevitabilism (in the sense of convergence on a whole description of the world), the employment of reliable methods would in fact guarantee that science continues to move closer to the truth about a portion of the world but non-necessarily on the whole truth. On the other hand, Allan Franklin discussed historical case studies in order to show that experiment provides constraints to the evolution of science (Franklin, 2008a, 2008b). He concluded that in many cases there is no room for reasonable alternative options.

I think it is unfortunate that so far the discussions about the contingency issue have not involved the concept of 'ways of thinking'. Indeed, it is important to note that this concept provides a notion in which the contingency issue could be fruitfully studied. Indeed, a given notion of the concept of 'ways of thinking' could imply precise claims about when and why certain ways of thinking have emerged and developed so to make certain results objective. And one could wonder what could have happened if a certain way of thinking had not emerged. As the theory of evolution provides the notion for a hypothesis of evolutionary relationships of species through time, so a theory about ways of thinking presupposes a precise order in which certain ways of thinking have evolved. The same mental experiment, described by Stephen Jay Gould (1941–2002) in *Wonderful Life* (Gould, 2000 chapter one), of replaying the 'life's tape' to see if the repetition would lead to the actual phylogenetic tree, could be conducted in the case of the evolution of ways of thinking. One could imagine pressing the rewind button and going back to the origins of the first way of thinking in order to see if a replay of the tape would lead to a science different from the actual one.

4.6 Styles of Reasoning, Relativism, Contingency

In the following chapters I shall address the questions concerning relativism and contingence raised above by focusing on the notion of 'style of reasoning'. Indeed, I believe that Hacking's notion of 'style of reasoning' is the best candidate for an investigation whose results have to be generalized to other notions. To start with, it represents the latest sophisticated version of a long tradition of notions of the concept of 'ways of thinking' that have been conceived in historical epistemology. In addition, while whether or not notions such as those of Fleck, Foucault, Kuhn and others involve relativism has been extensively discussed in the literature, in the case of Hacking the question has not yet received the attention it deserves. Although the latter did not want to provide criteria for the identity of styles, he characterized his notion by making precise claims that are philosophically interesting because they are suspected of relativistic implications: styles are characterized in terms of their bringing new objects into existence, of their self-authenticating character and their role in creating criteria of truth and falsehood. One can imagine to strip styles of certain features so to make them more similar to other notions and ask whether or not the conclusions previously drawn still hold true. Finally, the relativism issue in the case of Hacking's notion is particularly interesting and challenging: first he posed the problem of the relativistic implications of his notion ('I wish to pose a relativist question from within the heartland of rationality' (Hacking, 1982)) and then rejected the accusations of relativism without ever wanting to advance a systematic argument.

With regard to the contingency issue, once one accepts the idea that in the history of science there have been different ways of thinking, two important questions emerge: is the birth of a given way of thinking an inevitable circumstance? Is the endurance of a way of thinking inevitable? The project of styles of reasoning has much to say about these two questions and not only because of the self-authenticating character of styles. Styles have a lot more to do with human rationality than other notions. They are ways of tapping fundamental human abilities such as those involving the use of deductive or abductive reasoning. It seems inevitable that we will continue to use styles such as the deductive one or the historico-genetic. Furthermore, as I have shown in Sect. 4.3 (see also (Sciortino, 2021)), together with that of episteme, the notion of style of reasoning illuminates conditions of possibility that are crucial for the emergence of objectivity and lie at a deeper level than those considered by the other two notions. Also for this reason it is a good candidate for studying the issue of contingency.

The investigation on the implications of the notion of style of reasoning must necessarily start from the attempt, in the next chapters, to develop, correct and present in a more systematic way the styles project. One of the reasons is that, in order to tackle the relativism issue, we need to be clear about what can be considered a style of reasoning in the sense of Hacking. Take for example Kusch's criticism about Hacking's styles project. The latter reminds us that epistemic relativism is the claim that the properties picked out by the predicates 'rational' and 'justified' are relative to different epistemic systems (Kusch, 2010). For him this definition fits the case of Hacking's styles: a style-dependent sentence can be justified in its own style but is meaningless outside of it; therefore its justification is relative to its style. The problem is that this point only countenances a possibility: the relativism issue in the context of the styles project remains unsolved until it is found an example of a scientific claim that is justified in a style and unjustified in another. Kusch said that an example that involves two scientific styles in the Crombie's list would not do the job because 'familiar forms of epistemological relativism do not assume that one and the same individual possesses more than one epistemic system' (Kusch, 2010, p. 167). Therefore he brought in Christian theology and argued that it is a style of reasoning in the sense of Hacking; then he concluded that the sentence 'God told me that I should do his will from day to day, in humility and poverty' is rational and justified for a Christian but not rational and justified for him (Kusch, 2010, p. 168). Indeed, he considers the Christian style incompatible with his commitment to other scientific styles. Here we see that whether or not Christian theology can be considered a style of reasoning is crucial for the argument. If it were not a style of reasoning in the sense of Hacking, Kusch could not conclude from his argument that the former's project implies epistemic relativism. So, what is a style of reasoning exactly? This question bears on the problem I want to discuss.

Furthermore, I see at least two problems with Kusch's argument. First of all, he did not provide any example of a *scientific* claim that is justified in a style of reasoning and unjustified outside it, rather an example of a claim that, he thinks, is incompatible with one's commitment to science. Consequently, Kusch left a doubt: in absence of an example of *scientific* claim one does not know whether, within science, there is incommensurability or not. Indeed, think of two different communities A and B that at different times adopt different styles of reasoning; could B justify all the scientific claims made by A? This question remains unanswered. In sum, the lack of an investigation on the relation between relativism and the style of reasonings project, viewed as it is meant to be, that is a project on *scientific* ways of knowing, seems to me a serious gap in the literature.

Furthermore, Kusch noted that the same individual can adopt different scientific styles of reasonings but added that for him the Christian style is incompatible with them. So there is compatibility between two scientific styles of reasonings and there might be incompatibility between a non-scientific style of reasoning and a scientific style of reasoning. Kusch did not provide any argument to account for this asymmetry. In the absence of arguments one is led to thinking that the incompatibility between the Christian style and a scientific style of reasoning lies on the fact that what is considered true in the Christian style is not true in the scientific style. If this were the case one could push further his argument by saying that the style of reasonings project does not only imply epistemic relativism but also alethic relativism, a conclusion that, as I shall argue, is untenable.

4.7 Conclusions

In Sect. 4.2 I have brought to the fore the family resemblances of a set of notions that can be better understood within the context of historical epistemology. All these notions are different species of the genus 'ways of thinking', whose main character is historicized Kantianism. In Sect. 4.3 I have provided a model for studying objectivity that incorporates the notions of Fleck, Foucault, Kuhn and Hacking. Used together, these notions provide a more profound understanding of how objectivity emerges. I have also proposed a layered diagram that maps the differences between the notions of thought style, episteme, paradigm and style of reasoning. In Sects. 4.4 and 4.5 I have highlighted two philosophical issues, relativism and contingency, connected with the concept of 'ways of thinking' in all its different versions. In Sect. 4.6, I have argued that this issue is particularly controversial in the case of Hacking's notion. In the next two chapters I shall develop Hacking's styles project with the aim of discussing this case.

Bibliography

Bell, J. S. (1966). On the problem of hidden variables in quantum mechanics. *Reviews of Modern Physics, 38*, 447–452.

Brenner, A. (2006). Quelle épistémologie historique? Kuhn, Feyerabend, Hacking et l'école bachelardienne. *Revue de Métaphysique et de Morale, 1*(49), 113–125.

Brunschvicg, L. (1934). *Les âges de l'intelligence.* Paris Alcan.

Chimisso, C. (2008). *Writing the history of mind.* Ashgate.

Cushing, J. (1992). Historical contingency and theory selection in science. *Proceedings of the Biennial Meeting of the Philosophy of Science Association, 1*, 446–457.

Cushing, J. (1994). *Quantum mechanics: Historical contingency and the Copenhagen hegemony.* Chicago University of Chicago Press.

Daston, L., & Galison, P. (2007). *Objectivity.* Zone Books.

Elwick, J. (2012). Layered history: Styles of reasoning as stratified conditions of possibility. *Studies in History and Philosophy of Science.* https://doi.org/10.1016/j.shpsa.2012.07.004

Feyerabend, P. (1962). Explanation, reduction, and empiricism. In H. Feigl & G. Maxwell (Eds.), *Minnesota studies in the philosophy of science: Scientification explanation, space and time* (Vol. 3). University of Minnesota Press.

Feyerabend, P. (1988 [1975]). *Against method.* Verso.

Feyerabend, P. (1989). Realism and the historicity of knowledge. *Journal of Philosophy, 86*(8), 393.
Feyerabend, P. (1999). *Conquest of abundance*. The University of Chicago Press.
Feyerabend, P. (2016 [2009]). *Philosophy of nature*. Polity.
Fleck, L. (1979 [1935]). *Genesis and development of a scientific fact*. The University of Chicago Press.
Foucault, M. (1969 [1972]). *The archaeology of knowledge*. Pantheon Books.
Foucault, M. (1994 [1966]). *The order of things*. Vintage Books Edition.
Franklin, A. (2008a). Are the laws of physics inevitable? *Physics in Perspective, 10*(2), 182–2011.
Franklin, A. (2008b). Is failure an option? Contingency and refutation. *Studies in History and Philosophy of Science Part A, 39*(2), 242–252.
Gayon, J. (1996). De la catégorie de style en historie de sciences. *Alliage, 26*, 12.
Gould, S. J. (2000). *Wonderful life: The burgess shale and the nature of history*. Vintage.
Gutting, G. (1991 [1989]). *Michel Foucaults's archeology of scientific reason*. Cambridge University Press.
Hacking, I. (1982). Language, truth and reason. In M. Hollis & L. S (Eds.), *Rationality and relativism* (pp. p. 48–66 [Reprinted in Hacking 2002]). Blackwell.
Hacking, I. (1983). *Representing and intervening*. Cambridge University Press.
Hacking, I. (1988). The participant irrealist at large in the laboratory. *The British Journal for the Philosophy of Science, 39*(3), 277–294.
Hacking, I. (1992). The self-vindication of the laboratory sciences. In A. Pickering (Ed.), *Science as practice and culture* (pp. 29–64). Harvard University Press.
Hacking, I. (1996). The disunity of science. In P. Galison & D. J. Stump (Eds.), *The disunity of science: Boundaries, contexts, and power* (pp. 37–74). Stanford University Press.
Hacking, I. (1999). *The social construction of what?* Harvard University Press.
Hacking, I. (2000). How inevitable are the results of successful science? *Philosophy of Science, 67*, S58–S71.
Hacking, I. (2002). *Historical ontology*. Harvard University Press.
Hacking, I. (2005a). La philosophie de l'expérience: illustrations de l'ultrafroid. *Tracés. Revue de Sciences humaines*. Retrieved February 11, 2008, from http://traces.revues.org/264
Hacking, I. (2005b). Les philosophes de l'expérience. *Tracés. Revue de Sciences humaines*. Retrieved February 11, 2008, from http://traces.revues.org/201
Hacking, I. (2006a). Méthodes de Raisonnement.
Hacking, I. (2006b [1975]). *The emergence of probability*. Cambridge University Press.
Hacking, I. (2009). *Scientific reason*. NTU Press.

Hacking, I. (2014). *Why is there a philosophy of mathematics at all?* Cambridge University Press.
Kochen, S., & Specker, E. (1967). The problem of hidden variables in quantum mechanics. *Journal of Mathematics and Mechanics, 17*, 59–87.
Kuhn, T. (1987a [1978]). *Black body theory and the quantum discontinuity*. University of Chicago Press Edition.
Kuhn, T. (1987b). What are scientific revolutions? In L. Krüger, L. Daston, & M. Heidelberger (Eds.), *The probabilistic revolution*. Cambridge The MIT Press.
Kuhn, T. (1996 [1962]). *The structure of scientific revolutions*. The University of Chicago Press.
Kuhn, T. (2002 [2000]). *The road since structure* (J. Conant & J. Haugeland, Eds.). The University of Chicago Press.
Kusch, M. (2010). Hacking's historical epistemology: a critique of styles of reasoning. *Studies in History and Philosophy of Science, 41*(2), 158–173. https://doi.org/10.1016/j.shpsa.2010.03.007
Kusch, M. (2020). *Relativism in the philosophy of science*. Cambridge University Press.
Lévy-Bruhl, L. (2015 [1910]). *How natives think*. Martino Publishing.
Möβner, N. (2011). Thought styles and paradigms: A comparative study of Ludwik Fleck and Thomas S. Kuhn. *Studies in History and Philosophy of Science Part A, 42*(3), 416–425. https://doi.org/10.1016/j.shpsa.2011.02.001
Oberheim, E. (2005). On the historical origins of the contemporary notion of incommensurability. *Studies in History and Philosophy of Science, 36A*(2), 363–390.
Pickering, A. (1984). *Constructing quarks: A sociological history of particle physics*. Edinburgh University Press.
Sady, W. (2012). Ludwik Fleck. In E. N. Zalta (Ed.), *The Stanford encyclopedia of philosophy* (Summer 2012 ed.).
Sankey, H. (2008). Scientific realism and the inevitability of science. *Studies in History and Philosophy of Science Part A, 39*(2), 259–264.
Sciortino, L. (2021). The emergence of objectivity: Fleck, Foucault, Kuhn and Hacking. *Studies in History and Philosophy of Science Part A, 88*, 128–137. https://doi.org/10.1016/j.shpsa.2021.06.005
Soler, L. (2006). *Contingence ou inévitabilité des résultats de notre science ?, 33*(2), 16.
Soler, L. (2008a). Are the results of our science contingent or inevitable? *Studies in History and Philosophy of Science Part A, 39*(2), 221–229.
Soler, L. (2008b). Revealing the analytical structure and some intrinsic major difficulties of the contingentist/inevitabilist issue. *Studies in History and Philosophy of Science Part A, 39*(2), 230–241.

Soler, L. (2011). The incommensurability problem: Evolution, Current Approaches and Recent Issues. *Philosophia Scientiæ, 8*(1), 1–38.
Stump, J. B. (2001). History of science through Koyre's lenses. *Studies in History and Philosophy of Science, 32*(2), 243–263.
Trizio, E. (2008). How many sciences for one world? Contingency and the success of science. *Studies in History and Philosophy of Science, 39*(2), 253–258.
von Neumann, J. (1955). *Mathematical foundations of quantum mechanics.* Princeton University Press.

CHAPTER 5

Developing the 'Styles Project': Towards a 'Theory of Styles of Reasoning'

5.1 Introduction

In this chapter I shall be engaged in turning the styles project into a more comprehensive analysis of scientific thinking so as to be able to assess Hacking's notion. One of the points implied by his claims is that ways of thinking, the statistical and the laboratory style, share a set of features ('the characterizing properties of styles') presented in Sect. 3.4: 1) being ways of thinking and *doing*; 2) relying on a new kind of evidence; 3) introducing new candidates for truth-or-falsehood 4) being self-authenticating; 5) representing a sharp break in the history of Western thought.

The label 'style of reasoning' wants to suggest that, despite their differences, these ways of thinking belong to the same species. Still at issue is how we can further characterize scientific thinking: can we identify other styles of reasoning so as to provide a more accurate description of it? It is a safe guess to say that the spectrum of our ways of finding out about the world is quite broad: there might be other ways of finding out that possess most of the characterizing properties of styles of reasoning in addition to ways of thinking that do not satisfy them. Actually, I shall put forward two related theses.

My first thesis is that, besides the statistical and the laboratory style of reasoning, there are at least four other styles of reasoning, of which some are very important because they are based on the most fundamental forms of reasoning: deduction, induction and abduction. My second thesis,

L. Sciortino, *History of Rationalities*,
https://doi.org/10.1007/978-3-031-24004-1_5

which I shall discuss in the last section, is that the spectrum of scientific ways of thinking also includes 'spurious styles of reasoning', as I shall call all those ways of thinking that satisfy only few characterizing properties of styles of reasoning. To prove the first thesis, in the following four sections I shall argue that the algorithmic, the postulational, the historico-genetic and the taxonomic way of thinking possess the characterizing properties of styles of reasoning listed above and explained in Sect. 3.4. To prove the second thesis, I shall consider most of the candidates for being styles of reasoning presented in the literature and explain why they do not possess some important characterizing properties.

The brief comments on each of these four styles of reasoning just mentioned provided by Hacking are not sufficient to give a full account of their properties. However, in his discussions he has drawn the reader's attention to certain historical works that provide profound insights into the nature of these styles. I shall rely on these works, although they have not been written within the framework of historical epistemology. Nevertheless, other sources will receive my attention for their connections with my claims. In the following chapters, when it will be necessary to address philosophical questions related to the styles project, I shall often refer to the styles of reasoning discussed in this chapter and in Chap. 3 in order to make my points clearer. However, the conclusions I will draw will be independent from my own characterization of scientific thinking. They will follow from Hacking's characterization of what a style of reasoning is, i.e. from its characterizing properties.

5.2 The 'Algorithmic Style of Reasoning'

When in 1992 Hacking hinted that there is a style of applied mathematics distinct from the postulational, he called it 'algorismic'[1] (Hacking, 2002b [1992], p. 185). This term stems from 'Algoritmi', the transliteration of the name of the Persian mathematician al-Khwārizmī (780–850) whose works, translated into Latin in the twelfth century, introduced to Europe several algorithms. As the historian of mathematics Carl Boyer put it, '[al-Khwārizmī's work] is to algebra what Euclid's *Elements* is to geometry' (Boyer, 1968, p. 258): for Hacking it was with the introduction of

[1] To my knowledge, Hacking has never used the term 'algorithmic' in English. Though, in French he has used the term 'algorithmique' rather than 'algorismique'. As I shall explain below I prefer to call it 'algorithmic' (Hacking, 2006b, p. 7).

algebra that the algorithmic way of thinking flourished in Europe (Hacking, 2002–2003, p. 543).

The use of the term 'algorismic' instead of 'algorithmic' reminds us that this way of thinking has to do with following step-by-step procedures in *calculations* rather than in more general problems that do not involve numbers. Indeed, while in the Middle Ages the Latin name 'algorismus' denoted a genre of works on methods of calculations (such as *Carmen de Algorismo* by Alexander de Villa Dei (1175–1240) or *Algorismus Vulgaris* by John of Sacrobosco (1195–1256)) (Karpinski & Waters, 1928, p. 45), the term 'algorithm' emerged much later from 'algorismus' and took on the *generalized* meaning of a set of rules to be followed in problem-solving operations. At any rate, Hacking regarded the algorithmic way of thinking as a *mathematical* style of reasoning that, together with the postulational one, forms the 'style of mathematics' (Hacking, 2002–2003, p. 543): to adopt the algorithmic way of thinking means to follow methods of calculation, i.e. step-by-step list of rules that might be represented by a formula.

In his latest overview of the styles project (Hacking, 2012) Hacking did not even mention the algorithmic way of thinking. By contrast, I believe that this way of thinking cannot be left out from any account of scientific thinking since it occupies a vital role in all the branches of scientific knowledge. Besides the intimate connection of knowing and doing, a theory of scientific thinking should also highlight the fact that the sciences combine qualitative *and quantitative* methods of analysis. Mathematics does not only give a formal expression to disciplines such as physics or economics but also provides a way of seizing the measure of certain magnitudes. According to Michel Paty, from nineteenth-century science, 'measurable quantities [...] have meant uniquely, and perhaps restrictively, quantities *taking numerical values*, as it was most usually the case in classical physics' (Paty, 2003, p. 126).[2] The magnitudes to be calculated are not only those that already exist in the physical corpus, they are also introduced by scientists (and expressed by functions or constants): entropy in physics, affinity in chemistry, volatility in economics are some examples (Lacki, 2003, p. 274). The use of the algorithmic way of thinking is therefore ubiquitous in scientific research. Incidentally, in order to know the value of a physical constant it might be necessary to perform a measurement that consists of step-by-step procedures. As I intend to account for this aspect of scientific

[2] Paty provides a review of the conceptions of quantity or magnitude over the centuries and investigates how these conceptions have legitimated the mathematization of physics.

research too, I prefer to use the term 'algorithmic' rather than 'algorismic': I want to describe a style of reasoning that accounts not only for the use of standard procedures in calculations but also for the use of standard procedures to be followed in more general problem-solving operations that may involve numbers. So, to call this style 'algorithmic' seems to me more appropriate. In the following lines I shall describe the algorithmic thinking. Let us start by giving an example.

5.2.1 *Algorithmic Thinking: An Example*

The first known algorithms in history go back to civilizations more ancient than Greece.[3] In particular, they can be found in one of the Egyptian papyri that have survived the ravages of time, known as 'Rhind Mathematical Papyrus' or 'Ahmes Papyrus' in honour of Ahmes, the scribe who copied it down in around 1650 BC.[4] One of the sentences in the papyrus states:

> There are 7 houses, 49 cats, 343 mice, 2401 ears of grain

The scribe was providing an answer to this problem[5]:

> In each of seven houses
> there are seven cats
> each of which eats seven mice,
> each of which would have eaten seven ears of grain.
> Adding up houses, cats, mice, ears of grain, how many objects are there in total?

This riddle can be iterated adding as many objects as one wants, e.g. one can imagine that each of the seven ears of grain had seven seeds and each seed was made of seven items and every item contained seven objects and so on.

[3] I shall focus my discussion on the Babylonian and the Egyptian civilizations. However, my arguments could also be tested against examples drawn from the mathematics of Mayan people between the third and ninth centuries (see for example Magli, 2009 [2005]).

[4] The Egyptian history is usually divided into three large periods that cover a period from 3100 to 332 BC: the *Old Kingdom*, the *Middle Kingdom* and the *New Kingdom*. The Ahmes papyrus belongs to the Middle Kingdom, between about 2050 and 1600 BC.

[5] The Ahmes papyrus presents several arithmetical, algebraic and geometrical problems. For a review of the mathematics in the Ahmes papyrus see: (Boyer, 1968, pp. 12-23).

The simple answer to the riddle is that there are

7 +7x7+7x7x7+7x7x7x7= 7+49+343+2401 objects

Today, a mathematician would recognize at first glance that these numbers represent the first four terms of a geometric series of ratio 7 whose sum of the first n terms is given by a standard formula. By performing a standard procedure consisting of successive steps, the Egyptians were able to solve the riddle whatever the total number of the objects added by iteration was. For example, if the iteration stopped at n, the first step to take would have been to consider n terms of which the generic mth term of the series was 7 multiplied by itself m times. The other steps would have been to work out each of the n terms of the sum by performing the multiplications. According to Boyer (1968, p. 15), in Ahmes's day multiplication was performed by a method of successive doubling. Indeed, by using an abacus it is easy to add up, multiply and divide by 2. Suppose they would have to calculate the fourth term of the sum, which is 7x7x7x7=7x343=2401.[6] Since 7=4+2+1, they would have doubled 343 and got 686, then doubled 686 and obtained 1372, i.e. four times 343. Afterwards, they would have added up 1372+686+343 obtaining 2401. By using the abacus this is an easy task because to double requires a small number of operations. The Egyptians also had another very similar method to work out multiplications: it is today known as the Russian farmer's method and involves divisions and multiplications by two. Thus, in principle, they were able to 'check' a result by using different methods of calculation. To sum up, in order to provide the answer to the riddle, whatever the number n of iterations, the Egyptians used the algorithmic way of thinking: they followed a procedure expressed by a list of well-defined instructions, which involved the use of the abacus.

The Egyptians used algorithms to solve many arithmetical problems, such as that of finding solutions of linear equations (Boyer, 1968, p. 16) or that of calculating areas and volumes (Boyer, 1968, pp. 19–23). On the other hand, the Babylonians were able to calculate the correct values of many astronomical phenomena even though their geometry was not as developed as in the ancient Greece: for example, by measuring and implementing algebraic algorithms they managed to make predictions of the

[6] Of course the Egyptians did not denote numbers as we do. They had a system of numeration derived from base ten and did not possess a placed-valued system. When operations were performed they grouped together the symbols in no particular order.

phenomenon of lunar eclipses every 18 years, the so-called Saronic cycle (Mason, 1962, p. 19). The ancient Greeks made use of algorithms too: the Euclidean one for finding the greatest common divisor of two integers and the numerical solutions of algebraic equations in Diophantus's *Arithmetica*, which was read by the Arab mathematicians of the ninth century, are the examples that stand out.

In all the historical examples I have given above, to reason in the algorithmic style meant to follow a set of elementary rules in order to reach a certain objective. This way of reasoning is fundamentally distinct from that used in proving a theorem. I interpret Hacking in the passage below as making just this point:

> There is nothing perspicuous about calculation. There is no feeling of compulsion. There is no sense of understanding. (Hacking, 2000, p. 101)

Applying an algorithm to solve Ahmes's riddle is different from thinking in order to demonstrate that an angle inscribed in a semi-circle is a right angle. In the former case there is no feeling of understanding; in the latter, there is a strong feeling that the conclusion *must* follow from two facts: the base angles of isosceles triangles are equals and the sum of the angles of a triangle is equal to 180 degrees. Whereas the steps to be taken in order to solve Ahmes's riddle iterated *n* times are fixed, within a line of reasoning from hypotheses to thesis every statement must be *deduced* from the previous one.

Algorithm and demonstration are the two faces of mathematical thinking. A mathematician has to switch from the way of thinking of demonstration (the postulational style described in the next section), which allows her to prove a theorem, to the algorithmic way of thinking, which for example allows her to determine the result of an operation, a number.

5.2.2 *Techniques of Self-vindication*

As I have pointed out, the Egyptians were able to use different algorithms for solving the same problem; therefore, one of these algorithms could serve as a means for checking the result obtained. Examples are: the already mentioned method of successive doubling and the 'Russian farmer's method' for performing multiplications; the 'method of false positions' and the 'method of factorization' for solving linear algebraic equations (see Boyer, 1968, p. 17). In other words, the algorithmic thinking does not answer to any criterion except its own: the criterion to prove whether or not what has been found out is correct still relies on another algorithm. In this sense the algorithmic thinking possesses what Hacking would call a technique of self-vindication: 'the solution of the riddle iterated *n* is *x*' is what we conclude by using the algorithm A; and how do we know that we are correct? We know that because by using another algorithm, the algorithm B, we find *x*. That is: the algorithmic way of thinking is self-authenticating.

5.2.3 Mathematical Ways of Thinking in Egypt, Babylonia and Greece

Many of the 110 problems in the Rhind and Moscow papyri, from which we derive our knowledge of Egyptian mathematics, have a practical origin. Similarly, Babylonian cuneiform tablets do not present general formulations in geometry but deal either with specific practical problems or with more 'theoretical exercises' for the joy of pure mathematics, to be worked out according to a certain algorithm (Jens Høyrup, 1990, p. 65).[7] Referring to the contrast between Greek and Babylonian (or Egyptian) mathematics Jens Høyrup defined the former 'scientific knowledge' and the latter 'sub-scientific knowledge': while the ancient Greeks pursued scientific knowledge *systematically* and *for its own sake*, the Babylonians and the Egyptians acquired and transmitted only sub-scientific knowledge, i.e. a specialist knowledge pursued only *in view of its applicability* (Jens Høyrup, 1990 p. 64 and p. 71). The Egyptian and the Babylonian mathematicians did introduce problems with no practical application such as Ahmes's riddle but they were determined by methods already at hand. On the contrary, 'Greek mathematics [was] determined by *problems* for the solution of which new methods would have to be developed' (Jens Høyrup, 1990, p. 65).

The Babylonian mathematicians did not have an algebraic notation as transparent as ours. They represented their algebra by using a geometry of measurable segments, squares and rectangles and '[their] prescriptions describe[d] what is done in the geometric representation, just as we may describe in words what we are doing to the equation' (Jens Høyrup, 2010, p. 34). According to Donald Knuth, Babylonian algebra consisted of

[7] The Babylonian texts to which Høyrup refers belong to the Old Babylonian period, from 2000 to 1600 BC.

step-by-step lists of rules, i.e. of algorithms for representing algebraic equations and, ultimately, computing them (Knuth, 1972, p. 672). In brief, in Knuth's understanding the use of algorithms is the only relevant insight to read Babylonian mathematical texts. Høyrup stigmatized this view as a mistake built on the translations current at the time (e.g. Neugebauer & Sachs, 1945) and gave his own interpretation (Jens Høyrup, 2010): as we do, the Babylonian mathematicians presupposed the existence and the properties of the objects they were looking for and operated on geometric representation in order to find them. Not only the algorithmic thinking but also decision-making was involved: '[Babylonian texts] describe a particular type of geometrical manipulation, which like modern equation algebra is analytical in character, and which displays the correctness of its procedures *without being explicitly demonstrative*' (Jens Høyrup, 2010, p. 1 my emphasis).

These points lead to the conclusion that the algorithmic way of thinking had already settled down into a routine in the Egyptian and Babylonian civilizations. Importantly, by no means can we conclude that Babylonians did not have a 'reasoned' mathematics, as Høyrup explains:

> Procedures were described in a way which […] turns out to be as transparent as the self-evident transformations of modern equation algebra and in no need of further explicit arguing […]. (Jens Høyrup, 2012, p. 17)

Nevertheless,

> No surviving text suggests that all this was ever part of an explicitly formulated programme, nor do the texts we know point to any thinking about *demonstration as a particular activity.* (Jens Høyrup, 2012, p. 17)

This passage suggests we should think of the Babylonian mathematicians as possessing an *algorithmic* way of thinking but *not* a *demonstrative* way of thinking: as I shall explain in the next section, only as a consequence of a different historical and social environment did a style of demonstration ('postulational style of reasoning') emerge in ancient Greece.

5.2.4 Doing

By saying that 'mathematics is *embodied*' and that 'it is in the hands and in the arms that move around' (Hacking, 2011) Hacking reminded us that the first men who counted needed to perform precise acts with their hands. For example, to solve Ahmes's riddle meant to carry out automatically multiplications by manipulating pebbles from one side to another of a tablet. Similarly, as I have already said, the Babylonian algebra had a geometrical representation; and some operations that we might call 'addition' and 'subtraction' consisted of specific *actions*: they were *concrete* operations. To make few examples: 'to append' meant to *join* an entity to another one, which conserves its identity; 'to tear out' meant to *remove* an entity from another quantity of which it is part; 'to compare' meant to *observe* that a length or an area exceeds or falls short of another (Jens Høyrup, 2010, pp. 6–7).[8] The above-mentioned concrete operations upon objects, which were involved in the implementation of Egyptian and Babylonian algorithms, show that the algorithmic style is a way of thinking *and* doing.

5.2.5 The Emergence of the Algorithmic Style of Reasoning

Research shows that basic numerical competences are rooted in biological built-in resources that can be explored in animals and humans (for a review of this research see Nieder, 2005). Different species of animals have the sense of the cardinality of a set, although they have distinct upper-numerosity discrimination limits (Köhler, 1957) (see also Davis & Perusse, 1988). On the other hand, pre-verbal human infants possess a sense of the numerosity of a set of objects and can engage in basic additions and subtraction operations with objects (Wynn, 1992) (Feigenson, Dehaene, & Spelke, 2004); Amazonian populations that lack words for numbers beyond five can compare large sets of items (Pica, Lemer, Izard, & Dehaene, 2004). Furthermore, it seems that humans have an innate ability to follow recursive rules: procedures for finding the number of elements in finite sets such as finger counting were in use in human societies before the invention of writing. For example, according to the researchers who found it, a bone of a baboon marked with 29 notches and dated to approximately 35,000 B.C was used as a calendar stick (Bogoshi, Naidoo, & Webb, 1987). To record the passage of days it was sufficient to follow a simple rule: for each sunset a notch had to be etched into the bone.

[8] Interestingly, the Babylonian way of *doing* mathematics has gone extinct. This shows that, even if a style of reasoning is long-lasting, the ways of implementing it can die out.

However, although our capacity for number as well as the faculty of language relies on recursive rules (Hauser, Chomsky, and Fitch 2002), neither the existence of fixed procedures for counting nor the sense of numerosity implies the existence of an *abstract* concept of number. To acquire this concept it is necessary to put a step forward: to represent a particular value by a symbol or a word, i.e. to represent a feature of sets of objects that is independent from the characteristics of the objects themselves—this feature is the cardinality of a given grouping. In as much as the concept of number implies the idea of a magnitude to be represented, it requires the existence of a language that makes it possible to process numerical information with precision. When did this concept emerge? Certainly, when the first human societies started to make use of commodity money a concept of number had to come about: money is a *measurement* of value (as well as a medium of exchange and a store of value). Therefore, I maintain that the algorithmic thinking came to the fore when the concept of number came about, in turn a consequence of the emergence of complex social structures. We may think of the birth of the algorithmic style as a discontinuity in human history by considering that small groups of hunter-gatherers possessed only the ability of finger counting and had no symbolic representation of number (Rudman, 2007, p. chapter one). It was only with the Neolithic transition from a life-style of hunting and gathering to one of settlement and agriculture (about ten thousands years ago) that an abstract concept of number emerged. According to Peter Rudman, the concept of division and of prime number came to the fore after 10,000 BC (Rudman, 2007, p. chapter one). The Neolithic transition, also called the 'Agricultural Revolution', was very rapid, as archaeological and demographical studies have documented (Bocquet-Appel, 2011). It is plausible to say that it produced a sharp break between the way of thinking and doing of the hunter-gatherers and that of the first complex communities who computed land areas or granary volumes by using algorithms.

5.2.6 New Sentences, Evidence

Of course, there has been a time in which a sentence such as

The number of objects in Ahmes's riddle is given by multiplying the following three terms: 7x7, 7x49, 7x343 and adding them up to 7

had no truth-value. It was a time in which, in order to make sense of that sentence, people would have had to possess an *abstract* concept of number and share the algorithmic way of thinking and doing. Humans may well have an innate ability to follow step-by-step procedures, but only when the first complex societies came about did they use algorithms as evidence of the truth of certain (style-dependent) propositions. In order to make sense of the sentence above, those humans who lived before the Agricultural Revolution would have had to possess certain concepts, standards of truth, methods, in brief, a way of thinking and doing. For them it would not have been so important whether the sentence was written by using the Egyptian numeral system or our placed-value system: how to use a numeral system can be learned. The point is that only for someone who is part of a community that shares the algorithmic way of thinking and doing does the sentence above acquire a truth-value.

My conclusion is that the algorithmic way of thinking[9] and doing satisfies all the characterizing properties of styles and, therefore, deserves the label of 'algorithmic style of reasoning'.

5.3 The Postulational Style of Reasoning

5.3.1 Greek Mathematics

This section contains a broad characterization of the deductive way of thinking involved in the demonstrative proofs of geometry. As I have explained, according to Høyrup the deductive method emerged in ancient Greece and was not a Babylonian practice. Another scholar, the already mentioned Reviel Netz, in his book *The Shaping of Deduction in Greek Mathematics* (Netz, 2000) advanced the same thesis. Actually, Netz's reconstruction of the missing diagrams (i.e. geometrical figures composed of lines) in the Greek texts provides deep insights into Greek mathematics that I shall use to argue that the deductive way of thinking possesses the characterizing features of styles of reasoning.

Greek mathematical texts consist of lettered diagrams and words. The deployment of these two elements is what makes Greek mathematics a recognizable genre. As we shall see, Netz maintained that, on a logical plane, lettered diagrams and words combine in necessity-preserving ways to form deductive chains and to yield knowledge of general validity: not knowledge of a particular triangle but of all the triangles (Netz, 1999

[9] In (Sciortino, 2016), I have connected some Wittgenstein's thoughts with my considerations about the self-authenticating character of the algorithmic way of thinking.

chapters 5 and 7). On a historical plane, while lettered diagrams are the mark of a literate society, deductive proofs are idealized versions of oral arguments, the mark of a polemical society (Netz, 1999 chapter 7 especially p. 312). On a cognitive plane, Greek demonstrations are the result of an interplay between visual resources and *indices* (letters that signify a point by standing next to it) (Netz, 1999, p. 47).

The logical, historical and cognitive analysis offered by Netz provides fertile ground for my arguments. Indeed, first of all Netz gives sound reasons for claiming that the diagram is a necessary element in the reading of the text (Netz, 1999, pp. 26–27): for example, some assertions of Greek proofs must be deduced from the diagram (Netz, 1999, p. 47). This is a point that will be important for my argument that the postulational style can be viewed as a way of doing: as I shall show, assertions are mediated via specific actions.

Furthermore, in the last chapter of *The Shaping of Deduction in Greek Mathematics* Netz made another point that, as I shall point out, is germane to the question as to whether the early development of the postulational style was gradual. For Netz, the practice of the Greek mathematicians would not have come about without a culture of agonistic debate, the existence of an intellectual network, a tradition of oral and literate culture.[10] Finally, it is possible to make it clearer what Hacking meant when he maintained that the styles project is a study in 'cognitive history'. As I have said, 'cognitive history' is an expression that Netz used in the introduction to *The Shaping of Deduction in Greek Mathematics.* He reminded the reader that the foundations of the theory according to which some functions of mind are modules, i.e. task-specific capacities, were laid by the cognitive scientist Jerry Fodor (Fodor, 1983) (Netz, 2000, p. 5). For example, according to the latter, syntax is a module: a biological capacity hard-wired in the brain that allows human beings to perform syntactic computation without even relying on other knowledge. However,

> central processes such as the fixation of belief are not modular [...] we can never have a neat universal model of such functions as the fixation of belief. [...] For the historian, study starts where universality ends. It is clear why cognitive history is possible. While there are no general, universal rules concerning, for example, reasoning, such rules do exist historically, in specific contexts. (Netz, 1999, p. 5)

[10] See also (Sciortino, 2021).

In other words, processes such as vision and language are modular and single modules can be analysed by the methods of cognitive sciences. On the contrary, more complex cognitive processes, such as ways of thinking and doing, cannot be reduced to general rules: instead of a cognitive science we need a 'cognitive history' that searches for rules in specific contexts. 'Cognitive' because it is a study of the practices of knowledge and the way human cognitive structures come to interact; 'history' because the only rules that exist concerning the reasoning are not universal but historical.

Hacking generalized this point by conjecturing that, if the research programme of innate cognitive modules will develop, then 'no single module will correspond to exactly one style of reasoning. Each demands many and modules of different types' (Hacking, 2009, p. 39). So, each style of reasoning represents a particular way in which different modules interact with each other. In the case of the deductive thinking,

> diagrams [...] are the Greek mathematical way of tapping human visual cognitive resources. Greek mathematical language is a way of tapping human linguistic resources. (Netz, 1999, p. 6)

In his book Netz made no mention of Hacking's concept of style of reasoning, although he did use the term 'style' a couple of times (e.g. Netz, 1999, p. 9) in its common-sense usage to hint at a Greek 'style' of mathematics. Hacking on his part wrote that 'the styles project does not engage in historical research but it is delighted by new historiographical ideas that it can exploit. Netz provides one such' (Hacking, 2012, p. 5). In Hacking's mind, just as *Leviathan and the Air Pump* represents an analysis of the emergence of the laboratory style, so *The Shaping of Deduction in Greek Mathematics* represents a history of emergence of the postulational style (Hacking, 2012, p. 5). Though, these two books belong to different philosophical perspectives. Indeed, Netz played with the word 'shaping' in the title of his book in order to distance himself from constructivist stances:

> This book should not be read as if it were 'The Shapin of Deduction', an attempt to do for mathematics what has been so impressively done for the natural sciences. [...] I do not see 'deduction' as a sociological construct. I see it as an objectively valid form, whose discovery was a positive achievement. (Netz, 1999, p. 3)

Yet this *caveat* did not prevent Bruno Latour (1947–2022), who was recommended to read the book by Hacking (Hacking, 2009, p. 67), from considering *The Shaping* as belonging to the field of science studies:

> Netz's book does exactly what he says he does not want to do: it offers for the origin of formalism what Shapin and Schaffer have done for the origin of experimental science. (Latour, 2008, p. 442)

On the other hand, Hacking did not share Shapin and Schaffer's constructivism:

> In the case of Netz, my reading is closer to the author's intentions than is Latour's. In the case of Shapin and Schaffer, Latour's reading is certainly closer to the intentions of the authors than mine. Latour very much plays down the aspect of Netz indicated in the subtitle, cognitive history [...] For me, it is also the first account of the discovery of a fundamental human cognitive capacity, the ability to make demonstrative proofs. (Hacking, 2009, p. 71)

In the following lines I shall show that Netz's book can be interpreted as a study on the emergence and development of what Hacking has called the postulational style of reasoning.

5.3.2 *Postulational Thinking*

Euclid's *Elements*, which contains the concepts that constitute the foundation of Greek mathematics, was written in Alexandria in about 300 BC. The treatise consists of 13 books that collect definitions, axioms, postulates, propositions and mathematical proofs. As Thomas Heath explained, 1) an *axiom* asserts a self-evident truth that is taken for granted; 2) a postulate assumes the truth of a statement without proof but, unlike an axiom, is not self-evident and may assume that something exists; 3) a *proposition* may be a *theorem* or a *problem*. A theorem is a conditional statement; a problem, once solved, *proves* the existence of a geometrical entity (Heath, 1908, pp. 117–119).

Consider the problem represented by Proposition I Book I :

> On a given finite line to construct an equilateral triangle. (Heath, 1908, p. 241)

For a start Euclid states:

Let AB be the given finite straight line. (Heath, 1908, p. 241)

Then he provides the directive to construct the equilateral triangle on the straight line *AB*: consider the circle with centre A and radius AB; and the circle with centre B and radius AB. The circles intersect one another in two points on opposite sides of the line AB. Call C one of them and join with a straight line B and C and with another straight line A and C. AC =AB because they are radii of the same circle with centre A; and BC =AB because they are radii of the same circle with centre B. It follows that AC=BC=AB

Euclid's proof follows this schema:

(1) AC=AB and (2) BC=AB so (3) AC=BC

The assertions (1) and (2) follow from Definition 15 Book I that defines what a circle is:

A *circle* is a plane figure contained by one line such that all the straight lines falling upon it from one point among those lying within the figure equal one another (Heath, 1908, p. 153).

The implication (3) follows from Axiom 1 Book I according to which things that equal the same thing also equal one another (Heath, 1908, p. 155).

Netz maintains that the building-blocks of proof, which he calls the 'atoms' of necessity, are the *starting points*, i.e. all the assertions which are not subject to argument in the text, and the *arguments* (Netz, 1999, p. 168). To think in a postulational way means to combine in necessity-preserving ways these two atoms. For example, the proof above combines a definition and an axiom (starting points) with an argument (e.g. steps (1)-(3) constitute a *single* argument). We could add another atom of necessity, e.g. that (4) AB>DE, where DE is another straight line and conclude that (5) BC>DE and AC>DE. That is:

(1) AC=AB and (2) BC=BA so (3) AC=BC, (4) AB>DE so (5) BC>DE and AC>DE

According to Netz's terminology, (1)-(3) and (4)-(5) are different arguments of the same proof. Step (4) could be a starting point or the conclusion of another argument, which is introduced *at the required moment*. In this case the proof would involve two arguments ((1)-(3) and (4)-(5) plus three starting points. Notice that the postulational way of thinking requires not only to do deductions, such as in steps (3) and (5), but also postulations, such as when the existence of a line AB is postulated. In this sense proofs involve a way of thinking that is both *postulational* and *deductive*.

The Greeks' proofs can be much more complicated than my example: they can form long almost linear chains, i.e. chains that include asides and breaks such as when a mathematician sets aside the point and switches to another argument. Furthermore, the Greeks' proofs prove results that are both necessary and *general*: for example, proposition 1 above is true for all the straight lines AB. Netz maintains that '[necessity is explained] in terms of atomic necessity producing elements, which are then combined in necessity-preserving ways' (Netz, 1999, p. 240). On the other hand 'generality is the repeatability of necessity' (Netz, 1999, p. 270): generality derives from the possibility of repeating for any other case a proof about specific objects in specific diagrams.

5.3.3 Doing

In order to realize that a proof involves not only a way of thinking but also a way of doing we need to take note of the fact that the diagram is a *necessary* element in producing and reading a proof. Let us re-examine the proof of Proposition I above. First of all, consider the first opening sentence 'Let *AB* be the given finite straight line'. This is not an assertion but a sentence that invites an *action*: to individuate the letters and the line in the diagram. Consider next the point that two circles with centres A and B meet: this is an *implicit* assumption *based on the diagram* (Netz, 1999, p. 27). Besides, in the course of the proof, the line AB must be examined within its context: is it the radius of a circle? Does it intersect another line? Is it also the radius of another circle? *Thinking and doing*, mental and visual resources, interact in order to construct the equilateral triangle: the proof consists of drawing a line and two circles with a ruler and a compass (action); of individuating a point C and drawing the lines AC and BC (action); finally, of deducting (3) from (1) and (2). In Latin, the acronym used to indicate that the task set in propositions such as Proposition *1* Book *1* has been accomplished is Q.E.F., i.e. *Quod Erat Faciendum* – 'Which Was To Be *Done*'.

Once the assertion is proved, proposition *1* can be inserted in a chain of deduction so to prove another proposition. To sum up, the text points to objects via letters and leaves to the 'eyes' or the visual imagination the task of identifying them; the arguments are mediated through visual *actions.*

5.3.4 *Techniques of Self-vindication, New Evidence, New Objects*

Hacking made the point that the postulational way of thinking is self-authenticating because the result of a proof is always confirmed by other proofs:

> There is no standard of what is correct proof, than proof itself. Proofs are self-authenticating. Many a proposed proof has proven to be fallacious, but the standard of validity or fallacy is proof itself. (Hacking, 2009, p. 40)

This is a claim that can be better understood by noticing that the structure of mathematical proofs is completely autonomous: 'the necessity of assertions is either self evident (as in starting-points) or dependent on nothing beyond the immediate background' (Netz, 1999, p. 215). The immediate background is constituted, in Hacking's terminology, by the new objects 'introduced' by the postulational way of thinking, e.g. straight points, lines and planes.[11] When doing proofs a mathematician refers visually to these representations, which is to say that the diagram supplies the evidence to which the propositions refer—the diagram is *the* standard of evidence within the postulational style. To say that there is no other standard of what is correct proof amounts to saying that there is no way to establish that the diagram *is* a valuable standard of evidence. That is: the diagram represents the inescapable universe in which knowledge is produced. It is the postulational way of thinking that has 'introduced' the diagram and it is the diagram that supplies its universe of discourse.

[11] As I have pointed out in the first chapter, in his *The Crisis of European Sciences and Transcendental Phenomenology* (1970 [1936]) Husserl referred to Greek mathematics as 'a *style* which was *new in principle,* unknown to the ancients' (1970 [1936], p. 21 my emphasis) characterized by 'the idealization of empirical numbers, units of measurement, empirical figures in space, points, lines and the impressive idea of a systematically coherent deductive theory' (1970 [1936], p. 21).

5.3.5 New Sentences

According to Hacking,

> those very sentences used to express the geometrical *a priori* propositions could not have that sense unless they were embedded in the practice of geometrical demonstration. (Hacking, 1983a, p. 457)

Let us consider three examples of sentences:

Straight lines which are parallel to the same straight line and are not in the same plane with it are also parallel to one another. (Heath, 1908 Elements IX,9 p. 290)
Similar polygons inscribed in circles are to one another as the squares on the diameters. (Heath, 1908 p. Elements XII,1 p. 369)
Any prism which has a triangular base is divided into three pyramids equal to one another with triangular bases. (Heath, 1908 Elements XII,7 p. 394)

For Hacking the sense of these sentences is determined by the way in which we can reason for their truth-or-falsehood. To see why consider the first two propositions listed above. They could not have been thought or understood by, say, the Babylonian mathematicians, even if a definition of infinite planes, straight lines and similarity of triangles had been given to them. Indeed, terms such as 'point' and 'straight line' in those sentences are embedded in ways of doing unknown to Babylonian mathematicians. The Babylonian and the Greek texts reveal *different* ways of doing: the former refer to objects that were actually moved, the latter refer to diagrams; the former are dependent from the context, the latter are abstractions. A Babylonian would have had to learn to *act* as a Greek and to *think* independently from the context in order to make sense of those sentences above.

On the other hand, in the last sentence listed above, there are no terms that, once translated, would have had no sense for an Egyptian; but the sentence is not a candidate for truth-or-falsehood for the presence of the quantifier 'any', which gives generality to the proposition. Indeed, 'Babylonian mathematics is limited, compared to Greek mathematics, by being tied to the particular operation upon the particular case' (Netz, 1999 note 154): the quantifier 'any' would have made no sense for a

Babylonian because generality was a 'historical novelty' (Høyrup, 2005, p. 143) brought into being by the Greek mathematical thinking.

Finally, it is worthy of note that in the few lines that precede Hacking's passage above, Hacking reminds the reader that:

> empirical propositions, to be established from case to case by measurement are to be distinguished from *a priori* and demonstrable propositions of geometry.

In the language of the styles project, we can say that my examples of *style-dependent* propositions above must be distinguished from *style-independent* propositions such as:

> The side of this pyramid is one meter long
> or
> Two sides of this trapezoid are parallel

Indeed, for Hacking these sentences above have a truth-value that is *independent* from any way of thinking and doing.

5.3.6 *Discontinuity*

As Netz declared, his vision of Greek intellectual life has been shaped by Lloyd (Netz, 1999, p. 292). According to the latter, while in Mesopotamia the investigation of the heavens was a matter of state importance (Lloyd, 2004, p. 17), in Greece there were no intellectual institutions. Greek society was very argumentative because 'lacking state institutions that gave stable employment, they [the Greeks] were in open and more or less continuous competition with one another' (Lloyd, 2000, p. 3). As Jean-Pierre Vernant had already stressed 'speech became the instrument of the city's political life' (Vernant, 1982 [1962], p. 52). So, argued Netz, the development of rigorous arguments in philosophy and mathematics must be seen against the background of rhetoric: the fact that the latter failed to meet certain standards of incontrovertibility led to a form of argument that goes beyond mere persuasion. It was in Greece, thanks to this particular historical setting that proof emerged and came to be considered as the paradigm of what it means to settle an argument. Before the late fifth century there were no conditions for the emergence of proof.

Netz concluded that the origin of Greek mathematics has been a 'sudden explosion of knowledge': it is impossible to imagine that the early history of Greek mathematics was gradual and that there has been a rudimentary form of pre-Euclideanism:

> Whoever was capable of proving one Euclidean result was capable of proving most of them. I therefore suggest that the origin of Greek mathematics could have been a sudden explosion of knowledge [...]. No more than a generation would be required to find most of the elementary results of Euclid [...]. (Netz, 1999, p. 273)

This is exactly Hacking's point that the discovery of the possibility of demonstrative proof represents a sharp break in the development of mathematical reasoning. To conclude, the Greek style of mathematics possesses all the characterizing properties of styles and therefore can be called 'the postulational style of reasoning'.

5.4 The Historico-genetic Style of Reasoning

'Historico-genetic' is the locution by which Hacking referred to the last style of reasoning in Crombie's list. The few comments he made on Crombie's account of the historico-genetic style are insufficient to reconstruct the physiognomy of this style. First of all, we gather that while the postulational style systematically employs deductive logic, '[the historico-genetic style] employs one branch of logic, namely abduction or inference to the best explanation' (Hacking, 2012, p. 5). Hacking's description of the historico-genetic thinking can be summarized by these steps: 1) assuming that things stand in a certain way in some domain of interest; 2) devising an explanation of how things got to be that way; 3) inferring from the power of the explanation, and its superiority to any other available candidate, that it tells how things are (Hacking, 2012, p. 5). 'Darwin's long argument', mentioned by Hacking in passing, can be taken as an example of historico-genetic thinking.[12] Indeed, the explanation of adaption of living beings is considered the primary fact to be explained by the theory of evolution (Gould, 2002, p. 119) (step 1). The 'long argument', as Darwin calls his argument in *The Origin of Species* (1985 [1859]), provided an

[12] In (Vitti Rodrigues & Claus, 2021, pp. 1416–1418) can be found another example of the use of the historico-genetic style of reasoning.

explanation of this problem (step 2); William Paley (1743–1805) had proposed a different explanation according to which adaptation must have been 'designed' by God. As there are several facts that are not explained by Paley's argument (see Darwin, 1985 [1859] chapter XIV) but are explained by the 'long argument', one can conclude that Darwin proposed the best explanation (step 3). In the next subsection I shall provide a more detailed characterization of the historico-genetic thinking.

Hacking has not been very clear as to when the historico-genetic style emerged. Indeed, although he pointed out that '[the historico-genetic] genre of reasoning began long before what we call the sciences' (Hacking, 2012, p. 5), he also noted:

> Probability crystallizes in the time of Pascal. Then there is historico-genetic explanation, whose triumph is Darwin's explanation of the species by natural selection. (Hacking, 2012, p. 5)

In this passage he seems to suggest that Darwin's time represents the point in which the historico-genetic explanation became the standard method of scientific disciplines such as the life sciences. For my part, I shall argue that the historico-genetic style, a way of tapping abduction, a human form of reasoning, emerged in the eighteenth century.

5.4.1 *Historico-genetic Thinking Versus Laboratory Thinking*

After this brief presentation of Hacking's claims, I am going to better characterize the historico-genetic *thinking*. The best way to do this is to compare and contrast it with laboratory thinking.[13]

In his book *What Makes Biology Unique?* (Mayr, 2004) Ernst Mayr (1904–2005) explains that 'biology consists of rather two different fields, mechanistic functional biology and historical biology' (Mayr, 2004, p. 24). The former deals with the physiology of living organisms '[which] can be explained mechanistically by chemistry and physics' and does not need any historical analysis. The latter deals with 'all aspects of the living world that involve the dimension of historical time' (Mayr, 2004, p. 24).

[13] In my analysis I shall only deal with the general features of the historical and the experimental sciences. For a review of the methods deployed by Darwin see (Gould, 2002, pp. 103–116). A compendium of the methods of geology can be found in (Laudan, 1987, pp. 7-16).

Whereas in functional biology the *most frequently* asked question is 'how?', in historical biology scientists ask 'why?' most of the time (Mayr, 2004, p. 25).

This distinction can be extended to disciplines such as geology, cosmology and palaeontology: these disciplines are 'historical', although they can ask 'how-questions' like in chemistry and physics. For instance, geology has provided insights into the structure of the Earth by applying physics methods to the study of seismic waves, i.e. by asking questions about *how* waves propagate ('how-questions'); on the other hand, thanks to Alfred Wegener (1880–1930), geology has offered a historical explanation of continental drift, explaining a large number of features of the surface of the Earth, e.g. *why* continents' shapes fit together like a jigsaw ('why-questions').

In subsection 3.5, I have explained that one of the cognitive steps involved in laboratory thinking consists of establishing what is a matter of fact and what is a speculative hypothesis, which theory has to be embraced and which has to be rejected. Laboratory sciences postulate statistical or causal regularities among types of events; from the laws supposed to describe these regularities, predictions are inferred and tested in laboratory; and, importantly, often in the face of a failed prediction *ad hoc* hypotheses are added and separately tested in order to protect the hypothesized laws from false negatives (see also Cleland, 2002, pp. 476–480).

Furthermore, the essence of explanations in physics is to show that a phenomenon 'has to be so' given certain physical laws and a few additional facts. This is what Isaac Newton (1642–1727) did when he proved that Kepler's three laws of planetary motion could be deduced from his laws of motion and universal gravitation.

Conversely, as Mayr notes, laws have a *non-essential* role in historical biology (Mayr, 2004, p. 27) given the contingent character of many phenomena. In biology a result *R* we expect could fail to come about simply because our hypothesis *K* admits exceptions. Indeed, what really characterizes historical sciences is the postulation of a common cause for explaining different puzzling effects of past events. Often this common cause must be connected with its effects through a long chain of events, some of which are contingent: for example, on Gould's model of history of life the Cambrian explosion produced a wide range of phyla, of which many disappeared rapidly; for him, it was a matter of luck, and not of being 'fitter', if a worm-like organism called *Pikaia gracilens*, from which we probably descend, did not go extinct (Gould, 2000b). This example shows the absence of regularities in biology: according to Gould, if one could somehow turn the clock back and allow life to evolve again, very likely our species would not exist. It is for this reason that *historical explanations* have the character of a story that cannot be tested in a laboratory: there is no way to test experimentally a thesis about human origins.

Historical scientists try to find 'the best explanation', i.e. the particular story that unifies under a single explanation the largest number of facts to be explained. As often many explanations compete, scientists must hunt for what Carol Cleland called a 'smoking gun' (Cleland, 2002, p. 480), namely a trace or a body of traces that establish whether an explanation is more unifying than another. The search for a smoking gun plays a key role in historico-genetic research, as Cleland explained. A case in point is the hunting for a smoking gun that discriminates between the different explanations of the extinction of the dinosaurs. According to some of the first explanatory stories offered before the 1980s, dinosaurs went extinct about 65 millions of years ago because of the lack of immunity to a very infectious disease or because of a rapid change of climate. Then, following the discovery of unusual high concentrations of iridium in a layer of Earth's crust in 1980 (iridium is rare on our planet and abundant in meteorites) another explanation gained support: an asteroid impact determined a mass extinction. Finally, during the late 1980s, geologists found the Chicxulub Crater underneath the Yucatán peninsula: the smoking gun, as it were.

It is a fact that since the beginnings of the modern historical sciences there has always been a clear awareness of the distinction between how-questions and why-questions. So much so that when Martin Rudwick outlined the historical setting that lies behind the Great Devonian controversy (about the dating of certain puzzling rock strata) in the 1830s, he wrote:

> Causal analysis was central to dynamical geology, whereas for stratigraphical geology it was peripheral [...] Most geologists in the 1830s were *not* centrally concerned with causal questions at all [...]. The problems of geological dynamics were perceived as neither more nor less fundamental than those of stratigraphical geology: they were just different [...] [G]eologists [...] treated them as essentially separate realms. (Rudwick, 1985, p. 45)

In sum, my points so far provide an illustration, although idealized, of two *distinct* ways of acquiring knowledge: laboratory scientists infer predictions from laws, test them and control auxiliary hypotheses; historico-genetic scientists suggest explanations of puzzling traces of the past searching for smoking guns that rule out alternative explanations. We can say that the historico-genetic way of thinking is well portrayed by the following three mental operations. The first one consists of focusing on the historical aspect of the world. It is a change of perspective that is well described by Darwin:

> When we regard every production of nature as one which has had a history; [...] when we thus view each organic being, how far more interesting, I speak from experience, will the study of natural history become! (Darwin, 1985 [1859], p. 456)

Darwin's perspective is radically different from that in which the world is investigated *as it is*: compare a question about the orbit of a planet with that concerning its history from its formation to its orbital capture. The second mental operation consists of postulating causal histories to explain the effect of long-past events. Finally, the third mental operation is the search for the smoking gun, the evidence that discriminates between two or more historical explanations.

5.4.2 Doing

It must be added that, by hunting high and low for the traces of past-events, historical scientists do not only support their conjectures or discriminate between conflicting explanations. They also *build* their stories just as if they pieced together a jigsaw: sometimes they have no clear vision of how the puzzle image will look like; their final product is often unexpected and results from an interchange of *fieldwork* and theoretical research. A case in point is the reconstruction of the evolutionary history between *Homo sapiens* and the ancestral species we share with chimpanzees and bonobos. The hypothesis of a new species, now gone extinct, *Homo rudolfensis*, was put forward in the 1970s on the basis of a collection of early *Homo* fossils from Koobi Fora (Kenya). However, this hypothesis remained controversial because of the incomplete preservation of the remains. Only with the discovery of other fossils in 2012 by the paleoanthropologist Meave Leakey did it gain support (Leakey et al., 2012). Another example comes from geology: the Scottish geologist James

Hutton (1726–1797), searching for a proof of the idea that internal and external forces of Earth mould rocks and mountains, went by boat to Siccar Point, a Scottish promontory in which he found overlapping strata of different geological eras. That was the evidence that the Earth was considerably older than previously thought and the beginning of modern geology.[14] As a final example, by removing ice cores from the polar ice, climate scientists have been able to examine the air trapped inside. The data revealed information about the atmosphere over the millennia and helped the reconstruction of the history of the climate of our planet.

Fossil hunting, geological digging, ice cores removing represent examples of *fieldwork* in historical sciences. Rudwick pointed out that, in the early nineteenth century, working on foot, wearing outdoor clothing and being frequently soaked to the skin were not thought unusual by the geologists. This was the indispensable part of their activity that Deborah Cadbury has recently recreated in her scientific novel *The Dinosaurs Hunters*, a story of the rivalry between the palaeontologists Gideon Mantell (1790–1852) and Richard Owen (1804–1892). A hammer for collecting samples of rock and finding fossils, a magnetic compass for measuring the tilt of strata and other few items were 'all that the most eminent geologist required in the field' (Rudwick, 1985, p. 40). Fieldwork was an *activity* fundamentally distinct from the activity in laboratory. As Rudwick put it

> [the] major issues in early nineteenth-century geology, w[ere] argued over in such public arenas as the Geological Society and the British Association. But all parties [...] acknowledged publicly that fieldwork was the primary locus of encounter between the geologist and the phenomena of his science. (Rudwick, 1985, p. 37)

Fieldwork marks the *way of doing* that is peculiar to historico-genetic sciences: without *this activity* no stories can be told about the portion of world under study. The primacy of fieldwork in the historical sciences parallels the primacy of laboratory in the experimental sciences or the primacy of the use of diagrams in geometry: if a practitioner of geology did not *act* 'in the field' she would not have been accepted as part of the community of the earliest members of the first geological societies.

[14] More details on this example can be found in (Eldredge, 1998, p. chapter II).

5.4.3 *Discontinuity*

The historian of science Paolo Rossi (1923–2012) wrote that

> In the one hundred of years that separate in time the *Discourse on Earthquakes* (1668) of Hooke from the *Universal Natural History and Theory of Heaven* (1755) of Kant, the discussions about the history of the Earth and of cosmos radically change[d]. (Rossi, 2000, p. 256 my translation)

In the seventeenth century the minds of the natural philosophers were still impregnated with the origin myths of *Genesis* and the alleged *brief* history of our planet. A century later the dichotomy between a prescriptive science of the unchangeable laws of the universe and a descriptive science of the evolution of phenomena had already emerged. Historical sciences such as cosmology, biology, palaeontology and geology constituted themselves around a set of practices that shared the historico-genetic way of thinking. As Antonello La Vergata pointed out, 'only in the second half of the eighteenth century did the geological *practice* consolidate and did a consensus about terms, concepts and data emerge' (La Vergata, 1988, p. 165 my translation). The first geological societies were founded at the beginning of the nineteenth century: in the case of the Geological Society, '[its] membership more than doubled between 1825 and 1845' (Rudwick, 1985, p. 22). When the legitimacy of a 'historical' consideration about nature was established, several alternative theories as well as various models of history of Earth and universe emerged.

> The more [...] naturalists looked at the earth's structure, the more complex it seemed [...]. Soon theories based on extensive transformations of the earth's surface became to appear [...] [and] geology and palaeontology would eventually emerge. (Bowler, 2003 p. 28)

On the basis of these considerations, it is possible to say that the emergence of the historico-genetic way of thinking and doing I have described so far was not gradual. In their book *Cartographies of Time* (Rosenberg & Grafton, 2010) the historians Daniel Rosenberg and Anthony Grafton looked into how humans have drawn time in the course of history. I think that their conclusions corroborate my point. Today the timeline with a single axis is a ubiquitous figure in the representation of time and one of the most inescapable metaphors for representing history by a drawing.

Yet, Rosenberg and Grafton showed that the timeline is not even 250 years old. It is difficult to imagine a sort of rudimentary idea of the timeline that grows by micro-accumulation as if the first forms of such line had continuously developed into our timeline. Rosenberg and Grafton's account of 'the birth of modern historical thinking traces a path from the enumerated but not yet narrated medieval date lists called annals, through the narrated but not narrative accounts called chronicles to fully narrative forms of historiography that emerge with modernity itself' (Rosenberg & Grafton, 2010, p. 11). In their research they have not observed such a micro-accumulation. Until the mid-eighteenth century, what Rosenberg and Grafton call the Eusebian model—a simple matrix with kingdoms listed across the top of the page and years listed down the left or right hand columns—was dominant. Only in the middle of the eighteenth century did scholars start to communicate the idea of historical development. For Rosenberg and Grafton, the crucial event was the publication in 1765 of the *Chart of Biography* and in 1769 *The New Chart of History* by the scientist Joseph Priestley (1733–1804). The reader could carry her eye vertically and see the contemporary state of all the empires subsisting in the world, at any particular time; or, she could carry her eye horizontally and note empires rise, flourishing and decline as time goes by.

> Priestley's chart [...] provided an intuitive visual analogue for concepts of historical progress that were becoming popular [...]. In Priestley's chart, historical thought and new forms of graphic expression came into dialogue [...]. (Rosenberg & Grafton, 2010, pp. 19-20)

Priestley's charts became an essential part of a gentleman's library: for example Rosenberg and Grafton say that the novelist Maria Edgeworth (1768–1849) and the physician Erasmus Darwin (1731–1802) recommended the new ways of visualizing history as aids in the education of children. In his book *The Life and the Opinions of Tristram Shandy, Gentleman*, the novelist Laurence Sterne (1713–1768) drew lines to represent the narrative patterns and the digressions from the main story. In conclusion, just when the historico-genetic thinking became central in the history of thought the timeline emerged. Its sudden coming to the fore suggests that the historico-genetic way of thinking and doing, with the features I am going to describe, had sharp beginnings.

5.4.4 *A New Kind of Evidence*

Thinking in the historico-genetic way involved a 'new kind of evidence'. This expression must be understood in two senses, as I am going to explain. First of all, let us compare and contrast how the experimental and the historical sciences provide pieces of evidence and how they use them. Experimental scientists test a target hypothesis not so much in order to reject it on the basis of a failed prediction (as Popper thought) but to save it from misleading confirmations. For example, the astronomers Urbain Le Verrier (1811–1877) and John Couch Adams (1819–1892) 'saved' the gravitational law. When it was found that the orbit of Uranus differed from that predicted by the gravitational law, they suggested the existence of another planet (Neptune) whose position was theoretically calculated and confirmed by telescopic observation. Two points must be noted: firstly, experimental sciences' use of evidence is limited to a *single* hypothesis (Cleland, 2001, p. 988); secondly, the analysis of the hypothesis is often carried out by man-made machines such as Boyle's air pump that bring about the test conditions specified by the hypothesis.

Conversely, as Cleland pointed out, in the historical sciences most times evidence plays a different role since it is used to discriminate *multiple* competing hypotheses about past events (Cleland, 2001, p. 989). The new kind of evidence that comes about with the emergence of the historical thinking is the 'smoking gun'. The 'smoking gun' is a kind of evidence entirely distinct from the 'man-made evidence' of the laboratory style: the former discriminates between different historical accounts, the latter concerns a single hypothesis that must be falsified or 'saved' and it is, most of the time, elicited by apparatuses. For example, the Chicxulub Crater represents the evidence that an asteroid impact determined a mass extinction. So, it discriminates between different historical accounts of the evolution of dinosaurs. On the other hand the death of a bird in the air pump is a piece of evidence *produced* by a man-made machine in order to test the *single* hypothesis of the existence of void.

I am not claiming that there are no examples in which historical scientists use apparatuses in order to sharp a particular trace or to reveal a smoking gun. Nor am I claiming that 'the smoking gun' is a kind of evidence that was not in use before the emergence of the historical sciences. My point is that when the Geological Society invited its members to display on a table set the relevant specimens after a paper was read (Rudwick, 1985, p. 23), those pieces of evidence were being used in a way of which the entire community had become *aware*——they were used to discriminate between different *historical* accounts. The existence of a 'smoking gun' had become the new standard of truth within a new way of thinking and doing.

There is a more obvious sense by which it is possible to say that a new kind of evidence emerged: historico-genetic scientists turned their eyes to a *previously unnoticed* portion of the world, to objects such as fossils, minerals, rocks and geological layers. Before the emergence of the historico-genetic way of thinking this portion of the world had no role as evidence. In his book *I segni del tempo* (Rossi, 2002 [1979], pp. 21–45), Rossi explained that, until the seventeenth century and with the exception of Leonardo da Vinci (1452–1519)[15] and Bernard Palissy (1510–1590), Platonic and Aristotelian interpretations of the origins of fossils were prevalent: Platonic thinkers thought that fossils grew inside the Earth like organisms thanks to particular virtues (*virtus plastica*, *lapidifica*, *vegetabilis*); Aristotelian thinkers ascribed the existence of fossils to the action of solar heat and winds that blew in the bowels of the Earth. Only when fossils were thought of in terms of their origin could thinkers interpret those puzzling objects as remains of old organisms: nature became the reign of history, of which fossils represented the documents (Rossi, 2000, p. 258). Consequently, the emergence of the *historical* vision of nature is closely tied to the coming to the fore of a new portion of the world, which for the first time came to play the role of evidence.

5.4.5 *Techniques of Self-vindication*

The evidence on which historico-genetic scientists rely in order to infer the best explanation may not be determined independently from the historico-genetic way of thinking itself. For example, the evidence for the extinction of the dinosaurs is provided by the claim that the Chicxulub Crater was caused by an asteroid-impact dating from the end of the Cretaceous period, when the dinosaurs went extinct. In turn, this claim is an inference to the best explanation of different traces of past events such as: clay with quantities of iridium much higher than normal and concentration of iridium in asteroids. In other words, the very evidence that supports the historical account of how dinosaurs went extinct is still another

[15] An account of Leonardo's theories about fossils can be found in (Gould, 1998).

historical account obtained by using the historico-genetic way of thinking and doing.

Consider these questions and answers: why is the historico-genetic way of thinking and doing reliable? The reason is that it provides the best explanation. And why is what it provides the best explanation? The reason is that there is a smoking gun. And why is just that a smoking gun? Because it follows from an inference to the best explanation obtained by using the historico-genetic way of thinking. These questions and answers form a *circulus in probando*: the historico-genetic way of thinking is self-authenticating.

This point can be viewed from a different perspective:

> science advances our understanding of nature by showing us how to derive descriptions of many phenomena using the same patterns of derivation again and again, and, in demonstrating this, it teaches us how to reduce the number of types of facts we have to accept as ultimate. (Kitcher, 1989, p. 432)

That is: humans apply the abductive method again and again in order to build up a unifying picture of all the disparate phenomena. For example, the extinction of dinosaurs and the crater in the Yucatan peninsula were 'brute facts' until they were put in relation by applying twice the inference to the best explanation: an inference to the claim that the crater was caused by an asteroid-impact and an inference to the hypothesis of the extinction of the dinosaurs. By doing so those two brute facts are fitted together. In this sense to be self-authenticating for the historico-genetic thinking corresponds to the fact that we apply again and again the abductive method in order to provide a coherent historical account of our world; and we cannot 'escape' from abduction.

5.4.6 *New Sentences*

In their book Rosenberg and Grafton reproduce a section of a famous mediaeval manuscript called the *Annals of St. Gall*, which records events in the Frank kingdoms during the eight, ninth and tenth century:

> 709. Hard winter. Duke Gottfried died.
> 710. Hard year and deficient in crops.
> 711.
> 712. Flood everywhere

713.
714. Pippin, mayor of the palace died
715.
716.
717.
718. Charles devastated the Saxon with great destruction
719.
720. Charles fought against the Saxons
721. Theudo drove the Saracens out of Aquitaine.
722. Great crops

Rosenberg and Grafton say that although 'these annals breathe with the life of Middle Ages' they appear 'strange and antic to a modern eye, beginning and ending seemingly without reason, mashing up categories helter-skelter like the famous Chinese encyclopaedia conjured by Borges' (Rosenberg & Grafton, 2010, p. 11). Indeed, the *Annals of St. Gall*

> make no distinction between natural occurrences and human acts; they give no indication of cause and effect; no entry is given more priority than another. [...] there is no distinction among periods, and lists begin and end as nameless chroniclers pick up and put down their pens. (Rosenberg & Grafton, 2010, p. 12)

Let us focus on a single sentence of *Annals of St. Gall*:

> 712. Flood everywhere

There is no history behind that sentence, no real sense of a cause such as an alleged cyclone that brought about the flood. The sentence could be contrasted with sentences such as:

> The flood was a consequence of a warming of ocean waters and the formation of a cyclone that rotated counterclockwise and reached France in 712.

The meaning of the latter sentence hinges on scientific practices unknown in 712: no sense can be attributed to it until one comes to consider floods as having a history and grasps how different explanations are selected and how and where the evidence can be found and used. What is necessary to make sense of the latter sentence is a way of thinking and doing that was absent during those centuries. The same arguments could be applied to sentences such as:

Modifications to the lower parts of the limbs of the horse shows that evolution proceeds in the direction of increasing specialization

The order of formation of rocks can be read from the observed sequence

For us these sentences could be true or false, but in the ancient times no truth-value could be attributed to them. Myriads of sentences such as these two above, i.e. sentences that are candidates for truth-or-falsehood only within the historico-genetic way of thinking, have appeared after the eighteenth century. To conclude, although humans have always used the abductive method, it was only in the seventeenth century that a way of thinking with all the features described in this section emerged. These features characterize this way of thinking as a style, 'the historico-genetic style of reasoning'.

5.5 The Taxonomic Style of Reasoning

By 'taxonomy' it is generally meant a branch of science concerned with the classification of *organisms*. As Ernst W. Mayr and Walter J. Bock explained, classifications are attempts to arrange a diversity of entities into sets of classes according to a criterion of similarity (Mayr & Bock, 2002, p. 170): each class is constituted by individual entities that are considered similar. Classifications are a subset of 'ordering systems', a term that alludes to *any* attempt to arrange objects into particular categorizations. For example, the 'cladistic method', introduced by Willi Hennig (1913–1976) in the late twentieth century, is not considered a classification but an ordering system in that it groups species not according to their morphological similarity but according to their degree of evolutionary relatedness. 'The entities ordered in a cladification are not classes, but clades' (Mayr & Bock, 2002, p. 182), which may be so heterogeneous to include, e.g. mammalians and primitive reptiles.

Rather than depicting our ancestral ability to perform the dullest forms of cataloguing, I want to characterize the way scientists think when they classify the living world. Hacking explicitly said that Crombie's taxonomic style has to be counted as a style of reasoning, but he never provided his own characterization. However, it is possible to say that he pointed out the *possibilities* that this way of thinking opens up for human action and described the way it affects the people classified (e.g. see Hacking, 2002a chapter six) (Hacking, 1999 chapters four and five). For example, Hacking reflected on the fact that, although certain perversions may have always existed, the 'pervert, as a diseased person, [was] created in the late nineteenth century' by classifications based on a new functional understanding of disease (Hacking, 1999; 2002a, p. 100). This is only one of the ways explored by Hacking in which human beings and their perceptions of themselves are affected by our classifications. The way of framing these issues, which are beyond the purpose of this work, was influenced by Foucault, as Hacking acknowledged (Hacking, 2002a, p. 100).

Another fact to be noted is that Hacking has been attracted by a thesis put forward by the American anthropologist Scott Atran in the book *Cognitive Foundations of Natural History* (Atran, 1990). The latter argued that there are *universal* principles of biological classification and nomenclature employed by peoples everywhere. In particular, Atran maintained that the beginnings of modern biosystematics are to be traced to folk classifications of living beings: the classification of plants and animals made by Aristotle exhibits the same constraints as the classification of other coeval peoples in different ecosystems.

According to Atran plants and animals may change from a geographical region to another but every person recognizes the same classes in its environment and eventually comes to use the same taxonomic structure. For example, peoples who live in two different ecosystems will give to the same plant or to two distinct varieties of the same plant the same place in their classification.[16]

What attracted Hacking in Atran's work was the fact that it provides a basis for claiming that the taxonomic style of reasoning is grounded in the humans' innate dispositions: in fact Atran proposed the existence of a module for the naïve classification of living beings. However, for Hacking, Atran's theses only represent the first step towards an understanding of the taxonomic style. What he would like to see is a theory of the contingent social and institutional conditions that have made this style possible (Hacking, 1991, p. 18).

[16] There is no consensus about Atran's theses. Developmental psychology does not offer much support to them: for example, Spelke maintains that young infants appear to have initial domain-specific knowledge only in four domains: physics, psychology, number and geometry (Spelke, 1994, p. 433) (see also Carey, 1985). Lloyd pointed out some difficulties with Atran's studies (G. E. R. Lloyd, 2007, pp. 47-53). Also, by comparing the taxonomies of animals in ancient China with Aristotle's, Lloyd argued that there is no universal common sense by which to compare and contrast the solutions to some questions about the classification of animals (G. E. R. Lloyd, 2007, p. 54) (G. E. R. Lloyd, 2004).

5.5.1 *Taxonomic Thinking*

It is a fact that in the extant literature before Aristotle there is no comprehensive explicit classification of animals and plants. Aristotle introduced a major distinction between blooded and bloodless animals that roughly corresponds to our division between vertebrates and invertebrates. He subdivided the former class into live-bearing and egg-laying animals (and a third category of animals that shares properties with both groups); then, he subdivided the latter into four groups. Aristotle's aim was not to carry out a taxonomic project in the real sense of the word (Lovejoy, 1960 [1936], p. 56) (Louis, 1986, pp. 7–8) (Gohau & Duris, 1997, p. 8): he did not invoke a single principle to classify animals and switched from anatomical to physiological or behavioural criteria (see G. E. R. Lloyd, 1961). In spite of that, it was Aristotle who suggested to later thinkers the idea of a graded *scala naturae* in which the animals are arranged according to their degree of 'perfection' (see Lovejoy, 1960 [1936]) (Bynum, 1975). This suggestion was based on a criterion of *rank*: the grade of development reached by the offspring. On the basis of this criterion he conceived 11 grades, with man, the *ens perfectissimum* at the top and at the bottom the zoophytes, those animals that look like plants. For example, 'bloodless' animals were ranked below animals that possessed blood; animals such as birds were ranked over fish; and animals were ranked over plants for their ability to move and sense.

Grouping (forming taxa) and *ranking* (placement of those taxa in a hierarchy) represent two *distinct* mental actions that can be recognized in Aristotle's way of thinking about the natural sciences (Winther, 2012, p. 634) (Stevens, 1994, p. 10). For example, when Aristotle divided animals into two classes by using as a criterion the presence of blood in the body he was *grouping*, an action that remains at an organizational level and is distinct from that of ranking. On the other hand, when he suggested a criterion of rank to arrange all the classes in a single ladder he was *ranking*.

The taxonomist Graham Griffiths distinguished 'classification' from 'systematization': to classify means to order into classes, which in turn are constructed from *similarities*; to systematize means to arrange classes according to certain relations (Griffiths, 1974, p. 85) such as developmental or philogenetic. Griffiths's concept of systematization implies something more than just ranking: to put forward an explanation in order to interpret the ranking (Winther, 2012, p. 7). Systematizing makes the work of a taxonomist distinct from that of stamp-collector. To make the concept clearer let us come back to Aristotle. When he applied the categories 'high' and 'low' throughout the animal kingdom he just *ranked* different classes of living beings, e.g. according to his classification dogs are more advanced creatures than fishes, and fishes are more advanced than any invertebrate. This mental action is to be considered *distinct* from that of systematizing, i.e. interpreting the classification. For example, Aristotle claimed that the latter shows the static plan of nature: each species has always existed and will never become extinct. This was a systematization of the classification.

Later on, from the Middle Ages until the late eighteenth century, the result of Aristotle's interpretation was a conception of the universe as a 'Great Chain of Being', which was made, for the principle of continuity, of an infinite number of links in hierarchical order. The naturalist Charles Bonnet (1720–1793) took seriously this classification and put forward an anti-evolutionist systematization, i.e. an explanation according to which 'each species was an eternal element in the divine plan' (Bowler, 2003 p. 64). In fact, Bonnet provided a causal explanation to interpret a classification whose basic principles had been introduced by Aristotle. He did not group, nor did he rank, he just systematized, i.e. he provided an explanation of the classification by denying evolution. On the other hand Lamarck *reinterpreted* the chain of being by 'temporalizing' it, to use Lovejoy's term, i.e. by abandoning the fixity of species and proposing the theory of a linear evolution.

To suggest a causal explanation of the classification is often an action strongly interconnected with those of ranking and systematizing.

> Whether with Linnaeus we attend to form [...] [or] function, whether [...] we arrange the taxa by counting similarities, or whether we construct a taxonomy that is homomorphic to an evolutionary tree, we are in every case invoking a conception of living beings. (Hacking, 1993, p. 288)

Yet my point is that it is theoretically possible to disentangle these three mental actions, which are peculiar to the taxonomic *thinking*.

When from Aristotle we turn to the modern taxonomists, the mental actions of grouping, ranking and systematizing are even more clearly distinguishable. In order to *group* plants, the naturalist John Ray (1627–1705) focused on the distinguishing features that perpetuate themselves in propagation from seed and established as a principle that 'a species never springs from the seed of another' (quoted in Gribbin & Gribbin, 2008,

p. 24); in so doing he provided a first version of the definition of species. By taking into account every visible character of every species Ray tried to introduce a ranking: he grouped *species infimae* (our species) into *species subalternae* (our genera) and the latter into *genera* (our orders). Eventually he concluded that there is no way of knowing which is the most relevant character for a classification. Carl Linnaeus (1707–1778) opposed Ray's conclusion and focused on the reproductive organs of plants. He determined classes and orders by counting stamens and pistils of every flower; then he determined the species by a closer inspection of the structures of these plants. By doing so he introduced taxonomic *ranks* into the living world: species were grouped into a higher category called *genus* and the genera were grouped into classes of more basic resemblances. Linnaeus thought that his scheme of mapping biological relationships as smaller classes within successive large classes mirrored the God's permanent order, which ought to be discernible by the human mind. This can be viewed as a systematization, that is an interpretation of his classification, but also as an assumption that guided his research of a comprehensive classification of the living world——at any rate, Linnaeus's creationist account is to be distinguished from his mental actions of grouping and ranking.

The interpretation in creationist terms given by Linnaeus was not the only one possible. Linnaeus's classification is open to another systematization in terms of evolutionary theory, as Stephen Jay Gould explained: 'Linnaeus's scheme happens to be compatible with the interpretation that the interrelationships among organisms derive from a genealogical hierarchy built by evolutionary branching' (Gould, 2000a, p. 22). Ultimately, like Aristotle, Linnaeus carried out three independent mental actions: grouping, ranking and systematizing. His systematization was only one of the possible ones.

To sum up, the way of thinking of *taxonomists* from Aristotle to the modern taxonomists is characterized by three mental actions: grouping, ranking and systematizing. Grouping and ranking require criteria such as those based on similarities in Aristotle's and Linnaeus's classifications. Systematizing consists of a causal explanation to interpret the classification: for example, the fixity of species for the great chain of being or the evolutionary theory for the Linnean classification.

Another point to note is that taxonomists themselves consider their methods as ways of finding out. This is evident not only in Aristotle, e.g. who wanted to explain the nature of kinds by identifying their place in Nature (Hacking, 1991, p. 18), but also in the thought of later students of classification. In the mid-seventeenth century, 'to name means to know' was the *idée fixe* of taxonomists (Rossi, 2000, p. 280): for Linnaeus *denominatio* was one of the aims of botanic and classifying constituted a way for *finding out* the structure of the divine thought. In the mid-eighteenth century the idea that a classification should mirror the principle of common descent caught on. Classification helped to find out about the history of life: for example, to distinguish, within vertebrates, morphological types such as fishes, amphibians, reptiles, birds and mammals suggested that each of them descended from a common ancestor which possessed the same body plan.

5.5.2 Emergence and Discontinuity

For Hacking, Linnaeus represents the point in time in which the taxonomic thinking *suddenly* came to dominate the scientific milieu. Perhaps it is not by coincidence that a similar thesis can be drawn from Foucault's *The Order of Things* (Foucault, 1994 [1966]). In the latter work, Foucault told of three major epochs in Western thought: the Renaissance, the Classical Age (from 1650 to 1800) and Modernity. Each epoch was marked by an epistemic system, i.e. by particular conditions of possibility of knowledge and its discourse that Foucault tried to describe. In particular, until the mid-seventeenth century 'the historian's task was to establish the great compilation of documents and signs—of everything, throughout the world, that might form a mark' (Foucault, 1994 [1966], p. 130). It was only in the Classical Age that 'the sciences carr[ied] within themselves the project [...] of an exhaustive ordering of the world' (Foucault, 1994 [1966], p. 74). This entirely new way of connecting things is for Foucault 'the hidden network' (Foucault, 1994 [1966], p. XX) that determines, *inter alia*, how we deal with the things classified. For Foucault, the *épistème* of the Renaissance is the period in which 'the naturalist [becomes] the man concerned with the structure of the visible world and its denomination according to characters' (Foucault, 1994 [1966], p. 161). As I am going to show, Foucault's account not only supports the claim that the taxonomic style of reasoning crystallized in Linnaeus's time but also that it *suddenly* opened up a conceptual space for new candidates for truth and falsehood.

To start with, as Lloyd noted, no thinker in pagan Greek antiquity attempted a zoological classification on the scale of Aristotle (G. E. R. Lloyd, 2007, p. 54). Theophrastus, Aristotle's disciple, focused only on plants and rocks and Roman naturalists such as Pliny the Elder (23–79 AD) did not really add anything scientifically important to Aristotle's achievements (Gohau & Duris, 1997, p. 9). The important naturalistic works of Albertus Magnus (1193/1206–1280) faded soon into oblivion and in the Renaissance animals were not studied as such but for the signs that could be read in them (Foucault, 1994 [1966] chapter 5). Furthermore, no new systematizations were proposed until the end of the seventeenth century when the idea of chain of being finally lost support.

It is a fact that, between the mid-sixteenth and the mid-seventeenth century the amount of data increased exponentially: as Rossi reports, in the *Herbarum Vivae Eicones*, written by the botanist Otto Brunfels (1439–1534) and illustrated by Hans Weiditz (1495–1537), we find a list of 258 species of plants. Less than 100 years later, in 1623, the naturalist Gaspar Bahuin, in the *Pinax theatre botanici*, listed 6000 species. In 1686, Ray published the first volume of his *Historia plantarum* in which he described 18,000 species (Rossi, 2000). An interesting point made by Atran is that Aristotle dealt with a local fauna of just 600 or so local species, a number that does not strain our innate capacities to sort; by the time of Linnaeus and Georges Cuvier (1769–1832) the complexity of the known world of nature increased so much that our built-in capacity to sort plants became insufficient. Linnaeus is therefore the point of history——argues Atran—in which a 'scientific breakaway' took place, i.e. a sort of gradual disengagement of science from common sense. What triggered this explosion of discoveries? Why did people become obsessed with classification?

Foucault's analysis suggests the following answer to these questions: for the taxonomic thinking to crystallize in Linnaeus's time it was necessary for the students of 'History' (Foucault's terminology) to become *observers* of the natural world. This has not always been the case:

> […] though it is true that the historian, for the Greeks, was indeed the individual who *sees* and who recounts from the starting-point of his sight, it has not been so in our culture. Indeed, it was at relatively late date, on the threshold of the Classical age, that he assumed—or resumed—this role. (Foucault, 1994 [1966], p. 130)

The year 1657, date in which the naturalist John Jonston (1603–1675) published a *Natural History of Quadrupeds*, is considered by Foucault the threshold of the Classical Age. For him, until that moment all that existed was *just* histories: a *History of the Nature of Birds* by Pierre Belon (1517–1564) an *Admirable History of Plants* by Claude Duret (1570–1611), a *History of Serpents and Dragons* by Ulisse Aldrovandi (1522–1605). In their works, these authors gave room to allegoric interpretations, myths and legends about a given animal or a plant. Their classifications included mythological animals, literary considerations, possible uses of a certain animal and its edibility. Rather than coming from a direct observation of nature, Renaissance scholars usually depended for their reference material on information obtained from previous Classical authors. The naturalist Conrad Gesner (1516–1565), whose *Historiae animalium* was widely read in the Renaissance, is another case in point: in that work he showed the animals' places in history, literature and art and included mythical creatures and imaginary beasts, intermixed with animals brought back from the New World. Foucault explained his attitude with these words:

> Until the time of Aldobrandi, History was the inextricable and completely unitary fabric of all that was visible of things and of the signs that had been discovered and lodged in them [...]. (Foucault, 1994 [1966], p. 129)

The reason why, as I have just said, Foucault considers Jonston as the threshold of the Classical Age is that in his *Natural History of Quadrupeds* the latter eliminated almost completely the literary considerations still present in Aldobrandi. For Foucault, in the Classical Age naturalists eliminated fables and documents, began to take close examinations of the living beings and reported what they see in the most neutralized way. The ancient texts ceased to be the documents of history, replaced by herbariums, collections and gardens. The taxonomic thinking as *we* know it, described above by the mental actions of grouping, ranking and systematizing on the basis of the characters of animals and plants, was fully emerged. Some historians of science[17] have noticed that the discovery of New World species led to the disappearance of emblematics: it was impossible to interpret astrologically, or according to certain rules of sympathies new discovered plants or animals such as the armadillo or the anteater.

When Rossi (2000) says that in the Classical Age scholars got rid of the superfluous, he must be interpreted as saying not only that they eliminated legends, hearsay and literary considerations from their classifications but also that they *focused* their attention on a particular portion of the world:

[17] For more details see (Kwa, 2011).

certain aspects of plants and animals up to that moment 'unknown' . The stamens and pistils of flowers began to be enumerated, the sizes of corollas measured, the shapes of the flowers recorded. It was at that point that to group on the basis of the many emerging differences and similarities between animals and plants, to rank the many species just discovered in order to arrange them and, finally to systematize, became crucial.

5.5.3 *A New Kind of Evidence*

Foucault points out a sentence of Linnaeus that established what should be taken into account as fundamental in the study of a living being.

> As Linnaeus said [...] 'every note should be a product of number, of form, of proportion, of situation'. For example when one studies the reproductive organs of a plant it is sufficient but indispensable to enumerate the stamens and pistils [...], to define the form they assume, according to what geometrical figure they are distributed in the flower [...] and what their size is in relations to organs. (Foucault, 1994 [1966], p. 134)

As a matter of fact, number, form, proportion and situations were the standards of evidence shared by the new community of *naturalist* historians in the Classical Age. This is a standard of evidence entirely new: as we have seen in Sect. 2.2 in the Renaissance an *opinio* was supported by signs. Different scholars could well disagree on which characters to consider in order to group and rank different classes of living beings: Joseph Pitton de Tournefort (1656–1708) looked at corollas and the features of fruits whereas Linnaeus focused on sexual characters of plants. But they all agreed that number, form, of proportion and situation must represent the features necessary for reconstructing the place of a natural object in the natural world; they all agreed on trusting only their eyes and disregarding other organs of sense. In other words, they all agreed on certain *standards of evidence* for justifying what can be considered a sound classification. In doing so they established that to classify means to *group* and *rank* on the basis of *visible* similarities. The emergence of this way of thinking with these standards of evidence is a fact that characterizes Linnaeus's epoch.

Botanic gardens and zoological collections represented the institutions that reinforced the 'rules' shared by the naturalist historians. What came into being was a community of people that shared common standards for describing the living world. In order to classify the members of this community had to perform measurements and make comparisons, to carry out experiments and observe the capacity of interbreeding of animals and plants. More recently, measures and comparisons involve genetic sequencing. These physical actions necessary for comparing and contrasting characterize the taxonomic way of thinking.

5.5.4 *Meaningful and Meaningless Classifications*

The change between the way of thinking of the Renaissance scholars and that of the seventeenth-century taxonomists was so radical that the classifications made by the former appear alien to us. To give an example, Gesner grouped unicorns with rhinoceroses; Aldovrandi described snakes under headings such as: various meanings of the word 'serpent', synonyms and etymologies, differences, form and description, nature and habits, temperament as well as prodigies and presages, monsters, mythology, fables, allegories and mysteries, hieroglyphics, emblems and symbols, proverbs, heraldic signs (Foucault, 1994 [1966]). The juxtaposition of monsters and snakes, mythical creatures and rhinoceroses as well as that of habits of animals and proverbs and hieroglyphic is impossible for us to make sense of—it is a way of classifying and describing living creatures that seems to cross the boundaries of our rationality.

Indeed, Aldovrandi's and Gesner's lists are reminiscent of the essay *The Analytical Language of John Wilkins* by Jorge Borges cited by Foucault in a well-known passage of the preface to *The Order of Things.* Borges described a Chinese encyclopaedia in which it is written that animals are divided into:

1. Those that belong to the Emperor,
2. Embalmed ones,
3. Those that are trained,
4. Suckling pigs,
5. Mermaids,
6. Fabulous ones,
7. Stray dogs,
8. Those included in the present classification,
9. Those that tremble as if they were mad,
10. Innumerable ones,

11. Those drawn with a very fine camelhair brush,
12. Others,
13. Those that have just broken a flower vase,
14. Those that from a long way off look like flies.

Foucault commented:

> It is not the 'fabulous' animals that are impossible, since they are designated as such, but the narrowness of the distance separating them (and juxtaposing them to) the stray dogs, or the animals that from a long way off look like flies. What transgresses the boundaries of all imagination, of all possible thoughts, is simply that alphabetic series (a,b,c,d) which links each of those categories to all the others (Foucault, 1994 [1966], p. XVI). For the same reasons, the juxtapositions of Aldovrandi and Gesner live in a space that is unthinkable for the student of classifications who lived some dozens of years later: the mode in which animals and plants are grouped makes their writings fantastic, irrational, meaningless.

5.5.5 *Techniques of Self-vindication*

Hacking followed William James (1842–1910) in considering classification as 'the fourth' branch of logic together with deduction, induction and abduction (Hacking, 2006a, pp. 8–9). Hacking's point can be developed in order to explain in which sense the taxonomic way of thinking is self-authenticating. When a classification does not satisfy us (e.g. because it does not happen to mirror our current theories) we have still to rely on taxonomic thinking in order to provide more and more suitable classifications. It is *never* the case that the taxonomic way of thinking and doing is dismissed as inadequate or conducive to falsehood. What might be refused is the systematization of a classification, i.e. the theory we use to interpret the classification. But whether or not we keep on using the same classification or provide a more suitable one it is taxonomic thinking that we use. In other words, the taxonomic way of thinking and doing itself is immune to refutation, as deduction, induction and abduction are.

This point can be illustrated by historical considerations. As I have explained, Aristotle suggested that zoological forms could be arranged on a hierarchical scale, which reflected the grades of perfection of living beings. The organizing schema of a single linear chain remained substantially identical in the following centuries although it was enriched of always new forms of living beings. When the evidence that the world is forever changing became overwhelming, it was thought that the static linear chain could be converted into a sort of biological escalator, leading from the lowest organisms to man. Lamarck thought that, due to the pressures of adaptation to changing environment, the *scala naturae* had to make room for a certain number of branches. In other words: he kept on using the taxonomic way of thinking although he changed his theories. Eventually, the theories that interpreted the schema of a single linear chain were refused and Darwin's theory triumphed. It was the hierarchical order based on continuous branching and no subsequent joining of branches erected by Linnaeus that was compatible with the theory of evolution. Linnaeus, as it were, had changed the organizing device behind the classification but, again, he had still relied on the taxonomic way of thinking. He proposed not a linear chain but a logical tree in which we can reach a species by making successive twofold divisions (vertebrate and invertebrate, mammals and non-mammals within the vertebrates, carnivores and non-carnivores within the mammals and so forth). This type of logical tree that represented Linneaus's new classification was not a novelty: it was familiar at least since Plato and Aristotle and Ray had already used it before Linnaeus (Gould, 2000a, p. 23). However Linnaeus had interpreted his classification in creationist terms. His way of grouping and ranking was neither true or false: it just happened to mirror the topology of evolutionary systems (Gould, 2000a). It was his systematization that was wrong. Ultimately, while our systematizations can be refused, taxonomic thinking is inescapable: we have still to rely on it again and again when we want to know about the living world. It is in this sense that one can consider taxonomic thinking self-authenticating. To summarize, to define the taxonomic way of thinking as a 'style of reasoning' is appropriate because it satisfies all the characterizing properties of styles of reasoning.

We can conclude that there exist six styles of reasoning that share a set of common characterizing features. I have described two of them in the third chapter, following some of Hacking's suggestions. In this chapter I have argued for the existence of four other styles of reasoning by outlining their features. All the differences notwithstanding, striking parallels can be drawn between different styles of reasoning: they emerged suddenly at certain points in time, crystallized around communities that shared certain standards of evidence and represent basic and inescapable forms of reasoning in the sense I have discussed above.

5.6 Candidates for Being Styles of Reasoning

According to my account in the previous sections, there is a set of ways of thinking and doing that share a few precise features, although they differ in questions, methods and trajectories of development. I make no claim that they exhaust the set of all the styles of reasoning, the actual ones or those that might come about in the future. I only claim that, should we exclude one of the styles of reasoning I have presented, we would inevitably end up with a partial understanding of scientific thinking. However, for the account I am giving to be complete it is necessary to address an issue: ever since Hacking introduced his notion, scholars have argued that other ways of thinking should be counted as styles of reasoning too. I shall suggest that their conclusions are too hasty for reasons that differ case by case and that we should not count them as Hacking's styles rather as somehow 'spurious styles of reasoning' in that they lack some important features. However, this circumstance does not prevent some of them from having played a crucial role in the history of scientific thought.

5.6.1 *Are Religions Styles of Reasoning?*

In the course of his argument for the thesis that 'Hacking has failed to make progress as far as criteria are concerned' (Kusch, 2010, p. 170) Martin Kusch made the point that Christian-Catholic reasoning satisfies the characterizing properties of styles of reasoning. For him this way of thinking introduces:

> new types of objects (God and angels that I can directly experience), evidence (that God appears to me is *prima facie* evidence of his existence), sentences ('God told me that I should do his will from day to day in humility and poverty') modalities and possibilities (it is possible for God to be perceived by me). (Kusch, 2010, p. 170)

Furthermore, according to Kusch, Christian-Catholic thinking is self-vindicating for two different kinds of reasons. Firstly,

> Theorists of mystical perception have developed techniques of self-stabilisation. That God seems to talk to us is not good enough; for the experience to pass muster it must cause 'interior peace' rather than perturbation, trust in God rather than despair, or calm rather than impatience. (Kusch, 2010, p. 170)

Secondly,

> The authority principle and the principle of infallibility of the Pope in fundamental theological questions are further stabilization techniques. (Kusch, 2010, p. 170)

Now, notice that in principle his argument could be extended to any religion. One could say that, in general, religious thinking 'introduces' a variety of 'objects' (ghouls, ghosts, spirits), new truth-candidates ('God made the angels endowed with one, two, three and also four wings' (Koran 35:1)), new evidence (miracles, signs, states of mind), possibilities (it is possible for spirits to inhabit a wood); and possesses techniques of stabilization (e.g. the Koran proves the prophethood of Mahomet, then the Koran is the revelation of a prophet).

However, the notion of style of reasoning has been conceived to describe *scientific* ways of thinking. That religions can be counted as scientific styles is a result difficult to accept. The problem here is that Hacking, as we know, is unwilling to provide definitions. Before 2012, the only claim germane to this point he made is that what distinguishes scientific from non-scientific discourse is that there are no self-authenticating techniques for the larger part of morality and humanistic thought (Hacking, 1996, p. 74). Unfortunately, he never substantiated this claim, which is unfounded in the case of religions in the light of Kusch's example: Christian-Catholic thinking—argues Kusch—does not belong to the scientific discourse *and* is self-authenticating.

However, we have to consider that styles of reasoning are 'ways of finding out in the sciences', i.e.: 'finding out that "so and so"' and 'finding out what's true, and how to change things' (Hacking, 2012, p. 3). They can be understood in an 'ecological way' (Sect. 3.5.5), as ways of *altering* our physical world in order to make it our 'home', in the Greek sense of the word 'ecological'. These suggestions can be summarized by saying that to adopt a certain style of reasoning involves the use of our minds and bodies for fulfilling one or more of these aims: to know why a given fact happens; to predict certain events; to find out how to elicit or describe a phenomenon; to reveal how facts, entities or phenomena are connected with one another.

Now, religious ways of thinking may well involve ways of doing but never for fulfilling the aims just mentioned. Indeed, compare the ways of doing of religions with those of the styles of reasoning I have described. The monastic life certainly involves particular activities such as praying or introspecting. However, whether or not these activities can cause effects on our world, it is never the case that believers systematically act and intervene 'hands-on' in the world in order to reveal patterns of relations between facts or entities that appear *prima facie* unrelated or think in order to predict certain facts. On the other hand, to think and act in the algorithmic style of reasoning allows us to discriminate, compare or associate objects and events on the basis of their numerosity. Physicists use the visual resources involved in the postulational style of reasoning in order to reveal the relations between different phenomena, e.g. they use geometry to study the phenomena of light. Laboratory scientists build machines that induce certain effects in order to find out about the existence of theoretical entities, connect their properties with one another and predict their behaviour. Statisticians perform several actions, such as selecting random samples of voters, in order to find out how the probability of the election of certain candidates is related to certain events. The taxonomists of the seventeenth century conducted field research in order to find out about the structure of divine thought: do different morphological types descend from a common ancestor? Finally, by hunting high and low for the traces of past events, historical scientists *build* patterns of relations between different facts, as when they reconstruct the evolutionary history of *Homo sapiens*.

In brief, by thinking and *doing*, scientists reveal patterns of relations and give explanations of how and why natural phenomena occur. On the contrary, religions do not systematically intervene in the world in order to put in relation different physical phenomena; they are rather concerned about questions of ultimate meaning and moral value. Ultimately, religions are ways of thinking that do not satisfy one of the characterizing properties of styles of reasoning (feature 1 in Sect. 3.4.5): the property of being ways of thinking and doing in order to find out about phenomena. Thus, as a matter of fact, we can count religions as styles of reasoning only by disregarding this fundamental property deeply rooted in Hacking's philosophy of science.

5.6.2 *Internal and External Techniques of Self-vindication*

Hacking wrote that

> [E]ach style has its own self-stabilizing techniques, but [...] some are more effective than others. [...] in the case of statistics there is an almost too-evident version of self-authentication (the use of probabilities to assess probabilities). But that is only part of the story, for I emphasize the material, institutional requirements for the stability of statistical reasoning. (Hacking, 1996, pp. 73-74)

This passage implies that there are different self-stabilizing techniques and that, e.g. in the case of the statistical style of reasoning, material and institutional factors contribute to the stability of a style of reasoning. So Hacking seems to treat these very factors and the use of probabilities to assess probabilities alike: he counts both of them as 'techniques of stabilization'. It seems to me that the two things (social factors and use of probabilities to assess probability) must be kept distinct. To take as an example the postulational style of reasoning, one thing is to attribute its stability to the argumentative character of the Greek society and other historical and social circumstances, another thing is to explain its stability by noting that there is no way of assessing the demonstration of a theorem apart from that of using another demonstration (the use of proof to assess proof). In the first case, what provides stability to the postulational style of reasoning is something external to its cognitive methods: social factors. In the second case, as for other styles of reasoning I have described so far, to be self-authenticating means that the postulational style of reasoning does not answer to any external canon of truth independent of itself: the truth of a proposition whose sense hinges on the postulational style of reasoning can only be assessed by using that very style of reasoning. This very fact makes the postulational style of reasoning immune to refutation and therefore stable. Ultimately, in the first case the techniques are external to the style of reasoning, in the second they are intrinsic to it.

I shall distinguish 'internal techniques of self-vindication' from 'external techniques of self-vindication'. An *internal* technique refers to cases in which stability is provided by the characteristics of the way of thinking itself, which is self-reinforcing: in order to assess what has been found out by using a certain form of reasoning (e.g. abduction for the historico-genetic style of reasoning, deduction for the postulational style of reasoning, computation for the algorithmic style of reasoning) one has to use that very form of reasoning. Styles of reasoning are inescapable, e.g. it is not possible to say that counting is a 'wrong' method of thinking and doing on which the algorithmic style of reasoning relies: we can only assess a calculation by using another kind of calculation. On the other hand, an *external* technique refers to cases in which social, cultural and historical factors make a style of reasoning stable. As I have argued in the previous sections of this chapter, all the styles of reasoning I have described possess internal self-vindication techniques in addition to external: both internal and external techniques concur to giving stability to a given style of reasoning.

Hacking wrote:

> The apparent circularity in the self-authenticating styles [styles of reasoning] is to be welcomed. It helps explain why, although styles may evolve or be abandoned, they are curiously immune to anything akin to refutation. (Hacking, 2002b [1992], p. 192)

I would correct this passage by making it more precise: when styles of reasoning possess *internal* techniques of self-vindication they are immune to refutation; its style-dependent sentences cannot be refuted, neither in principle nor in fact. Of course, styles of reasoning might be *abandoned* should we find better ways of knowing or should we stop craving for knowledge, a circumstance impossible to imagine. Conversely, if a style of reasoning possesses *only* external techniques of self-vindication *in principle* it could be refuted, e.g. when the social context changes. I am not saying that this has to happen; I am saying that whereas a given style of reasoning that possesses internal techniques of stabilization cannot be assessed by other styles of reasoning, in principle it would be possible to use one or more styles of reasoning to assess a way of thinking that does not possess internal techniques of self-vindication. For example, it would be possible to use the styles of reasoning I have presented in the previous sections to assess other ways of thinking, but it would be impossible to assess the postulational style of reasoning by another style of reasoning.

The styles project as formulated by Hacking does not provide an answer to the question as to why styles of reasoning have endured. I suggest that a possible answer is that as they rely on internal techniques of self-vindication they can never be refuted (I will say more about this issue in Chap. 8). The examples I am going to give in the next subsections will clarify the points of this subsection (see also my parallel between Wittgenstein and Hacking in (Sciortino, 2016)).

5.6.3 A Scholastic-inquisitorial Style of Reasoning?

Barry Allen argued that the reasoning of the inquisitors of the early modern period (1490–1690) possesses the features of a style of reasoning (Allen, 1993). When examining cases of witchcraft, two works were often consulted: the *Malleus Maleficarum* by Heinrich Kramer (1430–1505) and James Sprenger (1436–1495), admired theologians and Dominican inquisitors, and the *Daemonolatria* by Nicolas Rémy (1530–1616), a secular magistrate who condemned 900 witches. For Allen, both these works proposed a debate about the existence of witches (the 'new objects'), contain candidates for truth-or-falsehood ('children can be generated by incubi and succubi' or 'witches are more common than in the past') (Allen, 1993, p. 115). Furthermore, the reasoning of demonologists is self-authenticating because of their attitude towards textual authority: 'each writer cited his predecessors and became an authority worthy of citation' (Allen, 1993, p. 112). Lastly, inquisitorial reasoning involves a new kind of evidence: confessions under torture, special signs in the body, taciturnity, and incapacity to recite the Lord's Prayer (Allen, 1993, p. 111). Allen concluded that the men who persecuted witches were not misogynistic monks with narrow minds but secular magistrates and highly educated lawyers who 'shared a [(style of reasoning)] which was [...] productive of reasoned discourse and knowledge' (Allen, 1993, p. 105). He called this way of thinking 'scholastic-inquisitorial'.

Notice that the attitude of appealing to the authority of predecessors, which provides stability of the scholastic-inquisitorial way of thinking, is an *external* technique of self-vindication: whereas the propositions of *Malleus Meleficarum* could well be assessed now by, e.g. the laboratory style of reasoning, which would reject its kind of evidence, other styles of reasoning are immune to refutation just because they possess an *internal* technique of self-vindication. Attributions of witchery might be mutually supporting but, as it is possible in principle to use another style of reasoning, e.g. the laboratory style of reasoning, to assess them, it cannot be said that the scholastic-inquisitorial way of thinking possesses *internal* techniques of self-authentication. Hacking is reluctant to state rules. However, it is important to note that *if* one required that *any* style of reasoning must possess *internal* self-vindication techniques then Allen's candidate would not be a style of reasoning. For sure, its techniques of self-vindication make it dissimilar to the styles of reasoning I have presented in the previous sections.

Another problem with Allen's candidate is that there is no relevant sense in which the former can be considered a way of finding out: demonologists used methods such as torture, interrogation and physical inspection in order to assess whether a person was a minion of Satan. These methods have a very narrow range of application: they only help to find out whether or not innocent human beings are witches.

5.6.4 *A Chemists' Style of Reasoning?*

Bernardette Bensaude-Vincent argued that both Crombie and Hacking wrongly located the emergence of laboratory style in Boyle's air pump (Bernadette Bensaude-Vincent, 2009). Indeed, in their *History of Chemistry* (1996), Bensaude-Vincent and Stengers had explained that the laboratory as a physical place was invented by the alchemists and then was adopted first by the chemists and only afterwards by the physicists (Bensaude-Vincent, B. and Stengers, I. 1996). Chemists had to fight to secure the identity of chemistry against other scientists, until it became an autonomous discipline. The conquest of this autonomy, they argue, was determined by the combined and changing action of three factors: professions, institutions and techniques of laboratory. For example, for Bensaude-Vincent the laboratory of chemistry must be conceived as distinct from the laboratory of physics (Bernadette Bensaude-Vincent, 2009, p. 370).

In Bensaude-Vincent's view the chemists' way of thinking possesses all the characterizing properties of styles of reasoning. For example, she noted that by *altering* materials chemists have been able to *know* the features of the substances. For her, chemistry is a way of knowing by *making*—only through technological processes are the features of the chemical elements revealed. She also considered the chemical atoms introduced by John Dalton (1766–1844) as the epitome of the 'new objects' introduced by chemists' style and the chemical formulas as a perfect example of 'new candidates for truth and falsehood'. As I interpret Bensaude-Vincent, different reasons explain why, despite 'all sorts of enterprises of reduction or refutation throughout their history, chemistry remain[ed] nevertheless relatively stable' (Bernadette Bensaude-Vincent, 2009, p. 376). Firstly, there is what I have called *external* techniques of self-stability: the combined role of institution and professions. Secondly, Bensaude-Vincent maintained that '[chemists] do not claim to provide a causal explanation, and their theory is closer to being a narrative than a fundamental explanation' (Bernadette Bensaude-Vincent, 2009, p. 372). For her chemists' practice is a sort of 'cookery' that does not involve theory books but recipes for manipulating substances in laboratory or mnemonic rules for dealing with chemical formulas (Bernadette Bensaude-Vincent, 2009, p. 371). Rather than representing the real world, chemical formulas are useful fictions to be manipulated in order to predict and characterize compounds. For Bensaude-Vincent there is no real effort for unveiling a deeper structure of the world: chemists mainly want to understand the principles behind particular classes of reactions, the changes undergone by matter under certain processes, the structure and the properties of certain compounds.

It is not clear whether by likening chemistry to cookery Bensaude-Vincent wanted to suggest that chemistry is self-authenticating. At any rate, on the basis of her claims it would be possible to argue for this thesis by noting that a chemical formula is correct because it follows certain rules, which are not really discussed and questioned by the chemists. For example, the reduction-oxidation reactions are worked out in reference to 'rules' imposed by the conservation of charge and mass. Chemists simply need to apply definite procedures in which only certain possible oxidation states of the elements are possible. And how can they double-check the correctness of a reduction-oxidation reduction? By looking at other chemical reactions in which it is found that a given element does get the same oxidation state number as in the reaction under study. In other words, the conclusion of the argument would be that chemistry could be thought as a set of procedures that are self-reinforcing. That is, according to this argument, chemistry would possess an internal self-authenticating technique (and therefore it would be comparable to other styles of reasoning).

If this is the essence of Bensaude-Vincent's point, it might be prone to the following objection. Suppose that a very large part of chemistry had been reduced to physics in the original sense of Ernst Nagel (1901–1985): a secondary science can be reduced to a primary science when the laws and theories of the secondary science are a logical consequence of those of the primary science (condition of derivability) and when the terms of the secondary science can be redefined in terms of the primary science (condition of connectability) (Nagel, 1961, pp. 336–397). Roughly speaking, that a science is reduced to another means that a set of propositions is derived from another set. In that case, there would be a sharp difference between the chemical way of thinking and the styles of reasoning I have presented.

Compare the postulational style of reasoning with the chemical thinking. The former is somehow irreducible in the sense that the truth-value of the propositions of geometry hinges on the very postulational style of reasoning: only demonstration can be used to check another demonstration. Conversely, in the hypotheses of reducibility, the propositions of chemistry would depend on those of physics and therefore the chemists' style would not be really autonomous from the laboratory style of reasoning. Otherwise stated, although chemistry is stable and autonomous *as a discipline*, its propositions would have an external justification——and therefore the chemical way of thinking would not be self-authenticating. In parenthesis, notice the difference between the chemists' way of thinking and the historico-genetic style of reasoning: as I have already argued, the latter has an internal self-stabilizing technique, and indeed biology, palaeontology and geology cannot be reduced to physics.

In conclusion, if chemistry can be reduced to physics it does not possess internal self-stabilizing techniques. Though, it must be said that it is still a *res controversa* whether or not chemistry can be reduced to physics. For different reasons Popper, Kuhn and Feyerabend were critics of reduction, while Paul Dirac (1902–1984) thought that the difficulty of reducing chemistry to physics was only technical (cited in Scerri, 1994, p. 162). Today it is widely recognized that the valence of an element can be calculated and explained by putting it in relation with the number of electrons in the outer orbit of atoms. However, according to Roald Hoffmann there are still in chemistry concepts such as PH or reactivity that resist reductionism (Hoffmann, 1995). More recently the philosopher of chemistry E.R. Scerri argued that the hope of any strict or exact reduction has been abandoned and that all that remains is the possibility of approximate reduction, a notion still vague because no criteria have been put forward (Scerri, 1994, p. 168).

Ultimately, the chemical way of thinking has played a fundamental role in the history of science and it is still crucial in many areas of scientific research. Since whether or not it is reducible to physics is a question for determination it is impossible to conclude that it satisfies all the characterizing properties: it might not possess internal techniques of self-vindication. We can only conjecture that if chemistry is in principle reducible to physics the chemists' mechanisms of reasoning are not so basic, fundamental, and inescapable as those of the styles of reasoning I have presented in the previous sections.

5.6.5 *An Astrological Style of Reasoning?*

As we have seen in Chap. 3, Hacking presented a sentence of Paracelsus as an example of style-dependent proposition. For him the way Paracelsus reasoned is entirely different from ours and exemplifies a 'Northern European Renaissance tradition of a bundle of hermetic interests: medicine, physiology, alchemy, herbals, astrology, divination' (Hacking, 1983b, pp. 70–71). I shall briefly discuss here only the astrological way of thinking, as similar points can be made about alchemy and divination.

Ancient peoples such as the Babylonians practised astrology and believed that many aspects of the skies had some importance in divination. However, it was with the conquests of Alexander the Great that astrology crystallized out of an amalgam of Babylonian, Greek and Egyptian elements. I shall focus on its elementary grammar, although astrology has greatly evolved beyond his apparent roots in Babylonian culture becoming a variegated group of doctrines.

To put it simply, astrological thinking is purely deductive: to think in the astrological way amounts to drawing some consequences from precise facts in the skies according to certain rules. For example:

> When Venus appears in Dilgan (Virgo), rains in heaven, floods on [earth] the crops of Aharru will prosper.

a sentence from a collection Assyrian omens called 'Enuma Anu Enlil' (Bobrick, 2005, p. 16).

However, the astrological way of thinking is more accurately described by three steps. The first step is an assumption: it is postulated that celestial bodies have an influence upon human affairs. To all intent and purposes, this is the *postulate* of astrology: 'astrology extrapolates factual influences by postulating that the positions of the Sun, Moon and 8 planets other than Earth with respect to the sky background, as well as with respect to each other, influence terrestrial events and human psychology and destiny' (Zarka, 2011, p. 420). I shall call this postulate 'the astrological correspondence principle' in that it is a principle that establishes a correspondence between mundane facts and cosmos. The second step consists of casting a horoscope, which is a map of the positions of Sun, Moon and eight planets drawn for a particular moment of time and read according to established rules. In many traditions this map is round: the circle is divided into 12 sectors, which are called the 'houses'; in the outer portion are drawn the signs of Zodiac, i.e. 12 stations or 'signs' along the ecliptic, the apparent path of the Sun around the Earth. Once the astrologer knows the date of birth, time and place of someone, he can draw up a horoscope. First, he calculates the ascendant; the point where the ecliptic intersects the horizon in the east at the place of birth; from that point, he draws the 12 sectors (houses) of the circle. Then he writes the signs along the round and, consulting an ephemeris, marks the location of the Sun, Moon and planets and their relative angles for that particular date and time. The third step consists of interpreting the horoscope, that is of drawing certain conclusions according to the rules of a certain tradition of astrology.

'Houses', 'signs', 'zodiac', and other conventions and reference frames can be considered the 'new objects' introduced by the astrological way of thinking. The horoscope represents a form of evidence for the believers in astrology: an astrologer will consider each of the possible relative angles between pairs of planets (i.e. Conjunction (0°), Opposition (180°), Square (90°), Trine (120°)), or how much an angle made by two points differs from the exactness, as *evidence* of the truth of certain sentences.

I think that two different *external* techniques of self-vindication made the astrological way of thinking endure. First of all, institutions, authorities and mass media have always been consistently woven into the discourse of astrology and astrologers cite their predecessors as authorities: the *Tetrabiblos*, Ptolemy's treatise on astrology, was considered the most authoritative astrological work of antiquity not only in the Medieval Latin West but also in the Islamic world. Finally there is the astrological correspondence principle, which is never questioned. Its role can be likened to the infallibility of Pope in Christian-Catholic reasoning.

The astrological way of thinking does not possess internal techniques of self-vindication because the astrological correspondence principle as well as sentences such as

> His condition quickly worsened because of the rapid motion of the moon (found in Bobrick, 2005, p. 137)

(written by the mathematician Gerolamo Cardano (1501–1576) in 1553) are obviously *not* potentially immune to refutation. The astrological thinking is not inescapable: one can reason in a different way to show that it is not a valid form of reasoning. But one cannot reason in a different way from the algorithmic style of reasoning to show that a calculation is wrong or in a different way from the historico-genetic style of reasoning to show that there is a better explanation for the extinction of dinosaurs.

The evidence for any astrological claim is provided through the horoscope by the positions of planets. No further investigations are carried out in order to know whether or not the positions of the planets exert an influence upon terrestrial events. This point lies behind Kuhn's complaint that 'astrology has rules to apply but no puzzles to solve and therefore no science to practice' (Kuhn, 1970, pp. 8–9) and Feyerabend's claim that '[astrology does not] attempt to proceed into new domains and to enlarge our knowledge of extra-terrestrial influence' (Feyerabend, 1978, p. 96). Therefore, even the astrological way of thinking cannot be likened to the styles of reasoning I have presented in the previous sections because it is much interested in making predictions and very little in finding out why and how. I suggest that we should consider the ways of thinking discussed in this sections as a variety of styles of reasoning that do not possess certain important characterizing properties such as that of possessing internal techniques of self-vindication. They are not self-authenticating in a strong sense. Moreover, apart from the chemists' way of thinking, they are not scientific, they are 'pseudo-styles' of reasoning. The main point of my discussion in this section is that scholars have implicitly compared ways of thinking as if they were all aiming to do the *same* thing: to produce knowledge of the world. A better approach would be to draw contrasts between their proposed 'styles of reasoning' and the styles of reasoning presented in the previous sections. Such an approach fits far better with one of the main points of Hacking's philosophy (see Chap. 8): the disunity of human knowledge.

5.7 Conclusions

There exist six styles of reasoning that share a set of common characterizing features. I have described two of them in Chap. 3, following some of Hacking's suggestions. In this chapter I have argued for the existence of four other styles of reasoning by outlining their features. All the differences notwithstanding, striking parallels can be drawn between different styles of reasoning: they emerged suddenly at certain points in time, crystallized around communities that shared certain standards of evidence and represent basic and inescapable forms of reasoning in the sense I have discussed above. In the last section, I have completed my characterization of scientific thought: I have argued that there are ways of thinking that differ from styles of reasoning. What I consider a relevant difference is that only styles of reasoning have strong techniques of self-authentication, techniques that are intrinsic to their forms of reasoning.

Bibliography

Allen, B. (1993). Demonology, Styles of Reasoning, and Truth. *International Journal of Moral and Social Studies, 8*(2), 95–122.

Atran, S. (1990). *Cognitive Foundations of Natural History: Towards an Anthropology of Science*. Cambridge University Press.

Bensaude-Vincent, B. (2009). The Chemists' Style of Thinking. *Berichte Zur Wissenschaftsgeschichte, 32*(4), 365–378.

Bensaude-Vincent, B., & Stengers, I. (1996). *A History of Chemistry*. Harvard University Press.

Bobrick, B. (2005). *The Fated Sky*. Simon & Schuster Paperbacks.

Bocquet-Appel, J.-P. (2011). When the World's Population Took Off: The Springboard of the Neolithic Demographic Transition. *Science, 333*(6042), 560–561.

Bogoshi, J., Naidoo, K., & Webb, J. (1987). The Oldest Mathematical Artefact. *The Mathematical Gazette, 71*(458), 294. https://doi.org/10.2307/3617049

Bowler, P. (2003). *Evolution: The History of an Idea*. University of California Press.

Boyer, C. B. (1968). *A History of Mathematics*. Wiley International Edition.

Bynum, W. F. (1975). The Great Chain of being. *History of Science, 13*, 1–28.

Carey, S. (1985). *Conceptual Change in Childhood*. The Mit Press.

Cleland, C. (2001). Historical Science, Experimental Science, and the Scientific Method. *Geology, 29*(11), 987–990.

Cleland, C. (2002). Methodological and Epistemic Differences between Historical Science and Experimental Science. *Philosophy of Science, 69*, 474–496.

Darwin, C. (1985 [1859]). *The Origin of Species*. : Penguin Books Ltd.

Davis, H., & Perusse, R. (1988). Numerical Competence in Animals: Definitional Issues, Current Evidence, and a New Research Agenda. *Behavioral and Brain Sciences, 11*, 561–615.

Eldredge, N. (1998). *The Pattern of Evolution*. W.H. Freeman & Company.

Feigenson, L., Dehaene, S., & Spelke, E. (2004). Core Systems of Number. *Trends in Cognitive Sciences, 8*, 307–314.

Feyerabend, P. (1978). *Science in a Free Society*. New Left Books.

Fodor, J. (1983). *The Modularity of Mind: An Essay on Faculty Psychology*. MIT Press.

Foucault, M. (1994 [1966]). *The Order of Things*. : Vintage Books Edition.

Gohau, G., & Duris, P. (1997). *Histoire des sciences de la vie*. Nathan.

Gould, S. J. (1998). The Upwardly Mobile Fossils of Leonardo's Living Earth. In *Leonardo's Mountain of Clams and the Diet of Worms*. : Harmony Books.

Gould, S. J. (2000a). Linnaeus's Luck? *Natural History, 109*(7), 18–29.

Gould, S. J. (2000b). *Wonderful Life: The Burgess Shale and the Nature of History*. Vintage.

Gould, S. J. (2002). *The Structure of Evolutionary Theory*. Belknap Press.

Gribbin, M., & Gribbin, J. (2008). *Flower Hunters*. Oxford University Press.
Griffiths, G. C. D. (1974). On the Foundations of Biological Systematics. *Acta Biotheoretica, 23*(3-4), 85–131. https://doi.org/10.1007/bf01556343
Hacking, I. (1983a). The Accumulation of Styles of Reasoning. In H. v. D. Heinrich (Ed.), *Kant oder Hegel? Über Formen der Begründung in der Philosophie* (pp. 453–465). Klen-Cotta.
Hacking, I. (1983b). *Representing and Intervening*. Cambridge University Press.
Hacking, I. (1991). Living Things. Review of Cognitive Foundations of Natural History: Towards an Anthropology of Science by Atran, S. London Review of Books. *London Review of Books, 13*(4), 17–18.
Hacking, I. (1993). Working in a New World: The taxonomic Solution. In P. Horwich (Ed.), *World Changes: Thomas Kuhn and the Nature of Science* (pp. 275–310). The MIT Press.
Hacking, I. (1996). The Disunity of Science. In P. Galison & D. J. Stump (Eds.), *The Disunity of Science: Boundaries, Contexts, and Power* (pp. 37–74). Stanford University Press.
Hacking, I. (1999). *The Social Construction of What?* Harvard University Press.
Hacking, I. (2000). What Mathematics Has Done to Some and Only Some Philosophers. In T. J. Smiley (Ed.), *Mathematics and Necessity* (Vol. 83, pp. 83–138). British Academy.
Hacking, I. (2002a). *Historical Ontology*. Harvard University Press.
Hacking, I. (2002b [1992]). 'Style' for Historians and Philosophers. In *Historical Ontology* (pp. 178-199). Cambridge, Massachussetts: Harvard University Press.
Hacking, I. (2002-2003). Des styles de raisonnement scientifiques - Résumé du cours. Retrieved from http://www.college-de-france.fr/site/historique/ian_hacking.htm#|p=./historique/ian_hacking.htm
Hacking, I. (2006a, 7/1/2013). Cognition. Retrieved from http://www.college-de-france.fr/media/historique/UPL32429_cognition.pdf
Hacking, I. (2006b). La Démonstration. Retrieved from http://www.college-de-france.fr/site/historique/ian_hacking.htm
Hacking, I. (2009). *Scientific Reason*. NTU Press.
Hacking, I. (2011). How Did Mathematics Become Possible? Retrieved from http://www.youtube.com/watch?v=PbKUsAR_8DY
Hacking, I. (2012). 'Language, Truth and Reason' 30 Years Later. *Studies in History and Philosophy of Science Part A, 43*(4), 599–609.
Heath, T. L. (1908). *The Thirteen Books of Euclid's Elements*. Cambridge Press.
Hoffmann, R. (1995). *The Same and Not the Same*. Columbia University Press.
Høyrup, J. (1990). Sub-scientific Mathematics: Observations on a Pre-modern Phenomenon. *History of Science, 28*, 63–86.
Høyrup, J. (2005). The Shaping of Deduction in Greek Mathematics: A Study in Coginitive History. *Studia Logica, 80*(1), 143–147.

Høyrup, J. (2010). *Old Babylonian "Algebra", and What It Teaches Us about Possible Kinds of Mathematics.* Paper presented at the ICM Satellite Conference 'Mathematics in Ancient Times', Kozhikode.

Høyrup, J. (2012). Mathematical Justification as Non-Conceptualized Practice: The Babylonian Example. In K. Chemla (Ed.), *History of Mathematical Proof in Ancient Traditions* (pp. 362–383). Cambridge University Press.

Husserl, E. (1970 [1936]). *The Crisis of European Sciences and Transcendental Philosophy.* : Northwestern University Press.

Karpinski, L. C., & Waters, E. G. R. (1928). A Thirteenth Century Algorism in French Verse. *Isis, 11*(1), 45–84.

Kitcher, P. (1989). Explanatory Unification and Causal Structure. In P. K. a. N. Salmon (Ed.), *Scientific Explanation.* University of Minnesota Press.

Knuth, D. E. (1972). Ancient Babylonian Algorithms. *Communications of the Association of Computing Machinery, 15*, 671–677.

Köhler, W. (1957). The Ability of Birds to "count". In N. J. R (Ed.), *The World of Mathematics* (Vol. 1, pp. 489–496). Simon and Schuster.

Kuhn. (1970). Logic of Discovery or Psychology of Research. In I. Lakatos & A. Musgrave (Eds.), *Criticism and the Growth of Knowledge.* Cambridge University Press.

Kusch, M. (2010). Hacking's Historical Epistemology: A Critique of Styles of Reasoning. *Studies in History and Philosophy of Science, 41*(2), 158–173. https://doi.org/10.1016/j.shpsa.2010.03.007

Kwa, C. (2011). *Styles of Knowing.* Pittsburgh University Press.

La Vergata, A. (1988). Geologia e paleontologia fra la metà del Settecento e la metà dell'Ottocento. In P. Rossi (Ed.), *Storia della Scienza* (Vol. 4). Unione Tipografico-Editrice Torinese.

Lacki, J. (2003). Styles of Physical Thinking Versus Mathematical Ones. *Synthese, 134*(1), 273–288.

Latour, B. (2008). The Netz-Works of Greek Deductions. *Social Studies of Science, 38*(3), 441–459.

Laudan, r. (1987). *From Mineralogy to Geology. The Foundations of a Science 1650-1830.* The University of Chicago Press.

Leakey, M. G., F, S, MC, D., CS, F., SC, A., C, K, & LN., L. (2012). New Fossils from Koobi Fora in Northern Kenya Confirm Taxonomic Diversity in Early Homo. *Nature, 488*, 201–204.

Lloyd, G. E. R. (1961). The Development of Aristotle's Theory of the Classification of Animals. *Phronesis: A Journal for Ancient Philosophy, 6*(1-2), 59–81. https://doi.org/10.1163/156852861X00080

Lloyd, G. E. R. (2004). *Ancient Worlds, Modern Reflections: Philosophical Perspectives on Greek and Chinese Science and Culture.* Clarendon Press.

Lloyd, G. E. R. (2007). *Cognitive Variations: Reflections on the Unity and Diversity of the Human Mind.* Oxford University Press.

Lloyd, G. R. (2000). Watching for the eclipse. *The Times Literary Supplement.*
Louis, P. (1986). *Histoire du vocabulaire scientifique.* Publications de l'Institut National de la Langue Française.
Lovejoy, A. O. (1960 [1936]). *The Great Chain of Being: A Study of the History of an Idea.* New York: Harper.
Magli, G. (2009 [2005]). *Mysteries and Discoveries of Archaeoastronomy: From Giza to Easter Island: From Pre-history to Easter Island.* : Copernicus Books.
Mason, S. (1962). *A History of the Sciences.* Macmillan Publishing Company.
Mayr, E. (2004). *What Makes Biology Unique? Considerations on the Autonomy of a Scientific Discipline.* Cambridge University Press.
Mayr, E., & Bock, W. J. (2002). Classifications and Other Ordering Systems. *Journal of Zoological Systematics & Evolutionary Research, 40*(4), 169–194.
Nagel, E. (1961). *The Structure of Science.* Harcourt, Brace and World.
Netz, R. (1999). *The Shaping of Deduction in Greek Mathematics: A Study in Cognitive History.* Cambridge University Press.
Netz, R. (2000). The Origins of Mathematical Physics: New Light on an Old Question. *Physics Today, 53*(6), 32.
Neugebauer, O., & Sachs, A. (1945) Mathematical Cuneiform Texts. In: *Vol. 29. American Oriental Series.* New Haven, Connecticut: American Oriental Society.
Nieder, A. (2005). Counting on Neurons: The Neurobiology of Numerical Competence. *Nature Reviews Neuroscience, 6*(3), 177–114. https://doi.org/10.1038/nrn1626
Paty, M. (2003). The Idea of Quantity at the Origin of the Legitimacy of Mathematization in Physics. In C. Gould (Ed.), *Constructivism and Practice: Towards a Social and Historical Epistemology* (pp. 109–135). Rowman & Littlefield.
Pica, P., Lemer, C., Izard, V., & Dehaene, S. (2004). Exact and Approximate Arithmetic in an Amazonian Indigene Group. *Science, 306*, 499–503.
Rosenberg, D., & Grafton, A. (2010). *Cartographies of Time.* Princeton Architectural Press.
Rossi, P. (2000). *La nascita della scienza moderna in Europa.* Feltrinelli.
Rudman, P. S. (2007). *How Mathematics Happened: The First 50,000 Years.* Prometheus Book.
Rudwick, M. J. S. (1985). *The Great Devonian Controversy.* The University of Chicago Press.
Scerri, E. R. (1994). Has Chemistry Been at Least Approximately Reduced to Quantum Mechanics? *Proceeding of the Biennal Meeting of the Philosophy of Science Association, 1*, 160–170.
Sciortino, L. (2016). Styles of Reasoning, Human Forms of Life, and Relativism. *International Studies in the Philosophy of Science, 30*(2), 165–184. https://doi.org/10.1080/02698595.2016.1265868

Sciortino, L. (2021). Caos e ordine: genesi e sviluppo dello stile deduttivo nell'Antica Grecia. *Informazione filosofica, 3*(2), 6–24.

Spelke, E. (1994). Initial Knowledge: Six Suggestions. *Cognition, 50*(1-3), 431–445.

Stevens, P. F. (1994). *The Development of Biological Systematics: Antoine-Laurent de Jussieu, Nature, and the Natural System*. Columbia University Press.

Vernant, J.-P. (1982 [1962]). *The Origins of Greek Thought*. : Cornell University Press.

Vitti Rodrigues, M., & Claus, E. (2021). Abduction and Styles of Scientific Thinking. *Synthese, 198*, 1397–1425. https://doi.org/10.1007/s11229-019-02127-7

Winther, R. G. (2012). Interweaving Categories: Styles, Paradigms, and Models. *Studies in History and Philosophy of Science, Part A, 43*(4), 628–639.

Wynn, K. (1992). Addition and Subtraction by Human Infants. *Nature, 358*, 749–750.

Zarka, P. (2011). Astronomy and Astrology. *Proceedings of the International Astronomical Union, 5*, 420–425.

CHAPTER 6

Styles of Reasoning and Relativism

6.1 Introduction

In this chapter I shall start to discuss a philosophical issue concerning the notion of style of reasoning. This issue boils down to the question as to whether Hacking's styles project implies epistemic relativism. My aim here is to bring it to the fore as well as examine and frame it. Only in the next chapter will I tackle this issue (henceforth 'relativism issue') by arguing that Hacking's claims do have relativistic implications. However, before starting my analysis in the second part of this chapter, first I wish to examine some preliminary questions that will also allow me to arrive at a more complete theory of styles of reasoning. In Sect. 6.2, I shall provide a historical example of a scientific problem to which two solutions in two distinct styles of reasoning have been given; also, I will clarify that styles of reasoning accumulated in the course of history. In Sect. 6.3, I shall examine the crucial question of whether or not members of a given community can make sense of a style-dependent sentence uttered by the member of another community that adopts a different style of reasoning. In Sect. 6.4, I shall frame the relativism issue and I will tackle it in the next chapter.

L. Sciortino, *History of Rationalities*,
https://doi.org/10.1007/978-3-031-24004-1_6

6.2 Styles of Reasoning in Action

6.2.1 *One Problem, Two Alternative Styles of Reasoning*

I now wish to make it clear that sometimes the same scientific problem has been tackled in two distinct ways based on *different* styles of reasoning. Consider this example. In the third century BCE, the Greek astronomer Aristarchus of Samos (310 BC–230 BC) calculated the distances of the Sun and the Moon from the Earth in units of the Earth's radius. His style of thinking was postulational: Len Berggren and Nathan Sidoli liken his treatise *On the Sizes and Distances of the Sun and the Moon* to mathematical texts of the early Hellenistic period, such as Euclid's works (Berggren & Sidoli, 2007). In fact, in order to work out the distance of the Moon and Sun, Aristarchus drew geometric diagrams of the Earth's, the Sun's and the Moon's respective positions during a half moon and a lunar eclipse, made few observations, formulated geometrical and computational hypotheses, used known geometrical theorems and finally derived the distance of these celestial bodies (Berggren & Sidoli, 2007, p. 215). However, in principle, the distance to the Moon can also be determined by adopting the laboratory style of reasoning through these steps: determination of acceleration g by measuring the period of a simple pendulum; measurement of the sidereal period of the Moon; calculation of the Earth's radius by measuring the gravitational constant G and applying the law of universal gravitation; finally, by equating the force between the Moon and the Earth to the Moon's centripetal force, the distance to the Moon can be expressed and calculated in terms of the acceleration g, the sidereal period of the Moon and the Earth's radius. In the case of Aristarchus the result is obtained by thinking in a postulational way: no laboratory thinking is involved, although Aristarchus had to make a few astronomical observations. In the second case, the laboratory thinking is necessary to reason in terms of physical laws and experimental apparatuses.

6.2.2 *The Accumulation of Styles of Reasoning*

Another point to note is that the project of styles of reasoning implies that 'styles of scientific reasoning tend to accumulate' (Hacking, 1983, p. 56). As styles of reasoning have emerged they have often been adopted together in order to solve single scientific problems: the algorithmic, the postulational, the statistical, the historico-genetic, the taxonomic, and the laboratory styles of reasoning are not mutually exclusive. Today scientists switch from the ways of thinking and doing, the methods, the objects, the type of evidence of a style of reasoning to another while they do scientific research. Algorithms, hypothetic deductive methods, distributions, inferences to the best explanation, taxonomies, analogical models and measurements as well as distinct ways of thinking and doing, take your pick: whatever scientific problem one considers, one can recognize the features of different styles of reasoning.

The simplest example of simultaneous use of different styles of reasoning that can be given has to do with the use of vectors to represent physical magnitudes such as forces and velocities. As vectors are often represented in the three-dimensional space as arrow-like objects, which can be multiplied by real numbers or added together, the simple rules that define their use rely on the postulational style of reasoning. Therefore, we can expect that in many problems of physics *both* the postulational style of reasoning and the laboratory style of reasoning play a crucial role. Consider for example three positive charges Q1, Q2 and Q3 arranged in a right triangle ABC. Suppose that one has to determine the force on Q2 on the point B, the vertex of the right angle. The solution of this problem is worked out by summing up two vectors: the one that represents the Coulomb force exerted by Q1 and that pushes Q2 away from it, and the one that represents the Coulomb force exerted by Q3 and that pushes Q2 away from it. By applying Pythagoras' theorem to the triangle formed by the two force vectors on Q2 and by their sum it is possible to calculate the magnitude of the resultant force on Q2. In principle, three styles of reasoning, the laboratory, the postulational and the algorithmic one, are necessary for working out the result. First of all, thinking and doing in the laboratory style of reasoning is necessary to put forward and test the Coulomb law; secondly, thinking and doing in the postulational style of reasoning is necessary to represent the forces in a plane and apply geometrical considerations: finally, the algorithmic style of reasoning is necessary to perform multiplications and divisions and obtain a number, the magnitude of the force.

6.3 Intelligibility, Understanding, Projectibility

6.3.1 *Truth-value, Intelligibility, Pure Nonsense*

In order to arrive at a complete theory of styles of reasoning I want to address a last question concerning the semantics of sentences whose truth-value hinges on their style of reasoning (style-dependent sentences). I shall elaborate on the claim that members of a community cannot make sense of a style-dependent sentence uttered by another community that adopts a different style of reasoning. More precisely, I shall define and examine the type of intelligibility that can be achieved for different kinds of sentences by communities that adopt different styles of reasoning.

True-or-false sentences are sentences that express true-or-false propositions. Henceforth, as I have done so far, I shall use the term 'sentence' rather than terms such as 'propositions', 'statements' or whatever the logicians use in order to designate assertive acts of thought: I want to call attention to the fact that I am referring to grammatically correct strings of words that come to be uttered by members of certain communities of people at specific points in time. Suppose that a style of reasoning crystallizes around a community of people. Its members will communicate to each other by expressing style-dependent and style-independent sentences. In that case, they will have three kinds of intentions: to utter those sentences, to confer meaning to them and to communicate them. The first intention is satisfied by using parts of the body to produce a sound; the second intention by expressing the sentences correctly and imposing truth-conditions on them. Once these two conditions are satisfied, a hearer could in principle recognize them and understand the sentence so as to satisfy the third intention of the speaker. Ultimately, putting aside the intention of uttering the sentence, the hearer must recognize the speaker's intention of 1) formulating the sentence correctly, i.e. according to the rules of the language and of 2) providing truth-conditions for that sentence. Points 1) and 2) express two different concepts: point 1) concerns the correct syntactical arrangement of the words that, by their meaning, will determine the *sentence meaning*, i.e. the conventional meaning of the sentence; point 2) concerns the *speaker meaning*, i.e. what the speaker means by expressing the sentence. This distinction between these two concepts has been brought to the fore by the philosopher Paul Grice (1913–1988): the notion of meaning involved when someone says that a *sentence (or a word) means* so-and-so is to be distinguished from the notion involved when someone says that by saying such-and-such *she means* so-and-so (Grice, 1968).

To give an example, if the sentence were 'the dog is barking', the hearer could recognize the intention of expressing a sentence correctly in English so as to attribute to it a conventional meaning. Then, recognizing this intention, the hearer could also recognize the second intention of providing truth-conditions: the sentence is true if, as a matter of fact, the dog is barking. Such being the case, the hearer could grasp the meaning of the sentence and the speaker could succeed in communicating it. Now, for the argument's sake, I shall imagine different cases in which speaker and hearer belong to different communities, even separated in time.

If the hearer does not belong to the community that adopts the same style of reasoning of the speaker the message conveyed from the latter may not get through. Take for example, this sentence in Middle English:

I ne have heer a fruyte

If the hearer (or the reader) belonged to the community of the Renaissance alchemists, she might not be able to recognize some words of that sentence and their syntactical arrangement—the conventional meaning of the sentence would be obscure. However, the hearer could investigate about the meanings of the words and the grammar rules employed by the modern scientists. She would find out that the translation is:

I do not have here a fruit

This sentence would not only have a conventional meaning but also a truth-value straightforwardly recognizable by the hearer—its conditions of truth do not differ from the conditions of truth she employs in her style of reasoning. In other words, *after investigation*, the hearer would be able to recognize that:

1) the sentence 'I ne have heer a fruyte' *means* 'I do not have a fruit' (sentence meaning)
2) by saying 'I ne have heer a fruyte' *the speaker means* that her sentence is true if and only if the speaker does not have a fruit (speaker meaning).

In these examples above, once the hearer finds out the 'sentence meaning' by virtue of the correct translation of the sentence, she has a straightforward access to the speaker meaning. However this is not always the case. Take for example, this passage of Paracelsus:

The Adamant is a black crystal called Adamant or else Evax on account of the joy which it is effectual in impressing on those who carry it. It is of an obscure and transparent blackness, the colour of iron. It is the hardest of all; but is dissolved in the blood of a goat. (Paracelsus, 2007, p. 90)

Suppose that the hearer belongs to a community of the eighteenth century that adopts the statistical style of reasoning. She would not find it difficult to recognize the correct arrangement of the words and therefore the conventional meaning, but that would not be of much help—the access to the speaker's meaning would not be straightforward anymore. To the hearer it would not be clear whether or not the speaker wants to attribute truth-conditions to that sentence and which ones they could be. The historical discontinuity between the way of thinking of the Renaissance thinkers and the statistical style of reasoning has not really turned words such as 'black', 'transparent', 'goat' into something else; but it has concealed from sight the conditions of truth given by the speaker. The hearer might want to ask how Paracelsus's sentences could be proven true or false; or, alternatively, she might decide that no condition of truth has been attributed to it and that the sentence has to be deemed *nonsensical*.

If the hearer opted for the former alternative she would need to investigate about the truth-value of the sentences in the passage. She could find out that 'the body of the Emerald is derived from a kind of petrine Mercury' and that 'it receives from the same its colour, coagulated with spirit of Salt' (Paracelsus, 2007, p. 90), but still struggle to understand the passage above. She could then turn to a study of the system of relations of similarity or equivalence between things posited by the Renaissance thinkers and become aware of the existence of criteria that provide a justification for sentences such as Paracelsus's. She could also learn that metals have virtues and essences and that planets sign metals, understand why crystals can be black and dissolve in animals' blood, acquire skills in following the logic chain of Paracelsus's reasoning. But still she would need an awareness of the ways of doing that enables one to bring to light the signs and the virtues or an awareness of the laws that connect them with other signs and lead the speaker to certain conclusions.

All these circumstances, from the belief in analogies and sympathies between heavenly bodies and earthly things to Paracelsus's chain of reasoning, from his criteria of evidence to his way of doing constitute *presuppositions* that were unknown to the hearer. As it is against these presuppositions that Paracelsus's sentence acquires a truth-value for the Renaissance community of thinkers, the hearer would need to be aware of them in order to be able to attribute to the speaker the intention of conveying a message. In other words, only when the hearer has gained an insight into all those *presuppositions* can she recognize the intentions of the speaker, i.e. can she recognize that the speaker has imposed conditions of truth on the sentence. When this is the case, the sentence becomes *intelligible*: by this term I mean that the hearer can recognize both the conventional meaning and the conditions of truth of the sentence—how it would be validated in a cognitive frame different from her own.

By the same token, one can consider the following style-dependent sentence in the postulational style of reasoning:

> In any triangle the sum of any two sides is greater than the remaining one. (Euclid, Book I prop. 20, Heath, 1908)

Suppose that this sentence were translated from the Greek spoken by Euclid in around 300 BC into Archaic Egyptian. Although, one can imagine, the translated sentence would have a conventional meaning because of the correct syntactical arrangement of the words, its truth-value, imposed by Euclid, would be concealed from a hypothetical ancient Egyptian. Indeed, according to the theory of styles of reasoning, it was only with the abrupt emergence of the postulational style that the methods of reasoning and standards of evidence necessary for validating that sentence came about. An ancient Egyptian would find it impossible to examine *any* triangle and establish that the sum of two sides is greater than the remaining one. In order to recognize the intention of imposing a truth-value to that sentence an ancient Egyptian would need to come to see the presuppositions necessary to validate the sentence: among them, the existence of the postulational method and the different steps of the postulational way of thinking. This point is better expressed in terms of a verificationist account of meaning: whether or not Euclid's sentence above is a candidate for truth-or-falsehood for a community of people depends on whether or not that community possesses the methods necessary to assess the truth of the sentence. Since the ancient Egyptians did not know the methods of postulational style of reasoning they could not assess Euclid's sentence. It is in this sense that this sentence would not have been meaningful for an ancient Egyptian. Notice that whereas we cannot understand what Paracelsus was saying, an Egyptian would arguably have understood Euclid's sentence, although she would not have had the tools to verify it. To conclude, only when the hearer comes to see the presuppositions of the postulational style of reasoning, and in particular its methods, does Euclid's sentence above become intelligible to her. It is also important to note that the hearer would need to know the kind of questions at the root of the Greek inquiries. The questions asked by the Greeks were radically different from those asked by the ancient Egyptians: the latter asked questions about single geometrical figures; the former were interested in underlying principles besides practical problems. Ultimately, the hearer cannot recognize the speaker's meaning without being aware of another kind of presupposition: the kind of questions that were at the root of the postulational style of reasoning.

6.3.2 Questions and Candidates for Truth-or-Falsehood

The points I have just made can be developed by considering what the philosopher Robin George Collingwood (1889–1943) wrote in the fifth and seventh chapters of his autobiography:

> You cannot find out what a man means by simply studying his spoken or written statements [...]. In order to find out his meaning you must also know what the question was [...] to which the thing he has said or written was meant as an answer. (Collingwood, 1979 [1939], p. 31) (Collingwood 1979 [1939], p. 31)

According to Collingwood, truth and falsehood do not belong to sentences in their own right but to the sentences as answers to questions. In this sense, what the ancient Egyptian would lack is to have a bearing on the kinds of abstract questions that are asked in the community of Euclid.

Questions about the first principles were not 'real' for the ancient Egyptian society, to use a term introduced by the philosopher Nicholas Jardine in his book *The Scenes of Inquiry* (1991). For Jardine, a question is *real* in a community if there is a disposition to acknowledge the relevance of evidential considerations to that question. Evidential considerations are considerations that are potentially evident under appropriate circumstances: for example, they might be empirical but also theological, aesthetic or self-evident. A consideration is relevant just in case it is taken by the community to favour an answer to the question over the others (Jardine, 1991, chapter 3). For example, for Jardine a question that is not real in *any* community is:

> Why is there something rather than nothing? (Jardine, 1991, p. 56)

On the other hand, he gives examples of questions that were real *only* in a given community:

> Is the organic the image of the planet? (Jardine, 1991, p. 56)

This is a question that the naturalist Lorenz Oken (1779–1851) addressed himself when he set out on his classificatory enterprise. Oken's aim was to rethink the natural history by following the programme of Friedrich Schelling (1775–1854), who thought that nature realizes an 'original ideal' through the entire sequence of types of organization. There is no consideration that one can take in order to argue for a possible answer. By the same token, from Hacking's account it follows that Euclid's question about the entire class of prisms would be unreal for an ancient Egyptian: there is no consideration that she could have taken to answer questions about the properties of entire classes of geometrical figures.

In *The Social Construction of What?* (Hacking, 1999) Hacking espoused Jardine's ideas: '[Jardine] shows, by means of compelling historical examples, how questions that make sense in one scientific framework are unintelligible in another' (Hacking, 1999, p. 165). Moreover, Jardine gave examples of sentences to which the comments I have made in the case of Paracelsus's sentence apply. For example, he quoted Oken's sentences such as:

> The nose is the thorax repeated in the head grown. (Jardine, 1991, p. 64)

> The organic must be a vesicle because it is the image of the planet. (Jardine, 1991, p. 51)

Jardine explained that much of the rationale necessary to understand those sentences is not explicit in Oken's work. A modern reader would need to put Oken in the context of Schelling's views and then learn how the latter explained the relations between different forms of living beings in the light of an interplay of natural forces. Only then would Oken's ways of defending his claims come to light and the sentences above become *intelligible.*

6.3.3 *Intelligibility Versus Understanding*

Both Hacking and Jardine use the term 'intelligibility' rather than other terms such as *understanding*. I think that this choice deserves a close examination. When the hearer comes to see the presuppositions of another community that has a *different* style she can recognize the truth-values imposed by the speaker. However, those presuppositions are not shared by her community, are not *her* presuppositions. As a consequence (in Hacking's terminology) the style-dependent sentences of a different style of reasoning have no truth-value, although they are in principle intelligible. Or, alternatively, in Jardine's terminology, one may say that the *questions* of different cognitive standpoints can become intelligible, yet they remain unreal. So, what should we reserve the term 'understanding' for? Suppose that a speaker uttered Herschel's sentence:

> The heat which has the refrangibility of the red rays is occasioned by the light of those rays

Hacking claims that, if the hearer were Archimedes, he would not be able to assign a truth-value to this sentence, even if she had learned Herschel's physics from the ground up. However, not any hearer who belongs to a community that has the laboratory style would grasp the meaning of this sentence straightforwardly. Yet, any hearer in the style of the laboratory, the same as Herschel's, could well *understand* the sentence, provided that she has learned some rudiments of physics. Indeed, she does not need to perform any hermeneutic work in order to come to see presuppositions such as questions, methods and standards of truth: these presuppositions are peculiar of her community and are 'real' for her—and I have used the verb 'to understand' to underline this very circumstance. To be clearer, understanding does not require an effort for gaining an insight into the presuppositions of a different style. Whereas Archimedes can only achieve intelligibility, any member of the style of the Laboratory can achieve *understanding* of Herschel's sentence.

It is important to bear in mind that, *for us*, understanding is in principle possible for all the style-dependent sentences of the six styles that I have presented in Chaps. 2 and 3. Indeed, as I have explained, they have emerged and still endure. That is to say that the presuppositions of those six styles, i.e. their criteria of evidence, methods, ways of reasoning and questions, are *our* presuppositions.

In his *Representing and Intervening*, Hacking wrote:

> There is no way to match what Paracelsus wanted to say against anything we want to say [...] we cannot assert or deny what is being said. (Hacking, 1983, p. 71)

What Hacking means can be interpreted in the light of my discussion so far: Paracelsus's sentences only 'live' in his community, i.e. only when there exist the presuppositions by which they can be validated. By virtue of a hermeneutic attitude Paracelsus's sentences do become intelligible in our community but, if we want to use them in order to argue, think and do we need to give up our own styles, i.e. our presuppositions: either we become members of Paracelsus's community or we cannot say what we want to say by using Paracelsus's sentences. By the same token, it is not possible, e.g. to defend a style-dependent sentence of the statistical style without relying on its presuppositions, e.g. the evidence of things and, say, the way of selecting a sample or dealing with errors. In other words, if a sentence such as

> The adult height for one sex in an ethnic group follows a normal distribution (Sect. 3.2)

has to be used to express claims of knowledge it cannot be separated from the presuppositions of the statistical style by which it can be validated. This claim implicitly relies on a verificationist thesis: the sentence is not meaningful for a community that has no methods to assess its truth. The methods and criteria of evidence of, say, the taxonomic style cannot validate the sentence above—either we switch from the taxonomic style to the statistical style or we are not able to make use of the sentence in order to know and find out.

6.3.4 Projectibility

This point can be expressed in a different way. One of Kuhn's ideas was that after a change in paradigm scientists 'live' or 'work'—he used both of these verbs—in a different world. In his paper *Working in a New World* (1993) Hacking used the term 'projectible' (introduced by the philosopher Nelson Goodman (1906–1998)) to restate Kuhn's thesis of untranslatability between different paradigms. He made it clear that the scientific terms of an earlier paradigm might well be intelligible to the members of the new paradigm, but the latter could no longer use these terms in a projectible way. By this term he means:

> a class of terms is [projectible when it is] used by a community for making law-like statements, forming general conjectures, picking and making things with expectations about what they will do and how they will work. (Hacking, 1993, p. 295)

For Hacking scientific names refer to scientific kinds within a paradigm and the notion of projectibility expresses their untranslatability into those of another paradigm. By analogy with Hacking's use of the term 'projectible' I say that style-dependent sentences are *not* projectible: they cannot be used to work and think in a different style. Very importantly, while scientific terms are *not* projectible because of an issue of untranslatability between paradigms that use different scientific kinds, style-dependent sentences are not projectible because of a difference in presuppositions between different styles. A style-dependent sentence is not projectible into a different style because it needs the presuppositions of its own style to be meaningful: it would be impossible to form conjectures, think and know.

Although in the case of Kuhn, projectibility is referred to untranslatability, Hacking's use of the term 'projectible' for scientific terms is compatible with my extension to style-dependent sentences:

> To call a term projectible is not to say that generalizations made with it […] can be justified in the community. Projectibility defines the class of possibilities envisioned or capable of being taken seriously by a science at a time. (Hacking, 1993, p. 296)

In other words, projectible terms are terms that can be used for forming statements that can be justified *in a given community*. Therefore, to study the projectibility of terms also includes the study of the possibility of justifying statements in a given community—just the possibility I have been discussing.

To sum up, what Hacking maintains can be couched in these terms: there are a number of presuppositions that must be known by a hearer in order to come to see the intention of the speaker of attributing truth-values to certain style-dependent sentences. The concept of intelligibility concerns the circumstance in which the speaker and the hearer have different presuppositions. The presuppositions of style-dependent sentences in the style of the speaker are recognized only thanks to the hermeneutic attitude of the hearer, who has a different style. The concept of understanding relates to the situation in which the speaker and the hearer share the same presuppositions but the hearer needs to acquire some background knowledge in order to grasp the meaning of the style-dependent sentence. The society we live in adopts all the six styles I have presented so that style-dependent sentences in those styles are in principle *understandable* by us. The term *presuppositions* refers to criteria of evidence, methods, ways of reasoning, questions. In other words, the presuppositions that need to be known in order to attribute truth-values to a style-dependent sentences coincide with some of the features of the styles they belong to. Ultimately, understanding has to do with sentences that have truth-value in a given style; intelligibility has to do with style-dependent sentences that lack truth-value according to the presuppositions of a given community that has a certain style.

Hacking's styles project also implies the impossibility of projecting style-dependent sentences into another style. For example, certain sentences of Paracelsus are not projectible into our styles: there is no way of justifying them according to our standards; there is no possibility of *doing* things as he does because that would be to hold Paracelsus's criteria of evidence. It would mean to drop out of our community. On the other hand, the style-*independent* sentences have truth-value in *any* style. They are always meaningful because they rely on the evidence of senses, which is not a presupposition of only one given style. Suppose that the following declarative sentences were translated from Greek to ancient Egyptian or vice versa:

Two sides of this trapezoid are parallel

Any hearer would recognize the truth-value that the speaker has imposed on these sentences straightforwardly: their being true or false depends on the evidence of senses.

6.3.5 *Taxonomy of Declarative Sentences: A Summary*

In this section I have deduced a classification of declarative sentences in correct grammatical form. The rules by which I have built this taxonomy are entirely derived by Hacking's claims in his styles of reasoning project. *From the point of view of a given style of reasoning* there are sentences that are *understandable* (in Hacking terminology: *meaningful* or sentences that have a truth-value), sentences that are *intelligible* and sentences that are *nonsensical.* Understandable sentences are either style-dependent sentences, i.e. sentences that have a truth-value only within the style of reasoning or style-independent sentences, which have a truth-value that does not hinge on a particular style of reasoning. Intelligible sentences have a truth-value only by virtue of the presuppositions tacitly assumed by a community that has a different style of reasoning. Nonsensical sentences are sentences that do not have truth-value in *any* style of reasoning—they are nonsense. Given a nonsensical sentence, we could well find out that a past style of reasoning has made it objective. By the term intelligible, I mean that, *in principle*, it is possible to come to see the *presuppositions* on which the truth-value of the style-dependent sentences is based on. Methods for finding out, standards of evidence, ways of thinking, questions constitute the presuppositions that need to be known in order to recognize the conditions of truth that have been imposed on the sentence. Presuppositions of a certain style of reasoning are constituted by some of its features. In other words, an appreciation of the fundamental features of the style of reasoning is necessary to obtain an insight into why a given style-dependent sentences is meaningful for a community of people. When this appreciation is gained, the style-dependent sentence becomes intelligible outside its style of reasoning.

Intelligible sentences are not projectible: even if certain style-dependent sentences have become intelligible, one cannot use them in one's own style of reasoning in order to know—they cannot appear as part of a body of sentences that express claims of knowledge, conjectures, hypotheses or law-like statements of another community that has another style of reasoning. Style-dependent sentences receive their sense, their being true or false only in their own style of reasoning by virtue of the presuppositions of that very style of reasoning. To use them in order to know and find out in another style of reasoning would mean to give up being part of one's own community and think, act and justify as the members of another community do. In my examples, this means that Laplace could have made Paracelsus's sentences intelligible by studying his thought but he could not have acted, thought or justified as Paracelsus did, and vice versa. Similarly, an ancient Egyptian could have learned about the postulational methods and the standards of truth of geometry but she could have not thought and acted as Euclid did when he proved the propositions in his *Elements*.

6.4 The 'Relativism Issue'

6.4.1 Trust and Presuppositions

Let us return to the Boyle-Hobbes dispute. When the laboratory style emerged, it raised the question (to which Hobbes answered in the negative) as to whether to consider the gentlemen of Boyle's community as trustworthy informants. A similar question might have been asked, *mutatis mutandis*, when other styles emerged: which informants should one accept? Which presuppositions? This is not an abstract dilemma that vanishes when we come to concrete examples. Suppose that a speaker that belongs to a given community that adopts a certain style utters a style-dependent sentence. A listener that belongs to a different community with a different style can come to see the presuppositions (criteria of evidence, ways of doing, methods etc.) that make that sentence true or false in the speaker's style. The sentence would become intelligible to the listener but she could ask: are those presuppositions right or wrong? At that point, the listener would require independent presuppositions to provide an answer to this question. Do these independent presuppositions exist? Is what is true or false for a certain community true or false for *any* community? And if so, how are we to justify this claim? In this section I shall give a precise identity to the philosophical issue I am discussing. Then I shall argue that Hacking does not provide any clue about how to address it.

It is important to stress once again that, at those points in time in which a given style emerges (according to Hacking styles have clear beginnings), whether or not to accept it can be a live option: for example, according to Hacking, Hobbes did not accept the standards of evidence of the emergent laboratory style of reasoning. Furthermore, new styles may come about in the future and whether or not to accept them will be a live option. For instance, one could argue that mathematical modelling and simulation *on computers* constitute a new style of reasoning. After all, the articulation by computers of existing mathematical models is a historical novelty, a new kind of experimental activity which enjoys a distinct status compared to ordinary experiments (Hacking makes a similar point speaking of 'computer-generated concepts and proofs' in mathematics (Hacking, 2002, p. 186)). The list of sciences that make extensive use of simulations is enormous, from climate science to astrophysics. However, whether and under which circumstances we should accept simulations on computers as a new form of evidence is still under debate.

6.4.2 *What Kind of Relativism?*

At first glance, it seems that the issue I am discussing resembles those incommensurability problems in which it is asked whether there is a 'common measure', that is a universal standard for comparing the merits of competing paradigms or theories (see Chap. 4). This intuition is correct although, as I have explained in Chap. 4, the kind of incommensurability issue posed by the styles project is different in nature from that discussed by Feyerabend and Kuhn. First of all, the styles project does not imply semantic incommensurability but only that there are certain sentences that are meaningful in certain styles of reasoning (e.g. because there are methods to assess their truth) and meaningless in other styles of reasoning. There is no change of meaning of sentences from a style of reasoning to another. The problem is another: although style-dependent sentences are intelligible, i.e. their presuppositions can be identified and understood, they cannot be projected into a different style. Nor it is always possible to switch from a style to another as when, as Kuhn imagines, one switches from the Keplerian system to the Ptolemaic. For Hacking, a Renaissance scientist could well learn how Boyle justified his sentences but in order to use all his sentences she would have to reject the presuppositions of her community and espouse Boyle's. Intelligibility is not enough. The Renaissance scientist would have to think, do and justify her conjectures as Boyle did—and she would not be able to perform these mental and physical actions *while preserving the presuppositions of her community*. Of course it is true that, e.g. today it is possible to switch from the probabilistic style to the taxonomic style: their presuppositions are *our* presuppositions. So, the suspicion of incommensurability of styles of reasoning comes from the fact that there seems to be no common measure, i.e. no universal set of presuppositions, valid for all the theoretical sentences expressed by communities that have different styles of reasoning.

Similarly, as I have suggested in Chap. 4, the styles project does not posit that advocates of different styles have different perceptions, as in the case of Fleck's, Kuhn's and Feyerabend's notions. In reference to earlier scientific conceptions such as Paracelsus's, Hacking said:

> [...] understanding is not enough. We cannot use, project, work in a former body of organized thought while we preserve our own. (Hacking, 1993, p. 298)

So there is a crucial difference between Kuhn and Hacking: for the former a Gestalt switch leads to understanding an ancient system of thought, for the latter even if we understand its presuppositions we cannot project its sentences.

At this stage of my argument, I cannot conclude that Hacking's characterization of styles of reasoning implies epistemic incommensurability. Indeed, although *some* style-dependent sentences of different styles hinge on distinct presuppositions, it is also true that all the styles share a criterion for assessing the truth-or-falsehood of style-*independent* sentences. This circumstance could make the scales tip in favour of the possibility of a universal criterion for assessing style-dependent sentences. Furthermore, Hacking could claim that certain scientific claims can be expressed in terms of style-*independent* sentences, which can be assessed by using the evidence of senses. To get convinced of the fact that we cannot conclude yet that the styles project implies incommensurability, one should look at the role of theory-ladenness in Kuhn's and Feyerabend's epistemologies. Whereas Hacking claims the existence of 'sense-datum statements', which are basic, foundational and out of history, and that do not require any reasoning to be assessed or false, Fleck, Kuhn and Feyerabend believe that all the observational propositions are theory laden. Thanks to this difference the latter thinkers are able to defend incommensurability. To sum up, the styles project leaves open what I have called the incommensurability issue.

Chris Swoyer has classified relativism by using the following schema:

Y is relative to X.

Each form of relativism is provided by: replacing Y by different concepts such as rationality, truth, standards of rationality, justification etc.; replacing X by what leads to the differences in the value of Y, e.g. cultures, paradigms, styles of thinking, languages etc.; and stating what 'relative' amounts to (Swoyer, 2008).

In order to obtain an economical classification, Maria Baghramian has proposed to classify relativism by simply considering that Y can refer to *cognitive*, *moral* and *aesthetic norms* (Baghramian, 2004, p. 6). For my purposes, I just want to point out that within the category of cognitive relativism it is possible to distinguish *alethic relativism*, which claims that the truth of statements is relative to an individual or social group and *epistemic relativism*, e.g. the claim that justification is relative to styles. Epistemic incommensurability between styles implies epistemic relativism for, if there is no universal and independent justification, justifications con only be relative to different styles. In particular, a proposition might be justified in a style and meaningless in another style, which would amount to saying that facts about justification are not absolute. Obviously, epistemic relativism implies epistemic incommensurability so we can conclude that the incommensurability issue is a relativism issue.

Notice that an alethic relativist claim about styles would consist of this double claim:

1) different styles lead to different claims, sometimes incompatible, about the world
2) the justification for each of these claims is relative to the style of thinking in which they are embedded.

The epistemic incommensurability of styles I am discussing has nothing to do with points 1) and 2). Indeed, think of two styles *S1* and *S2*. For Hacking, what might happen is: that there exists a proposition *C* such that *C* is claimed true in *S1* and is neither true nor false in *S2*; and that *C* is justified in *S1*. This is different from saying that *C* can be claimed true in *S1* and false in *S2*, which would entail alethic relativism.

6.4.3 What Hacking Says

My contention in this section will be that Hacking has not offered any conclusive argument that helps to solve the incommensurability issue. Let us examine the passages relevant to this issue. In his first paper on styles he wrote:

> Consider Hamlet's maxim, that nothing is either good or bad but thinking makes it so. If we transfer this to truth and falsehood, it is ambiguous between (a) nothing which is true is true, and nothing which is false is false, but thinking makes it so, and (b) nothing is either true-or-false but thinking makes it so. It is (b) that preoccupies me. My relativist worry is that the sense of a proposition *p*, the way in which it points to truth and falsehood,

> hinges on the style of reasoning appropriate to *p*. (Hacking, 2002 [1992], p. 180)

In this passage Hacking highlights that the key distinction to bear in mind is the difference between truth-and-falsehood and truth: point b) would imply alethic relativism; point a) only states that the truth-or-falsehood of style-dependent sentences is relative to styles. I think that Hacking is correct when he says that the styles project does not imply alethic relativism. Indeed, if the styles project implied that truth has a history then the concept of knowledge would have a history too, in contrast with what I have concluded. Ultimately, Hacking only manages to dissociate himself from *alethic* relativism—point b) is still compatible with epistemic relativism. The incommensurability issue remains open.

In his paper of 1992 Hacking touched on another important point: what is the reason of the disappearance of a style? Arguably if there were a universal and atemporal criterion to assess style-dependent sentences it might allow us to give up a style because it does not get at the truth. Hacking made no mention of such a criterion claiming that, although particular theories might be refuted a style, thanks to its self-authenticating techniques, endures. In his words:

> It is our knowledges that are subject to revolutions, to mutation, and to several kinds of oblivion; it is the content of what we find out, not how we find out, that is refuted. (Hacking, 2002 [1992], p. 192)

In more recent writings Hacking distanced himself from constructionism (Hacking, 1999, 2000). He distinguished *form* from *content* of knowledge, a distinction that is relevant to the styles project. By *form* of a branch of scientific knowledge Hacking means 'sentences that can be true or false, together with techniques for finding out which ones are true and which ones are false' (Hacking, 1999, p. 170). In the terminology of the styles project: style-dependent sentences as well as methods and criteria of evidence, which are used to determine the truth-values of those very style-dependent sentences, constitute the form of a given style. The content of a style is what is found out, e.g. the Maxwell equations.

After having distinguished form from content, Hacking replied to constructionist thinkers:

> the answer to a clear question about some aspects of the world is determined by how the world is. […] when the question is a live one, and there is a context in which there are ways of addressing the question, or even methods of verification for possible answers, then aspects of the world determine what the answer is. (Hacking, 2000, p. S 69)

In the terminology of the styles project 'live questions' are the questions that a community that adopts certain styles asks in a certain historical period: for example, the questions about the evolution of man became relevant within the historico-genetic style. We may say that certain questions are relative to styles: they become lively when a style emerges; they are condemned to oblivion when the style withers away. Hacking's point is that, although questions may be relative to styles, the correct answers to them are not relative to anything:

> the answers to live questions about the natural world have nothing to do with us. (Hacking 2000, S70)

By putting forward these ideas Hacking also wanted to distance himself from those social constructionists such as Shapin and Schaeffer who spoke of the social construction of the actual answers to well-asked questions (Hacking, 2000). However, his anti-relativist turn is all the more evident in writings unrelated to science studies, e.g. in his rereading of the concept of style inspired by the philosopher Bernard Williams. In his review of *Truth and Truthfulness* (Williams, 2002), Hacking insisted that the key premise of the styles project is that truth is external to history: the truths discovered by reasoning in a certain way are independent of how we have found them. In other words, although only if we reason in a particular style can we attribute a truth-value to certain sentences, their being true or false has nothing to do with the fact that we reason in a particular style.

I want to highlight two of Hacking's central claims: 1) there are *correct answers* to live questions 2) these *correct answers* are found when there are ways of addressing the question. Now, think of an answer expressed in terms of style-dependent sentences found by a community that adopts a certain style. The problem is that we might not be able to establish that this answer is correct until we find an atemporal and universal criterion to justify style-dependent sentences. If the ways of addressing the questions and the methods of verification are changeable how could one ever know whether the answers provided by a certain style are to be trusted? Hacking's arguments can only oppose alethic relativism but are a blunt spear against epistemic relativism. My argument in the following sections will support the thesis that the styles project, as conceived by Hacking, does have relativistic implications despite his recent anti-relativistic remarks.

I believe that, if Hacking wished to offer such a justification or, more in general, give an example of a 'correct answer' that can be expressed in terms of style-independent sentences and therefore justified independently of the styles in which is found, he could bring up his arguments in favour of realism about entities. In the next chapter I shall argue that there cannot be such an atemporal and independent justification. In his latest paper on styles Hacking implicitly disagreed with those who draw relativistic implications from his styles project but he postponed his reply to another occasion:

> Relativism, or reductio ad absurdum of the styles project, take your pick. This is a sufficiently focused question to which the project can reply. For reasons of space, it will have to do that elsewhere. (Hacking, 2012, p. 10)

My argument in the following two chapters will support the thesis that the styles project, as conceived by Hacking, does have relativistic implications despite his recent anti-relativistic remarks.

6.4.4 Solutions Proposed to the Relativism Issue

In Chap. 4 I have already explained why my assessment of Hacking as not successfully avoiding epistemic relativism is philosophically important. My main point was that whether or not the styles project invites relativism can be considered a sub-problem of a wider issue hotly debated: is relativism a philosophical consequence of the historicization of epistemology? My hope in the following chapters is to clarify this sub-problem in order to shed light on the general problem itself. However, I should add that to assess the notion of style of reasoning is important for another reason: the relativistic implications of the styles project might be at odds with Hacking's entity realism (i.e. the epistemic claim that we *know* about unobservable entities). Like other historical epistemologists, Hacking is very much aware of the relativism issue posed by his historicization of scientific reason. It is also this worry, and not only a mere attempt to distance himself from social constructionism, that forces him to allay any suspicion of relativism in his more recent writings. More importantly, his

Representing and Intervening (Hacking, 1983) makes it impossible to sidestep more abstract epistemological debates concerning the relativistic implications of the styles project. Indeed, in that book, where the notion of style is explicitly mentioned three times (Hacking, 1983, pp. 56, 71, 127), he makes the *epistemic* claim that the best kind of evidence for the existence of a theoretical entity is that we can manipulate it in the laboratory. By making this claim Hacking raised the doubt that, if his *characterization of styles is correct*, his justification strategy based on experimental realism would be no viable option for a member of a community that does not adopt the laboratory style. Indeed, in that case, one might argue that the existence of a theoretical entity would be justified for an advocate of the laboratory style and unjustified for someone, like Hobbes, who refutes to think and *do* in that style. Aa a matter of fact, there has really been a debate about the existence of void between Hobbes and Boyle: it was a debate, so to speak, external to the laboratory style as to whether to accept the new evidence used by Boyle. The death of a bird in the air pump is a piece of evidence *produced* by a man-made machine in order to test the *single* hypothesis of the existence of void. Hacking stresses that Hobbes did not want to accept this piece of evidence and the new kind of evidence of the laboratory style *tout court*:

> Hobbes saw exactly what Boyle was doing, and hated it. He foresaw that laboratory apparatus for generating phenomena was radically new. He was dead against it. This was not a quarrel about the relative weight of empirical evidence as against deductive proof. The question was more profound and more consequential. What shall be evidence? Is it to be what we find among us, bring home from abroad, chart in the skies – or is to be what we make with laboratory apparatus? (Hacking, 2009, p. 114)

Ultimately, the point I am making is that to ask whether or not the styles project has relativistic implications is crucial for assessing the internal coherence of Hacking's philosophy of science. Indeed, the styles project might imply that the existence of unobservables is justified for a community that adopts the laboratory style and unjustified for a community that does not adopt it, in contrast with the claim that we *know* about unobservable entities. In other words, the relativistic implications of the styles project might be at odds with his entity realism (i.e. the epistemic claim that we *know* about unobservable entities). This doubt concerning the internal coherence of Hacking's philosophy of science makes it unavoidable to ask whether or not epistemic relativism as such is a feature of the styles project.

With regard to the relation between the styles of reasoning project and relativism some scholars have proposed a different thesis from mine, others have presented arguments that, in my view, do not really hit their target. Newton-Smith claimed that the point in (Hacking 2002 [1982] -a) that a proposition can be determined as true by a style of reasoning for which there is no external justification implies relativism (Newton-Smith, 1982). Although Newton-Smith added that whether or not Hacking thinks that there are external justifications is not really clear to him, he suggested that one should describe Hacking's ideas as a form of relativism, which he called 'muted relativism'. However, this point did not prevent Newton-Smith from saying that it would be possible to find a way out of the relativism issue by formulating an independent criterion to justify styles of reasoning in terms of style-independent sentences ('observational statements' whose sense does not hinge on their style of reasoning). However, Newton-Smith did not exhibit this criterion nor did he discuss how it could be found—one can suspect that this criterion does not exist.

Jardine mooted the possibility that one could show the superiority of the methods of a given style of reasoning *M* against those of a past one *M'* (Jardine, 1991). He imagined the case in which *M'* is ancestral to *M* and the local methods and the practice of *M'* have gone through successive stages of modifications in the body of local practices: the outcome of these successive refinements, displacements and replacements of methods is *M*. Under these circumstances, Jardine concluded, it is possible to consider *M* to be more reliable than *M'* because derived from *M'* by a sequence of reliability enhancing steps (Jardine, 1991, p. 167). It will be clear later that Jardine's point cannot be accepted in the light of Hacking's insistence that the evolution of methods and criteria of evidence are discontinuous. A case in point is that of the laboratory style of reasoning. According to Hacking, styles of reasoning may be characterized by methods that have not been calibrated against previous ones. He follows Shapin and Schaeffer in thinking that the air pump is the emblem of a new way of providing evidence: to create effects that did not exist in isolation. There is no possibility of considering this new way of doing as derived from that of another style of reasoning through successive refinements.

Baghramian, as Newton-Smith did, pointed out that Hacking does not propose a form of alethic relativism: what varies from culture to culture is the availability of propositions, not their truth-value (Baghramian, 2004).

For her, Hacking's account is an example of relativism because for him what counts as evidence is internal to a given style of reasoning. However, even without an independent criterion and by using only the norms internal to one's style of reasoning, she suggested, it is possible to become convinced that certain questions are better addressed by the methods of an alternative style of reasoning. I consider this suggestion unjustifiably optimistic, as it will be clear.

In Chap. 4, I have explained that Kusch argued that the styles project implies epistemic relativism. However, I have also pointed out that he did not provide any example of a *scientific* claim that is justified in a style of reasoning and unjustified outside it, rather an example of a pseudo-style of reasoning that, he thinks, is incompatible with one's commitment to science. In the absence of an example of *scientific* claim it is not clear whether, within science, there is incommensurability or not. Otávio Bueno disagreed with Kusch by claiming that Hacking has resources to avoid relativism (Bueno, 2012). For him, statistical analysis could provide the relevant common standard to assess certain propositions of the laboratory style of reasoning. In personal conversation he clarified his point by maintaining that the laboratory and the statistical styles of reasoning have common standards. I disagree with this conclusion and the reason will be clear in Chap. 7.

Recently, Adam Carter and Emma Gordon presented a strategy for what they claim to be Hacking's line of argument for 'a radical denial of the possibility of objective epistemic reasons for belief' (Carter & Gordon, 2014). In fact, they ignored the evolution of Hacking's thinking on styles over the last 30 years and took it for granted that he self-identifies as a relativist. The most serious thing, however, is that for them 'differences in objectual understanding' (Carter and Gordon, 2014 p. 8) suffice to explain the epistemic incommensurability between different styles. By 'objectual understanding' they mean understanding achieved when informational items are pieced together. Essentially, for them the epistemic incommensurability between the style of Paracelsus *P* and another style *S* can be explained by a difference in objectual understanding: an advocate of *S* fails to appreciate how the propositions taken as true in *P* are interrelated and the reasons taken in *S* for believing a certain proposition cannot have force for her. By relying on these points Carter and Gordon argue that the relativistic implications of the notion of style can be blocked.

Their diagnosis of epistemic incommensurability between styles is misconceived as my analysis above shows.[1] Differences in objectual understanding are not the reason of epistemic incommensurability. Hacking has been clear that they can be overcome: he remarked that, although the way Paracelsus proposed and defended propositions is alien to us we can reconstruct Paracelsus's way of thinking and even talk as Paracelsus talked (Hacking, 2002 [1982]-a, p. 170). On the contrary, as I have explained in the previous sections, the problem is that, even if we came to see the presuppositions of Paracelsus's style (its criteria of evidence, methods, questions, ways of thinking and doing, including the way these things are interrelated) they would not be *our* presuppositions. *Understanding would not be enough.* We would not be able to 'project' Paracelsus's style-dependent sentences into our styles of reasoning. We could not think, *act* and justify our conjectures as Paracelsus did—*while preserving the presuppositions of our community.* The epistemic incommensurability issue amounts to the following doubt: are the presuppositions of a style which is alien to us right or wrong? Are there independent presuppositions to provide an answer to this question?

6.5 Conclusions

In the first half of this chapter I have discussed some preliminary questions in order to arrive at a more complete theory of styles of reasoning. I have argued that, while in other cases different styles of reasoning are used together to solve the same scientific problem, in other cases either of two styles of reasoning can be used alternatively. Afterwards, I have addressed the question as to whether or not members of a given community can make sense of a style-dependent sentence uttered by a member of a community that adopts a different style of reasoning. My main point has been that there are intelligible sentences which are not projectible: one cannot use them in one's own style of reasoning in order to know.

In the second part of this chapter, I have argued that from the styles project emerges the question whether or not Hacking's ideas imply epistemic relativism. Despite Hacking's dissent, most of the commentators claimed that the styles of reasoning project does imply a moderate form of relativism. However, a few maintained that within Hacking's project there is room for a universal criterion for comparing different styles of

[1] See also (Sciortino, 2016).

reasoning. Unfortunately, none of them put forward such a criterion. Consequently, the literature lacks a thorough investigation about its existence. Still worse, Hacking has been elusive on the core topics of justification. Of those who claimed that the styles of reasoning project implies epistemic relativism none has provided examples of *scientific* claims that are justified only within a certain style of reasoning. Notably, an in-depth analysis of how scientific claims are justified should have a role in any theory that claims the existence of different ways of finding out. This is for me another good reason to undertake this task. *Inter alia*, in the next chapter I shall show that there cannot exist any universal and atemporal criterion of justification within the framework of Hacking's characterization of styles of reasoning.

Bibliography

Baghramian, M. (2004). *Relativism*. Routledge.

Berggren, J. L., & Sidoli, N. (2007). Aristarchus's On the Sizes and Distances of the Sun and the Moon: Greek and Arabic Texts. *Archive for History of Exact Sciences, 61*(3), 213–254.

Bueno, O. (2012). Styles of Reasoning: A Pluralist View. *Studies in History and Philosophy of Science Part A, 43*(4), 657–665.

Carter, A., & Gordon, E. (2014). A New Maneuver Against the Epistemic Relativist. *Synthese, 191*(8), 1683–1695.

Collingwood, R. (1979 [1939]). *An Autobiography.* : Oxford University Press.

Grice, P. (1968). Utterer's Meaning, Sentence-Meaning, and Word-Meaning. *Foundations of Language, 4*(3), 225–242.

Hacking, I. (1983). *Representing and Intervening*. Cambridge University Press.

Hacking, I. (1993). Working in the New World: the Taxonomic Solution. In P. Horwich (Ed.), *World Changes. Thomas Kuhn and the nature of Science* (pp. 275–310). Bradford Book.

Hacking, I. (1999). *The Social Construction of What?* Harvard University Press.

Hacking, I. (2000). How Inevitable Are the Results of Successful Science? *Philosophy of Science, 67*, S58–S71.

Hacking, I. (2002). *Historical Ontology*. Harvard University Press.

Hacking, I. (2002 [1982] -a). Language, Truth and Reason. In *Historical Ontology* (pp. 159-177). : Harvard University Press.

Hacking, I. (2002 [1992]). 'Style' for Historians and Philosophers. In *Historical Ontology* (pp. 178-199). : Harvard University Press.

Hacking, I. (2009). *Scientific Reason*. NTU Press.

Hacking, I. (2012). 'Language, Truth and Reason' 30 years later. *Studies in History and Philosophy of Science Part A, 43*(4), 599–609.

Heath, T. L. (1908). *The Thirteen Books of Euclid's Elements.* Cambridge Press.
Jardine, N. (1991). *The Scenes of Inquiry: On the Reality of Questions in the Sciences.* Clarendon Press.
Newton-Smith, W. (1982). Relativism and the Possibility of Interpretation. In M. Hollis & L. S (Eds.), *Rationality and Relativism* (pp. 106–122). Blackwell.
Paracelsus. (2007). *The Hermetic and Alchemical Writings of Paracelsus* (A. E. Waite, Trans.): Forgotten Books.
Sciortino, L. (2016). Styles of Reasoning, Human Forms of Life, and Relativism. *International Studies in the Philosophy of Science, 30*(2), 165–184. https://doi.org/10.1080/02698595.2016.1265868
Swoyer, C. (2008). Relativism. In E. N. Zalta (Ed.), *The Stanford Encyclopedia of Philosophy* (Vol. 2008).
Williams, B. (2002). *Truth and Truthfulness. An Essay in Genealogy.* Princeton University Press.

CHAPTER 7

The Incommensurability of Styles of Reasoning: The Case of the Existence of Theoretical Entities

7.1 Introduction

In the previous chapter, I have argued that an important issue underlies the styles of reasoning project. I have called it 'the incommensurability issue': the question as to whether there exists a universal and atemporal justification for the claims made by a community that adopts a particular style of reasoning. In this chapter, I shall present a case study in which a claim made in the laboratory style of reasoning has no universal and atemporal justification. As I shall explain, Hacking *justifies* his belief that unobservable entities exist on the ground that they can be *regularly* manipulated by experimenters in order to find out and produce various phenomena (experimental realism); he also maintains that no belief that our theories are true is required in order to be realist about unobservable entities. I shall argue that his justification is *relative* to the laboratory style of reasoning. In particular, my point will be that, *if Hacking's characterization of styles of reasoning is correct*, his justification strategy based on experimental realism is no viable option for a member of a community that does not adopt the laboratory style of reasoning—the existence of a particle would be justified in the laboratory style of reasoning and unjustified outside it. Ultimately, from this case study I shall draw the important conclusion that the styles of reasoning project implies epistemic relativism.

L. Sciortino, *History of Rationalities*,
https://doi.org/10.1007/978-3-031-24004-1_7

In the next section I shall present Hacking's entity realism and theory anti-realism and put them in relation with the styles of reasoning project. In the third section, I shall first discuss the consequences of theory anti-realism for the possibility of comparing different styles of reasoning. Afterwards I shall deploy my argument that Hacking's entity realism is not justified outside the laboratory style of reasoning. The argument will be divided into two parts: whether or not we grant Hacking that experimentation is nearly theory-free, the justification for the existence of unobservable entities is relative to the laboratory style of reasoning.

7.2 Hacking's Experimental Realism

The greater part of Hacking's ideas about scientific realism is expounded in his book *Representing and Intervening* (Hacking, 1983), published one year later than the first paper on styles of reasoning. This book mentions the notion of style of reasoning three times (Hacking, 1983, pp. 56, 71, 127) but never refers to the notion of 'laboratory style of reasoning' as the latter was introduced in the review of 1991 to *Leviathan and the Air Pump* (Hacking, 1991).

Hacking distinguishes between two types of realism: entity realism and theory realism (Hacking, 1983, p. 37). Generally speaking, *entity realists* deny some or all of the beliefs held by the *theory realists* (such as theories aim at the truth; if a theory is true then the fundamental terms of the theory denote real entities). However, unlike *entity anti-realists*, entity realists do consider fundamental entities as real and as part of the causal structure of the world. Hacking describes himself as an entity realist: an anti-realist about theories and a realist about entities. To start with I shall explain his theory anti-realism, then I shall focus on entity realism.

7.2.1 Entity Realism

In the paper 'Experimentation and Scientific Realism' Hacking acknowledged that '[his] own approach owes an enormous amount to Nancy Cartwright's parallel developments, which have often preceded [his] own' (Hacking, 1984, p. note n. 3 p. 171). For her part, the philosopher Nancy Cartwright wrote that her book *How the Laws of Physics Lie* (1983) is a complement to *Representing and Intervening* (Cartwright, 1983, p. 20). Hacking espoused Cartwright's distinction between phenomenological laws and fundamental laws, the former being laws that merely describe what happens in specific situations. In Cartwright's example, Airy's law of Faraday's magneto-optical effect is a characteristic phenomenological law: the physicist George Airy (1801–1892) added some *ad hoc* further terms (first or third derivatives) to the equations of light (which are to be interpreted as fundamental laws) in order to represent analytically the rotation of the plane of polarization of a beam of light sent parallel to the lines of a magnetic field inside a piece of material. What Airy obtained is, in Cartwright's terminology, a phenomenological law. The wave equations of the theory of light are fundamental laws that Airy modified in order to *describe* the *particular* situation under study: the passage of light through a definite piece of material. Ultimately, Cartwright argued, as compared with fundamental laws, phenomenological laws do not describe idealized situations. They are faithful representations of certain aspects of the phenomena as they appear, considering all the circumstances in which they occur.

Building on these ideas, Cartwright argued that, although the fundamental laws of physics provide us with great explanatory power, they do not describe reality but rather idealized situations; as the title of her book states, the laws of physics lie, i.e. they do not get the facts right. On the other hand, when they are amended to be true by transforming them into phenomenological laws, they lose their fundamental explanatory force. Indeed, for her fundamental laws are actually *ceteris paribus* laws—for example the gravitational law is to be expressed: '*If* there are no forces other than gravitational ones at work, *then* two bodies exert a force between each other which varies inversely as the square of the distance between them, and directly as the product of their masses' (Cartwright, 1983, p. 58). So, for Cartwright it is wrong to say that for *any* two bodies the force between them is given by the law of gravitation. If the two bodies are charged, they will not behave just as the gravitational law states. '[The gravitational law] is irrelevant to the more complex and interesting situations' (Cartwright, 1983, p. 39). To come out true the gravitation law must be reinterpreted either as *ceteris paribus* (that is, by including the antecedent 'if there are no forces other ...') or as describing a component force (that is, as one of the forces that act on the body in addition to, say, frictions, Coulomb's law and so on). But for her to interpret a fundamental law *ceteris paribus* means to loose descriptive adequacy since *ceteris paribus* conditions never hold in reality but only in ideal situations. On the other hand, she considers the addition of forces as an operation *we* perform when we do calculations; nature does not add forces.

Hacking also espouses Cartwright's model of explanation. According to the *covering law model*, proposed by the American epistemologist Karl Hempel (1905–1997), to explain means to provide a law of nature that 'covers' the phenomenon, namely to show that the phenomenon had to be so given the circumstances in which it occurred and the existence of certain laws that always act in a certain way. Cartwright sets up against this model what she calls the 'simulacrum account': to explain means to construct models, within a theory (to which they may not be consistent), which are formal representations of the phenomenological laws. Indeed, she sees models as simulacra of things, i.e. mental pictures or even hold-in-your-hand models that mimic features of the world. As Cartwright says '[the simulacrum] account says that we lay out a model, and within the model we "derive" various laws which match more or less well with bits of the phenomenological behaviour' (Cartwright, 1983, p. 161). Once a model is deployed, phenomenological laws describe those aspects of the phenomenon that are syphoned off by the model.

Hacking stresses that, as there are many models, often mutually inconsistent and embedded in the same theory, there can be no single explanation:

> The covering-law account supposes that there is, in principle, one 'right' explanation for each phenomenon. The simulacrum account denies this. (1983, p. 17)

For the same reasons given by Cartwright, Hacking downplays the role of explanation:

> [...] explanation may play a less central role in scientific reasoning than some philosophers imagine. [...] *Explanations are relative to human interests.* [...] Explaining is a feature of the historical or psychological circumstances of a moment. (Hacking, 1983, p. 53 my emphasis)

The hypothetical construction of analogical models is, together with measuring and exploring relations between phenomena, part of the laboratory thinking and doing. According to Hacking's ideas above we should conclude that thinking in the laboratory style of reasoning by propounding models of phenomenological laws within a given theory does not provide any valid argument for scientific realism. There is no auto-corrective mechanism that simplifies models and makes them converge on better and better descriptions of reality: we are condemned to the local truths of phenomenological laws.

7.2.2 *Experiments-with, Experiments-on*

There is an important point made by Cartwright that helps us understand how Hacking defends entity realism. She considers the following inference: 'if *x* explains *y* and *y* is true then *x* is true'. For her it would be invalid if *x* were a fundamental law but it would be valid in the case of causal explanation:

> Suppose we describe the concrete causal process by which a phenomenon is brought about. That kind of explanation succeeds only if the process described actually occurs. To the extent that we find the causal explanation acceptable, we must believe in the causes described. (Cartwright, 1983, p. 4)

To illustrate Cartwright's claim let us consider the example of a Geiger counter: a cylindrical capacitor with a processing electronics that is filled with gas. When a α-particle passes through it, it ionizes some of the atoms of the gas and produces a current pulse that can be measured. In some Geiger counters the pulse produces audible clicks associated with the number of α-particles. To provide a *causal* account of why a Geiger counter produces a click, a physicist would implicitly resort to a certain number of fundamental physical laws and eventually would describe the principle of operation in terms of phenomenological laws such as: 'when the gas is so-and-so and the ions travel with such-and-such speed then they create an avalanche described by this equation'. In Cartwright and Hacking's view, Maxwell's equations lie but, to the extent that we find credible the explanation that the *cause* of the clicks is a flux of particles, we must believe that, as a matter of fact, a flux of particles enters the Geiger counter. It is in this sense that Hacking's anecdote about what convinced him of the existence of electrons ('if you can spray them on a niobium ball they are real'; Hacking, 1983, p. 22) must be understood. Like in the case of the Geiger counter, if the existence of electrons is the only plausible cause of the effect of spraying, it must be true that they are real. Ultimately, causal reasoning provides good grounds for concluding that unobservable entities exist.

In order to be confident about the existence of a cause of a given effect—stresses Hacking—causal reasoning must be integrated with experimenting. He distinguishes two types of experiments: experiments-with and experiments-on (Zeidler & Sobczynska, 1995/1996; Hacking, 1995/1996). For example, when in 1908 the physicist Robert Millikan (1868–1953) determined the charge of the electron by experimenting *on* electrically charged oil droplets he was performing experiments-on. Conversely, electron diffraction, which makes it possible to infer the structure of a crystal by firing at it electrons and observing the interference pattern, is a kind of experiments-with. Experiments-on are experiments that aim to measure the properties of the putative entity. Experiments-with are manipulations of an alleged real entity in order to find out something else or create a new phenomenon such as the electron diffraction. Hacking directs the reader's attention to experiments-with and to the fact that a one-off experiment would not dispel our doubts about the most likely cause of a given effect. He maintains that when we *regularly manipulate* a putative entity in order to investigate other aspects of nature, that is when we routinely perform experiments-with, we are *justified* in thinking that the entity exists:

> Experimental work provides the strongest evidence for scientific realism. This is not because we test hypotheses about entities. It is because entities that in principle cannot be 'observed' are regularly manipulated to produce new phenomena and to investigate other aspects of nature. (Hacking, 1983, p. 262)

I wish to stress two points: the first is that Hacking is claiming that the experimental realist *knows* that theoretical entities exist:

> My realism about entities implies both that a satisfactory theoretical entity would be one that existed (and was not merely a handy intellectual tool). That is a claim about entities and reality. It also implies that we actually *know*, or have good reason to believe in, at least some such entities in present science. That is a claim about *knowledge*. (Hacking, 1983, p. 28 my emphasis)

The second point is that, since experimenting-with involves the 'creation' of new phenomena, it must be considered part of the laboratory way of doing. So we can say that it is by doing in the *laboratory style of reasoning*, and in particular by using causal reasoning and experimenting in order to elicit new phenomena, that experimenters find the most likely cause of a certain effect.

7.2.3 *Causal Effects as Style-Independent Sentences*

Consistently with his theory anti-realism, for Hacking experimentation is free from theory as far as it has the power to justify our beliefs in the existence of theoretical entities. He argues that an important part of experimentation is theory-free by making different points, which are relevant to the styles of reasoning project, as I shall show here and in the next section. First of all, he points out that there is no theory of a theoretical entity to which all the experimenters are realistically committed (Hacking, 1983, p. 264). Within the same experiment *different* models coexist and each of them enables us to make calculations relevant to *only* one particular aspect of the particle. To the question 'is there not a common *core* of theory, the intersection of all the different theories?' Hacking answers: '[T]here is a common lore, not a common core' (Hacking, 1983, p. 264). Furthermore,

> [w]e design apparatus relying on a modest number of home truths about electrons, in order to produce some other phenomenon that we wish to investigate. (Hacking, 1983, p. 265)

The use of colloquial terms such as 'common lore' or 'home truth' is meant to downplay the theoretical analysis of experimentation: one of the aims of *Representing and Intervening* was to urge philosophers, too much concerned with theories, to focus on practical aspects of experiments. Hacking's point is that, although groups of experimenters 'may hold different and mutually incompatible accounts of electrons' (Hacking, 1983, p. 264), they share a way of doing and a tacit knowledge of the behaviour of apparatuses, of various effects given certain conditions and of '[causal properties] derivative on properties like spin, mass, and charge [which] are the best candidates for home truths' (Morrison, 1990, p. 20). Sentences such as 'a beam of electrons is halted by an aluminium plate' or 'electrons with features such-and-such spiralling in a magnetic field emit radiation so-and-so' are examples of the shared language of experimenters not too much infected by theory, so to speak.

More importantly, I wish to remind the reader that for Hacking 'there have been important observations in the history of science which have included no theoretical assumptions at all' (Hacking, 1983, p. 176). To illustrate this point he gives the example of Herschel's discovery of the infrared radiation of sunlight (Herschel, 1800b) (Herschel, 1800a, 1800c). As I have explained, Hacking is adamant that the corpuscular theory, in which the physicist Herschel believed, did not infect several statements of Herschel's research on radiant heat. For example Herschel wrote:

> When I used some of them [the filters] I felt a sensation of heat, though I had but little light, while others gave me much light with scarce any sensation of heat. (Herschel, 1800b)

Commenting on this sentence, by which Herschel described the sensation of heat caused by light rays coming through different filters, Hacking wrote that '[there is no] better sense-datum report than this in the whole of physical science' (Hacking, 1983, p. 176).

Herschel's sentence is to be considered style-*independent*. Indeed, as I have explained in Chap. 3, Hacking contrasts it with another Herschel's sentence, which for him is style-*dependent*: 'The heat which has the refrangibility of the red rays is occasioned by the light of those rays'. Ultimately, the distinction between these two classes of sentences (style-independent and style-dependent sentences) mirrors the distinction theory-observation put forward at the level of his argument for realism: a sentence infected by theory (e.g. 'the heat has the refrangibility ...') is supposed to be style-dependent on the laboratory style of reasoning, whereas a theory-free sentence 'my skin is warmed' is supposed to be style-independent.

Notice now that the sentence 'my skin is warmed' can be also viewed as a sentence that describes a *causal effect* of the light rays. Therefore, in the case of Herschel's study of radiant heat, a causal effect is described by a style-independent sentence. As I am going to argue, it is possible to say that this is a general fact: for Hacking, the causal effects of entities are described by style-independent (observational) sentences. For example, one can imagine that during the most important experiments that mark the history of the electron these style-*independent* sentences, which describe *causal effects*, have been pronounced:

(a) 'The glass tube glows' (Thomson's experiment)
(b) 'The oil drops are in mechanical equilibrium' (Millikan's experiment)
(c) 'There is a track in the photographic plate (Wilson's experiment)

Experimenters of different groups communicate by using sentences that can be more complex than these above and that can describe, in addiction to effects, causes and properties of causes. For example, one can imagine these alterations of the sentences (a), (b) and (c):

'Electrons emitted by the cathode make the glass glow.'

'The value of the charge on the droplets in mechanical equilibrium is always a multiple of the charge of an electron.'

'That is the track of a positron in a photographic plate.'

These sentences refer to causes (entities) and properties of causes (charge) in addiction to effects. For example, the first sentence states that the cause of the glowing is an entity called electron; the second that the droplets are dragged upward by the field because they are charged with particles called electrons; the third that the track is the effect produced by an entity that have opposite charge to the electrons.

Hacking is to be interpreted as maintaining that the sentences (a), (b) and (c), which describe causal effects, are style-independent (observational). Indeed, speaking of the sentence 'That is the track of a positron in a photographic plate', Hacking writes:

> There is a tendency to infer from stories like that of the positron that anyone who reports, on looking at a photographic plate, 'that's a positron' is thereby implying or asserting a lot of theory. I do not think is so. (Hacking, 1983, p. 179)

Hacking is implicitly expressing his disagreement with those philosophers, such as N.R. Hanson, Kuhn and Feyerabend, who put forward theses of theory-ladenness by claiming, for example, that observational reports assert theoretical presuppositions. The passage above clearly supports my interpretation: for Hacking sentences which describe causal effects are to be considered observation sentences and, therefore, style-independent. After all, to characterize the causal interactions of unobservable entities with other parts of nature in terms of observation sentences is crucial for Hacking's argument for realism. Indeed, only if causal effects can be characterized without theory can someone be anti-realist about theories *and* believe in the existence of an entity only on the basis of its causal effects.

Hacking's point that sentences such as 'that is the track of a positron in a photographic plate' are observation sentences contrasts with what many scholars think. For example, this is what the historian of science Gerald Holton writes:

> [O]ur naked eyes would see only an unconvincing curlicue; but the mind's eye sees, through the use of a Feynman diagram version of the same phenomenon, that a neutrino scatters an electron without any change of charge. (Holton, 1996, p. 92)

Holton's point is in fundamental conflict with Hacking's. Indeed, Holton is claiming that Feynman diagrams are necessary to 'see' a scattering between particles instead of a squiggle. Feynman diagrams are *models* that both help to visualize the relevant aspects of particle interactions (in which strong, weak and electromagnetic forces are involved) (see Wüthrich, 2012) and help to calculate their more interesting properties. They were introduced in the late 1940s by the physicist Richard Feynman (1918–1988) for simplifying complex calculations in quantum electrodynamics (QED), the theory that explains the electromagnetic force at the quantum level. In other words, Holton is claiming that certain models are the presuppositions held by investigators who deal with the causal effects of electrons, i.e. they infect those elements of experimentation that Hacking considers free of theory.

To summarize the content of the last two subsections, Hacking claims that we *know* that unobservable entities exist. Regular manipulation is part of the way of doing of the laboratory style of reasoning and for Hacking it is the strongest evidence for the existence of an unobservable entity. What the expression regular manipulation of an entity wants to convey is the idea that we perform experiments *with* that putative entity in order to produce macroscopic causal effects. These effects are described by style-*independent* sentences, which may be used for forming more complex sentences, not so much infected by theory, which include causes and properties of causes. Consequently, it is possible to say that an important part of experimentation, which *in primis* includes causal effects, is theory-free. By making this point Hacking is able to defend his realism about entities while being anti-realist about theories. Finally, since he is anti-realist about theories regular manipulation is also *the only one* justification he gives for realism about theoretical entities.

7.2.4 *Putnam's Theory of Reference and Experimental Realism*

In *Representing and Intervening* Hacking asks:

> [...] is not the substance of the theory about positrons among the truth conditions or truth presuppositions for the type of utterance that we may represent by 'that's a positron'? (Hacking, 1983, p. 179)

His answer confirms what I have explained in the previous section, i.e. that 'that's a positron' is an observation sentence:

> Possibly, but I doubt it. The theory might be abandoned or superseded by a totally different theory about positrons, leaving intact what had, by then, become the class of observation sentences represented by 'that's a positron'. (Hacking, 1983, p. 179)

However, the way Hacking replies also points to another problem: how can we be sure that scientists holding competing or successive theories are referring to the same entity, the positron? This difficulty is closely related to the notorious issue of incommensurability between different theories debated since the 1960s: as I have explained, thinkers such as Kuhn and Feyerabend maintained that certain terms take on different meanings if used in two incommensurable theories. In order to overcome this difficulty Hacking wedded himself to Putnam's version of causal theory of reference[1] because, as I am going to explain, it suggests that scientists having competing theories about the positron 'may still be talking about the same thing' (Hacking, 1984, p. 157). In 1970s the philosophers Hilary Putnam (1926–2016) and Saul Kripke (1940–2022) proposed causal accounts of how terms acquire referents. They attacked the descriptivist theory of meaning according to which meanings of names coincide with the descriptions associated with them, whereas their referents consist of the objects that satisfy these descriptions.

As the descriptions of a name change (e.g. 'water is a colourless liquid' or 'water is that substance made of atoms of hydrogen and oxygen') the descriptivist view seems to imply that its reference changes as well (as if, by the two descriptions of water, we did not refer to the same thing).

Since Hacking claims that a positron is something in the world, regardless of the theories or descriptions we have of it, the descriptivist view does not provide the support he needs.

According to Putnam the meaning of a word has four components: the word's syntactic marker, the semantic marker, the stereotype and the referent. The *syntactic marker* states the grammatical role played by the word, e.g. subject or predicate; the *semantic marker* points to the category of things the word applies, e.g. the word 'horse' applies to mammals and quadruped; the *stereotype* is the conventional idea people have about a

[1] Putnam's theory has been criticized by many scholars (see for example Dupré, 1996, pp. 17–36). Hacking is aware of these criticisms. However, he has made it clear that, although he does not believe Putnam's theory literally, he is happy to employ his account (Hacking, 1984). He has extensively written on Putnam's theory of reference in Hacking (1983, 2007).

certain thing, that is the description commonly accepted, e.g. ‘zebra’ is thought of as a striped animal; the *referent* of a word is the thing or the class of things it refers to, e.g. the reference of ‘water’ is a certain kind of stuff individuated by the chemical formula H_2O. Only when the reference of a term has varied is it legitimate to state that the meaning has changed. For example, the concept ‘natural kind *all* of whose members live under water, breath through gills, etc.’ does not fit the natural kind ‘fish’ because some fish do not breath through gills. However, according to Putnam it would be wrong to conclude that the concept ‘does not *correspond* to the natural kind Fish’ (Putnam, 1975, p. 196). If anything, the concept of fish changes as our knowledge of the organisms develops. Yet Putnam maintains that it still corresponds to the same natural kind.

To argue for the invariance of reference Putnam put forward the concept of *introducing event*: everyone who uses a term is connected by a certain kind of causal chain to a situation in which a description of that term is given, e.g. an event that singles out ‘positron’ as the entity with certain features and responsible for certain effects. For example someone could tell me that the positron is that object that makes a Geiger counter produce a click, leaves a track in a gas chamber and has the same mass of an electron but different charge; hence, each of my later uses will be connected with this event. Furthermore, even the use of any other person who has been told about the term ‘positron’ by me will be causally linked to its being in my vocabulary and, therefore, to my original introducing event (Putnam, 1975, p. 200).

Kripke argued that when the name of a referent is fixed by an act of naming it becomes a ‘rigid designator’. By this expression he meant that, for example, the name George Washington refers to the same person in all the possible worlds, whereas the expression ‘the man who was the first president of the United Stated’ could in principle refer to another person. If two rigid designators are found to designate the same object in the actual world it must be true—argued Kripke—that they designate the same object in all possible worlds. In other terms, if it is true that two rigid designators are found to refer to the same object then it must be necessarily true that they designate the same object. He famously gave the following example: suppose that Hesperus is the name that is given to a star seen in the evening and Phosphorus is the name given to the same star seen in the morning. Suppose also that astronomers find out that Hesperus and Phosphorus designate the same star, Venus. Kripke concluded that it is necessarily true, i.e. true in all possible worlds that Hesperus=Phosphorus (Kripke, 1972, pp. 213–215).

If we return to the initial question of this section: if a theory about the positron is abandoned or superseded by a totally different one, are we always referring to the same object, the positron? The causal theory of reference suggests that the answer is in the positive. Indeed 'positron' is a rigid designator: everyone who has the minimal linguistic information about 'positron' and uses the word in a way that is causally connected to an introducing event will refer to the positron, even if she advances new theories or comes to possess additional information about the properties of the positron. As in the case of the star Venus, if two groups of researchers refer to same object, the positron, at a certain point in time they will always refer to it in all possible worlds, for example in cases in which one of both of them will advance radically new theories about positrons based on new findings. There is no possible situation in which 'positron' designates something else. In conclusion, we now have a deeper sense of why Hacking considers Putnam's casual theory of reference as the logical skeleton for his experimental realism view. He needs a notion of reference that is not defined in terms of our theories and conceptions.

7.3 The Incommensurability of Styles of Reasoning

Now that I have offered an account of the position Hacking attempts to defend, I return to my original question: whether there is a criterion independent of any style of reasoning by which to judge the claims of knowledge made by a community that has a particular style of reasoning. Until it is found an example of a claim in a particular style of reasoning whose justification is relative to that very style of reasoning, it is not possible to conclude that the styles of reasoning project implies epistemic relativism. I think that the experimental argument is a good candidate for representing such an example: Hacking maintains that we ***know*** about the existence of unobservable entities, in particular we are justified in believing so; I shall argue that his justification is ***not*** atemporal and universal but relative to the laboratory style of reasoning. Differently stated, my point will be that Hacking's experimental argument might ***justify*** realism about theoretical entities ***in the laboratory style of reasoning*** (if one agrees with Hacking); however, outside the laboratory style of reasoning realism about theoretical entities is ***unjustified***. Before focusing on Hacking's entity realism and arguing for this thesis I want to point out that Hacking's theory anti-realism makes it impossible to rank different styles of reasoning in terms of their capacity to explain why certain phenomena happen.

7.3.1 Incommensurability and Theory Anti-realism

Consider the example I have given in Sect. 6.2: the distance of the moon from the Earth can be calculated by using the geometrical style of reasoning or the laboratory style of reasoning. As far as the *calculation* of the moon distance is concerned, there is no much difference between the postulational style of reasoning adopted by Greeks and our laboratory style of reasoning—one has no strong reasons to conclude that a style of reasoning is 'better' than another. However, if we take a *realist* attitude, unlike Hacking, and interpret the gravitation law as a law that is approximately true then, as far as the study of the moon is concerned, the laboratory style of reasoning can be considered 'superior' to the geometrical style of reasoning. Indeed, in that case not only does the former provide the distance to the moon but also explains why the value of the distance is what it is; on the contrary, the geometrical style of reasoning provides no explanation of why the moon rotates at a certain distance with its specific period.

In contrast to theory realists Hacking's view towards the gravitational law is instrumental: the law 'lies' but represents a useful devise that helps to calculate the distance to the moon. It does not provide *the* explanation of the phenomenon. So Hacking would conclude that the laboratory style of reasoning does not get any closer to the truth than the postulational style of reasoning—for him, as far as the calculation of the distance to the moon is concerned, there can be no independent criterion for ranking these two styles of reasoning.

The crux of the matter is that Hacking espouses Cartwright's view that 'rendered as descriptions of facts, fundamental laws are false; amended to be true, they lose their fundamental explanatory force' (1983, 54). Consequently, if Cartwright is right, we are left with no answer as to why the distance to the moon is what it is: the gravitational law has to be amended in order to be used for the calculation of the distance to the moon; therefore, it does not explain. Alan Chalmers raised a similar point when he remarked that if we deny an explanatory role to the fundamental laws we are at loss to say what governs the world outside experimental situations. His example concerns the orbit of Halley's comet:

> The first sighting of the comet, by a Voyager spacecraft, enabled the predicted orbit to be *corrected*, thereby facilitating later terrestrial sightings. The regularity predicted on the basis of Newton's laws lied. Nevertheless,

> Halley's comet did return very nearly as scheduled. To echo Cartwright, if Halley's comet is not governed by Newton's laws, then I do not know why it returned as and when it did. (Chalmers, 1999, pp. 8–9)

To conclude, Hacking's theory anti-realism leads to a view in which it is impossible to rank different styles of reasoning in terms of their capacity to explain phenomena. On the other hand, Hacking's experimental realism represents an interesting example of a claim that *might* be universally and atemporally justified. I am going to show that this idea is fallacious.

7.3.2 The Incommensurability of Styles of Reasoning: First Argument

As I have done in Sect. 6.3, for the argument's sake I shall now imagine a situation in which a 'speaker' tries to communicate with a 'hearer' who has a different style of reasoning. Speaker and hearer are to be imagined as belonging to different communities, which may be separated in time. Suppose that the speaker, who adopts the laboratory style of reasoning, expressed the sentence: '[T]he muon is real'. Can a hearer who belongs to a community that does not adopt the laboratory style of reasoning *understand* that sentence? One problem is how the hearer could attribute a meaning to the term 'muon'. Putnam's causal account of reference can be applied to this problem. Although the hearer would have no knowledge of the working of experimental apparatuses, she could well be able to perceive the sound of a Geiger counter or see the effects of the track in a photographic plate; she could also acquire some information about other properties of the muon. On Putnam's view, this can be defined as an introducing event, i.e. an event in which an adherent of a different style of reasoning would grasp that the term 'muon' designates an entity responsible for certain effects and, for example, capable of motion and negatively charged. From that moment on, both the speaker and the hearer would be connected by a causal chain to the act of naming the muon. In sum, Putnam's account fulfils our desideratum—it makes the reference of the term 'muon' independent of the style of reasoning in which it is used.

Suppose now that the speaker justified the existence of the muon as Hacking does: she could say that the muon exists because it can be regularly manipulated in order to produce causal effects. Hacking assures that she would also be able to describe the causal effects in terms of *style-independent sentences*: for example, 'that's the track of a muon in a photographic plate'. Ultimately, if the speaker made the claim 'the muon is real' and justified it by using Hacking's experimental argument, the relevant sentences that describe the causal effects would have a truth-value in a style of reasoning different from the laboratory style of reasoning. Then, is the hearer justified to believe in the claim that the muon is real? My answer is in the negative. Indeed, the meaning of the term 'muon' may not depend on the presuppositions of the laboratory style of reasoning, Putnam account implies, but the criteria for the existence of a muon do depend on the presuppositions of the laboratory style of reasoning even if the causal effects can be described in terms of style-independent sentences. I shall explain. What the speaker proposes is to assess the existence of the muon by using devices that create new phenomena and involve the way of doing of the laboratory style of reasoning. Whether or not the speaker uses style-independent sentences to illustrate his justification, the hearer finds herself in rather an awkward situation that may well be likened to that described by Shapin and Schaffer. Indeed, a hearer coeval with Hobbes could object to the speaker that God has given us enough phenomena and that we should not get involved in dubious ones. The speaker might well describe the causal effects by using style-independent propositions but this would not make her argument plausible for the hearer—the presuppositions of the laboratory style of reasoning would be implicit in her argument. Indeed, when illustrated in detail, the speaker's point would amount to say that, by adopting the *way of doing* of the laboratory style of reasoning and its *methods for finding out*, it is possible to perform experiments *with* the muon in order to *create new phenomena*; these new phenomena, which are represented by macroscopic effects described in terms of observational sentences, are *evidence* of the existence of the muon. However, the speaker's way of doing and her methods for finding out would be no presuppositions of the hearer's style of reasoning, nor could creating new phenomena represent evidence of the existence of an entity. I have already quoted Hacking as saying that 'the substance of the theory about positrons [is not] among the truth conditions or truth presuppositions for the type of utterance that we may represent by "that's a positron"'. This is beside the point. If anything, the crux of the matter is that, echoing Hacking, the 'substance *of the laboratory style of reasoning*' (i.e. its presuppositions) is among the truth presuppositions for the utterance 'that's a muon'. It is true that 'the object called "Geiger counter" made a click' is a style-independent sentence but it is the very practice of using the Geiger counter for producing new phenomena that does not belong to the presuppositions of styles of reasoning different from the laboratory style of reasoning. To use a Geiger counter in order to elicit a phenomenon does not count as evidence of anything for a member of a community that does not have the laboratory style of reasoning. Sentences such as 'the object called "Geiger counter" has made a click' are not really projectable in a different style of reasoning: they cannot be used as they are used in the laboratory style of reasoning. In the laboratory style of reasoning these sentences can be used to illustrate an argument for the existence of the muon; but that argument implicitly assumes a criterion of evidence that is not part of the presuppositions of different styles of reasoning. It is as if the sentences that describe the causal effects of the muon can be admitted into the hearer's world but nothing significant can be claimed about them. They are projectable in the ordinary language of the hearer for describing the consequences of certain actions performed by others but cannot be employed to present the argument of experimental realism: Hacking's argument is relative to the laboratory style of reasoning. In sum, if the styles of reasoning are characterized as Hacking does, the knowledge of unobservable entities in the laboratory style of reasoning has no foundations apart from the very presuppositions of the very laboratory style of reasoning.

7.3.3 Supporting the First Argument: An Historical Example

To substantiate my argument I shall provide a historical example concerning Galileo's discovery of Jupiter's satellites. The example will show that, although the 'speaker' (Galileo) used style-independent sentences to deploy his argument, the 'hearers' (Aristotelian scholars) rejected it on the basis that the speaker's kind of evidence and way of doing were not acceptable.

The last part of Galileo's *Sidereus Nuncius* (Galilei, 2008 [1610], pp. 299–319) reports that on 7 January 1610 Galileo 'realised that close to Jupiter there were three stars, little in truth, but very bright [...] they were lying in a straight line and parallel to the ecliptic' (Galilei, 2008 [1610], p. 299). He went on observing in the successive nights until early March 1610 and detailed the relative positions by the simple insertion of linear drawings embedded within his published text. These numerous observations of Jupiter did not conform to the principles of Aristotelian cosmology but supported the Copernican sun-centred hypothesis of the world system in which Galileo believed. However, although Galileo did have Copernicanism in mind when he discovered the four moons, he did not mention this theory at all in his treatise, since it was forbidden; he just wanted to offer to the reader a visual experience.

Three points can be noted. Firstly, even if it had turned out that Copernicus' model was wrong, this fact would not have called in question Galileo's detailed observations. From Hacking's perspective this fact could be expressed by saying that, even if theories change, different scientists can still refer to the same things, represented by the drawings in the *Sidereus Nuncius.* Secondly, the relative positions of Jupiter and its moons are described in terms of what Hacking would call observation sentences: in many cases they could be viewed as captions for his drawings. From Hacking's perspective this second point suggests that the claim for the existence of an *observable* entity can be substantiated by pure observation sentences. Finally, as Mario Biagioli stressed,

> To Galileo [...] the evidence that counted was not a snapshot of individual luminous dots around Jupiter, but the 'movie' of their motions. It was a long chronological perspective that linked his string of observations and turned the luminous bodies near Jupiter into satellites, not fixed stars. (Biagioli, 2006, p. 103)

In other words, Galileo's observations presupposed *a commitment to conducting observations over several days* so as to differentiate the periodic motions of the moons from other visual patterns. In this sense his evidence was inherently historical. Anyone who had looked through a telescope *for only a few minutes* might have legitimately believed that the bright points were artefacts of the telescope, rather than satellites. But anyone who had observed the bright points for many days would have concluded that they followed over time the exact pattern of rotating satellites and that, therefore, they were not artefacts such as double images and colour fringes (which were present in early telescopes). This third point implies that one needs *to commit oneself to a way of doing*, as it happens in my imaginary discussion between a speaker who adopts the laboratory style of reasoning and a hearer who adopts a different style of reasoning: a one-off observation is not sufficient to realize and find out that four bright dots are satellites rather than artefacts; one needs a great incentive to take time to learn how to see through a telescope and to observe for a long period. Above all, one needs to accept that the telescope, an instrument rarely used before to observe the sky, represents a plausible way of finding out.

Ultimately, these three points above suggest that, the differences notwithstanding, there are striking analogies between my imaginary example in the previous subsection and the concrete case of the discovery of Jupiter's moons. What I want to show now is that just as the imaginary hearer could not accept the speaker's experimental argument on the ground that it was based on presuppositions that were alien to her so Galileo's opponents could not accept the claim that Jupiter's moons exist on the ground that they did not accept his presuppositions.

Few months after the publication of the *Sidereus Nuncius* Galileo moved to Florence where his discoveries were met with disbelief by the court scholars. One of them, the Aristotelian philosopher Cesare Cremonini (1550–1631), did not want to look through the telescope. At that time it was thought that the physics of sky was different from the terrestrial physics, in particular that the laws of light interaction with terrestrial objects could not be extended to the sky (see Feyerabend, 1988 [1975], p. Chapter 10). Consequently, Cremonini was sceptical about the idea that the telescope could provide evidence for the existence of celestial objects. He had little incentive to invest time in Galileo's observations. As Biagioli explains, '[A]n observation would have been a very unwise investment of time and resources for someone [i.e. Cremonini] of his disciplinary affiliation and professional identity' (Biagioli, 2006, p. 113). In other words, Galileo's methods of finding out, the evidence he provided and his way of doing were so different from Cremonini's that the latter would have lost his professional identity if he had invested time in *doing* what Galileo had done.

Unlike Cremonini other scholars did look through the telescope, but *only for a short time*. The brevity of the observation was fatal for them: for example, the mathematician Martin Horký saw minute spots close to Jupiter on 24 and 25 April in exactly the same configuration described by Galileo but he concluded that they were an optical illusion. That is not surprising. One needs to have Galileo's style of doing for committing oneself to refining telescopes and observing bright spots for months.

In *Life of Galileo* (Brecht, 2007 [1938]), the playwright Bertolt Brecht (1898–1956) presents a scene in which Galileo invites a philosopher and a mathematician, who act the part of the sceptical scholars, to observe Jupiter's satellites. The dialogue below is emblematic of the epistemic incommensurability between Galileo's way of thinking and doing and that of the two scholars. Although Galileo's arguments were not infected by theories, Hacking would say, the mathematician did not accept the kind of evidence offered by Galileo.

> GALILEO: If you gentlemen are agreeable, we shall begin with the inspection of the satellites of Jupiter, the Medicean stars.
> [...]
> THE PHILOSOPHER: [...] before we apply ourselves to your famous tube, we should like to request the pleasure of a disputation: Can such planets exist?
> GALILEO: Your Highness, would you care to observe those impossible and unnecessary stars through the telescope?
> THE MATHEMATICIAN: One might be tempted to reply that if your tube shows something that cannot exist it must be a rather unreliable tube.
> GALILEO: What do you mean by that?
> MATHEMATICIAN: It certainly would be much more to the point, Mr. Galilei, if you were to tell us your reasons for supposing that there can be free-floating stars moving about in the highest sphere of the immutable heavens.
> THE PHILOSOPHER: Reasons, Mr. Galilei, reasons!
> GALILEO: My reasons? When a look at these stars and my calculations demonstrate the phenomenon? This debate is getting absurd, sir.
> THE MATHEMATICIAN: If it were not to be feared that you would get even more excited than you are, one might suggest that what is in your tube and what is in the sky might be two different things. (Brecht, 2007 [1938])

In the passage above Galileo insists that for accepting his *reasons*, that is his *justification* for the existence of Jupiter's moons, it is sufficient to look through the telescope. The two scholars, though, insist in asking other reasons because they do not accept his kind of evidence. Even if Galileo can offer an argument in terms of observable sentences—and, indeed, observations—his opponents do not accept his claim because the kind of evidence he offers does not belong to their way of thinking and doing.

It took some dozen of years before Galileo's truth was accepted. Galileo's milieu could do little to favour the establishment of *new standards of truth*. Indeed, Biagioli wrote that 'lack of consensus about *style of argumentation and standards of evidence* as well as the scant interdependence among the members of this field hindered closure of the debate' (Biagioli, 2006, Italic is mine). Of course, had Galileo's discoveries been discussed within an intellectual milieu of astronomers, as Boyle's ones were discussed within an elite of people who explored nature in the same public place, his kind of evidence and his way of doing would not have been called into question.

I conclude by summing up my point. Hacking claims that we are justified in believing that unobservable entities exist when we regularly manipulate them. I counter that, *if his characterization of styles of reasoning is correct*, then his justification is relative to the laboratory style of reasoning. My argument has followed these steps: I have pointed out that regular manipulation is a way of doing that is not part of the presuppositions of another style of reasoning. Even if causal effects obtained in experiments-with can be described by style-dependent statements, the experimental argument implicitly assumes the presuppositions of the laboratory style of reasoning: methods of finding out, ways of thinking and doing and types of evidence. Therefore, the experimental argument is not a valid justification for the existence of a particle outside the laboratory style of reasoning. To substantiate my thesis I have given a historical example in which it is not sufficient that a claim is justified in terms of observational statements in order to be accepted. Galileo's claim is dismissed a priori by Aristotelian scholars because it relies on a kind of evidence alien to their mental horizon.

7.3.4 The Incommensurability of Styles of Reasoning: Second Argument

Up to now, I have argued that Hacking's argument for realism is relative to the laboratory style of reasoning but I have not criticized it: in my imaginary example I have supposed that the claim that unobservable entities exist can be defended as Hacking does. In this subsection, I shall use the historical account of the discovery of the muon provided by the historian of science Peter Galison in *How the Experiments End* (Galison, 1987) for assessing Hacking's experimental realism. My analysis will show that, in contrast with what Hacking maintains, theoretical style-*dependent* sentences in the laboratory style of reasoning are *necessary* to justify the claim that unobservable entity exists. Since style-dependent sentences in the laboratory style of reasoning have no truth-value for a community that adopts a different style of reasoning I shall conclude that, in this case too, the justification of an observable entity is relative.

The history of the discovery of the muon is intertwined with that of the research on the cosmic rays, high-energy particles of galactic and extragalactic origins, most of them protons, which reach the Earth's atmosphere. It was by investigating the compositions of cosmic rays that muons were detected for the first time. Today it is an undisputed fact that when cosmic rays enter the atmosphere, they collide with its molecules and produce a so-called shower of ionized particles, of which some decade into muons, able to reach the surface of our planet. A shower is a cascade of secondary particles with smaller quantities of energy than the incoming particle, each of them producing other cascades of billions and billions of particles. As a proton collides with an air molecule, it creates pions, protons and neutrons; pions decay in particles such as muons, which are *penetrating* particles, i.e. they are able to reach the surface of the Earth. Muons were unknown particles to physicists until the last years of the 1930s, when numerous experiments and theoretical considerations helped to settle the dispute about their existence.

The physicist Robert Millikan (1868–1953), one of the pioneers of this research, believed that the primary cosmic rays impinging the atmosphere were gamma rays, i.e. photons of high frequency penetrating the atmosphere isotropically from space. Gamma rays are the most penetrating of all radiations because they lose little energy in interactions with matter. In the 1920s Millikan persuaded that in the cosmic rays there were no very penetrating *charged* particles, measured the intensity of radiation (he was convinced of measuring photons) as a function of penetration depth for different energies and concluded that his hypothesis was correct: for him the penetrating particles were gamma photons arriving from outside the atmosphere. However, in 1929 Walther Bothe (1891–1957) and Werner Kolhörster (1887–1946) set up an experiment that challenged Millikan's conclusions. By using two Geiger tubes separated by a block of lead and measuring the *coincidences* (when the same particle crosses both the counters and make them produce a click) they found evidence for the passage of a single *charged* particle through the intervening lead (Galison, 1997, p. 441). Indeed, by measuring how the coincidences decrease as the thickness of the intervening lead is modified, it is possible to rule out the possibility that the cause of the coincidences is the passage of gamma rays. This was a conclusion based on the macroscopic effects of the investigated entity and on the knowledge of how gamma rays are attenuated by plates of different thickness. However, all Bothe and Kolhörster came to know was that the particles were charged; they acquired no knowledge about its real identity.

Later on the physicist Bruno Rossi (1905–1993) improved the apparatus and was able to confirm with more certainty Bothe and Kolhörster's result. In the following years three independent experiments (see Swann, 1961, p. 812) confirmed the existence of the so-called latitude effect: the variation of the intensity of cosmic rays with the latitude. This is the result of the Lorentz force generated by the terrestrial magnetic field on charged particles. This force has a full component when the speed of the particle is orthogonal to the Earth's magnetic field (at the equator) and is null when the speed is parallel to it (at the Poles). The latitude effect supported the idea that cosmic rays could not consist of gamma rays, because on photons the Lorentz force has no effect since they are not charged. To illustrate the point of the experiment, that it is impossible that the penetrating particles are not charged, it is necessary to use the Lorentz force, a fundamental law, or at least phenomenological laws derived from it.

The latitude effect was the reason of a long dispute between Millikan and Arthur Compton (1892–1962) (Bertolotti, 2013, pp. 85–95), who had conducted a large-scale geographical survey: in the beginning the former, who had made experiments himself, insisted that there was no latitude effect, then he tried to explain it under wrong assumptions, i.e. by claiming that primary photons could knock off some electrons from interstellar matter. Millikan also refused to accept the results of Bothe and Kolhörster's experiments as well as Rossi's on the basis that the internal processes inside the Geiger counter were misunderstood. As Galison reports, in 1934 Millikan and three colleagues published a paper on *Physics Review*, in which they asserted:

> [The counters' coincidences] cannot in general be due to the passage of one charged article through both counters and the intervening lead, but must rather be due to some mechanism but which a photon can release successively along, or in the general neighbourhood of, its path a number of different particles whose separate but practically simultaneous action on the two or more counters is responsible for the observed coincidence. (quoted in Galison, 1987, p. 95)

The way Millikan challenged the experiment is reminiscent of the criticism of Galileo's opponents. The analogy lies in this fact: the scholars of Galileo's time thought that the bright spots were not satellites but rather artefacts produced by the instrument, of which there was no accurate knowledge; similarly, Millikan and colleagues suggested that the coincidence was not the effect of a *charged* particle but that of a mechanism inside the counter produced by a gamma ray. In both cases, doubts about the internal processes of the instrument were a reason for not believing in the claim of the existence of an entity. Of course, there was an important difference between the two disputes: Galileo's opponents contested the use of the telescope as a form of evidence in principle and did not share with Galileo the same way of thinking and doing; on the contrary, Millikan and Compton shared the same style of reasoning and the former did accept the use of the Geiger counter as an instrument of investigation. This difference notwithstanding, in both the situations it was thought that the instrument was misleading. I shall return to this point later.

In the light of the experiments performed in the 1930s it became clear to many physicists that the primary cosmic rays impinging the atmosphere were protons and that the coincidences at the Earth's surface were caused by charged particles. Millikan and others were not convinced, though. About this time or immediately following it, several phenomena concerning the interaction between protons and the molecules of atmosphere as well as the production of new particles became clear. In 1934 Rossi recognized the phenomenon of air shower: the primary cosmic rays when impinge the atmosphere produce a cascade of different particles (Rao & Sreekantan, 1998, p. 5). However, whether or not a new kind of particle was the cause of the coincidence remained a mystery: it was thought by some that it was an electron.

The growth in complexity of theoretical physics became increasingly crucial for tackling the problem. New theoretical descriptions were important to find out the elementary processes that constituted the phenomenon of showers such as decays, processes of pair creations and radiation. For example, on the basis of these speculations the possibility that certain known particles could reach the surface of the Earth was ruled out. In particular, the theoretical physicists John Carlson and Robert Oppenheimer (1904–1967) modelled the showers by treating them as a diffusion process similar to the expansion of a drop of ink in a glass of water. In fact, they described the showers with phenomenological equations within the quantum theory. Their conclusion was that either these equations were inapplicable in the domain of cosmic rays energies or the actual penetration of these rays had to be ascribed to the presence of a component either than electron (and photon) (Galison, 1987, p. 95). Carl Anderson (1905–1991) and Seth Neddermeyer (1907–1988) made measurements of the energy loss of particles occurring in the cosmic ray showers that led to rejecting the first hypothesis: Carlson and Oppenheimer's theoretical estimates were applicable to cosmic ray energies; therefore, the coincidences could not be caused by electrons.

In 1937, Jabez Street and Edward Stevenson provided the first quantitative analysis of the mass of the penetrating particle by examining the track curvature of the particle in a double-cloud-chamber and subjected to a magnetic field. Indeed, by measuring the radius of curvature and applying the formula of Lorentz force they concluded that the value of the muon's mass is about 130 electron masses. Many experimental physicists were definitely persuaded of the existence of the muon by this remarkable experiment. However, as Galison explains, 'among theorists, especially, the acceptance of the new particle was greatly facilitated by an argument from the lofty heights of quantum field theory that had been absolutely irrelevant to the experimentalists' (Galison, 1987, p. 124). Indeed the muon had approximately the same mass of a particle predicted by a theory of nuclear forces put forward in 1935 by the theoretical physicist Hideki Yukawa (1907–1981). Nevertheless theoretical physicists were wrong: later on it became clear that the muon's behaviour did not conform to that predicted by Yukawa's particle, which actually describes the pion. It was only after the war that the muon and the pion were distinguished, thanks to an improved theory of quantum field, the discovery of new particles and the construction of new accelerators.

The story of the muon makes it clear that it is impossible to identify a 'moment of discovery', an event that convinces everybody of the existence and the identity of a particle: '[W]e should envision the ending of the muon experiments as a progressively refined articulation of a set of phenomena' (Galison, 1987, p. 127). Furthermore, Galison draws from the story of the muon some conclusions that strikingly contrast with Hacking's emphasis on experimentation. Hacking is adamant that high-level theories are not necessary to develop an argument for the existence of a theoretical entity. But Galison remarks that 'theory itself played a complex role in ending of the cosmic ray experiment' and adds that 'if the first role of theory in the ending of these experiments was *ostensive*, demarcating a domain of phenomena, the second was *constitutive*' (Galison, 1987, p. 129). For example, the quantum theory of electrodynamics drew the attention on the phenomenon of penetrating radiation, and theory helped to isolate a set of possible experimental techniques as important. Beside this ostensive role, there is a constitutive role: '[T]he acceptance of quantum theory of electrodynamics and its identification with shower particles was not separable from the acceptance of a new particle. These questions are complementary pieces of the same conceptual structure' (Galison, 1987, p. 129). Galison concludes:

> Theory enters first ostensively, acting in broad, qualitatively way: look at penetrating particles, look at showers. [...] But theory enters a second time, now not to point out types of phenomena but to provide a quantitative analysis that plays a constitutive role in conclusion. And at this level theory is often mediated by a model which instantiates features of the more general theory without invoking the generality of an overarching structure. (Galison, 1987, p. 131)

Bearing these thoughts in mind let us return to imagining an adherent of the laboratory style of reasoning (the speaker) who has to convince a member of a community who adopts a different style of reasoning (the hearer) about the existence of a muon. Among other things, the speaker would need to prove that the causal effects are not artefacts of the Geiger counter as Millikan thought. Hacking might be right that Galileo did not need much theory to make his point, but the example of the muon shows that it is not possible to isolate a group of observation sentences in order to characterize this particle and convince someone of its existence. Indeed, Galison explains that particle physicists did need fundamental theories of theoretical physics to become confident about the existence of the muon. Our belief in muons is intimately connected with our ideas about the counters, the Lorentz's force and the phenomenological laws that describe the showers. So, while observation style-*independent* sentences were all Galileo needed for arguing for the existence of Jupiter's moons, theoretical style-*dependent* sentences were necessary in order to argue that the Geiger counter was really counting muons. For example, in the case of the muon, the hearer would need to resort to theoretical sentences to express the phenomenological laws of Carlson and Oppenheimer or the theoretical interpretation of the latitude effect by Lorentz's force. These sentences would have no truth-value in a style of reasoning different from the laboratory style of reasoning although, since the laboratory style of reasoning is self-authenticating, they would be true by its standards of truth. Carlson and Oppenheimer's model, for example, is for Hacking not *the* explanation of the phenomenon of the shower but a theoretical characterization that is true to the result of an experiment. In turn, the result of the experiment is true to the model. To sum up, the speaker would need theoretical style-*dependent* sentences in order to deploy her argument and the hearer could not attribute any truth-value to them. Hence, that very argument is relative to the laboratory style of reasoning.

On the basis of the results of this section I conclude that whether or not it is possible to provide a justification for the existence of unobservable particle that is as theory-free as possible, Hacking's characterization of styles of reasoning implies that that very justification is *relative* to the laboratory style of reasoning. This conclusion has two important corollaries: the first is that Hacking's experimental realism is at odds with the relativistic implications of his styles of reasoning project. Indeed, on the one hand, Hacking's experimental realism is the claim that we *know* about unobservable entities, and, on the other hand, the styles of reasoning project implies that experimental realism is unjustified outside the laboratory style of reasoning. The second corollary is that by claiming that 'when the question is a live one [...] then aspects of the world determine what the answer is', Hacking does not really manage to distance himself from constructionism. Indeed, this claim amounts to saying that there is only one *correct* answer to a live question. But the answer to the question as to whether or not the particles posited by physicists exist cannot be considered as universally and atemporally 'correct': its being 'correct'—the styles of reasoning project implies—is unjustified outside of a given style of reasoning.

So far, my discussion of incommensurability has focused on the laboratory style of reasoning. However, in Sciortino (2016) I have offered a more general argument for the incommensurability of styles of reasoning that concerns any style of reasoning. In other words, my argument that the styles of reasoning are incommensurable in this chapter can be pushed further. By comparing and contrasting the styles project with Wittgenstein's later works such as *Philosophical Investigations* (Wittgenstein, 1997 [1953]) and *On Certainty* (Wittgenstein, 1995 [1969]), in the article just mentioned I pointed out that both the conceptions of meaning of the early and the later Wittgenstein can be recognized in the styles project and I discussed the philosophical consequences. Then relying on some analogies between Wittgenstein's notion of 'form of life' and the notion of style of reasoning, I argued that the styles project implies anti-foundationalism, i.e. the idea that each style is ungrounded, not epistemically justified. I concluded that, although in the case of Wittgenstein whether or not anti-foundationalism leads to epistemic relativism is at issue, in the case of Hacking this inference is *inevitable*. Indeed, while Wittgenstein does not posit the actual existence of different epistemic systems (see Coliva, 2010), the claim that there are different epistemic systems (i.e. styles of reasoning) is present in Hacking. Thus, we can conclude that his anti-foundationalism leads to epistemic relativism: the justification of a given sentence might be relative to one of the different epistemic systems of which he claims the *actual* existence.

7.4 Conclusions

To sum up, in this chapter I have analysed Hacking's experimental realism in the light of the styles of reasoning project. Hacking's theory anti-realism denies the possibility of ranking styles of reasoning in terms of their capacity to explain phenomena. Furthermore, Hacking is not successful in showing that his entity realism is justified for a community that does not adopt the laboratory style of reasoning. It follows that the styles of reasoning project is in conflict with the claim that we know about unobservable entities. My discussion of incommensurability has focused on the laboratory style of reasoning. In Sciortino (2016), I have offered a more general argument that concerns any style of reasoning.

Bibliography

Bertolotti, M. (2013). *Celestial messengers: Cosmic rays: The story of a scientific adventure.* Springer-Verlag.

Biagioli, M. (2006). *Galileo's instruments of credit.* The University of Chicago Press.

Brecht, B. (2007 [1938]). Life of Galileo. http://buehnenkunst.ohost.de/auditions_coming_up_files/Life%20of%20Galileo%20by%20Brecht.pdf

Cartwright, N. (1983). *How the laws of physics lie.* OUP.

Chalmers, A. (1999). Making sense of laws of physics. In H. Sankey (Ed.), *Causation and laws of nature* (Vol. 14). Kluwer Academic Publishers.

Coliva, A. (2010). Was Wittgenstein an epistemic relativist? *Philosophical Investigations, 33*(1), 1–23. https://doi.org/10.1111/j.1467-9205.2009.01394.x

Dupré, J. (1996). The disorder of things: Metaphysical foundations of the disunity of science. In U. O. C. Press (Ed.), *The disunity of science.*

Feyerabend, P. (1988 [1975]). *Against method.* Verso.

Galilei, G. (2008 [1610]). *Opere.* Mondadori.

Galison, P. (1987). *How experiments end.* University of Chicago Press.

Galison, P. (1997). *Image and logic: A material culture of microphysics.* University of Chicago Press.

Hacking, I. (1983). *Representing and Intervening.* Cambridge University Press.

Hacking, I. (1984). Experimentation and scientific realism. In J. Leplin (Ed.), *Scientific realism* (pp. 154–172). University of California Press.

Hacking, I. (1991). Artificial phenomena. *The British Journal for the History of Science, 24*(02), 235–241.

Hacking, I. (1995/1996). Comments on Zeidler & Sobczynska's paper. *Foundations of Science, 4*, 537-542.

Hacking, I. (2007). Putnam's theory of natural kinds and their names is not the same as Kripke's. *Principia, 11*(1), 1–24.

Herschel, W. (1800a). Experiments on the refrangibility of the invisible rays of the Sun. *Philosophical Transactions, 90*, 284–292.

Herschel, W. (1800b). Experiments on the refrangibility of the invisible rays of the Sun. [Abstract]. *Abstracts of the Papers Printed in the Philosophical Transactions of the Royal Society of London, 1*(1800–1814), 22–23.

Herschel, W. (1800c). Investigation of the powers of the prismatic colours to heat and illuminate objects. *Philosophical Transactions, 90*, 255–283.

Holton, G. (1996). *Einstein, history and other passions*. Addison Wesley.

Kripke, S. (1972). *Naming and necessity*. Harvard University Press.

Morrison, M. (1990). Theory, intervention and realism. *Synthese, 82*(1), 1–22.

Putnam, H. (1975). Explanation and reference. In H. Putnam (Ed.), *Mind, language and reality: Philosophical papers* (Vol. 2, pp. 196–214). Cambridge University Press.

Rao, M. V. S., & Sreekantan, B. V. (1998). *Extensive air showers*. World Scientific Publishing.

Sciortino, L. (2016). Styles of reasoning, human forms of life, and relativism. *International Studies in the Philosophy of Science, 30*(2), 165–184. https://doi.org/10.1080/02698595.2016.1265868

Swann, W. F. G. (1961). The history of cosmic rays. *American Journal of Physics, 29*, 811.

Wittgenstein, L. (1995 [1969]). *On certainty* (D. Paul & G. E. M. Anscombe, Trans., 11th ed.). Blackwell.

Wittgenstein, L. (1997 [1953]). *Philosophical investigations* (G. E. M. Anscombe, Trans.). Oxford Blackwell.

Wüthrich, A. (2012). Interpreting Feynman diagrams as visual models. *Spontaneous Generations, 6*(1), 172–181.

Zeidler, P., & Sobczynska, D. (1995/1996). The idea of realism in the new experimentalism and the problem of the existence of theoretical entities in chemistry. *Foundations of Science, 4*, 517–535.

CHAPTER 8

Styles of Reasoning, Contingency and the Evolution of Science

8.1 Introduction

By 'contingency issue' it is meant the question as to whether the history of a particular branch of our science could have taken a different route and provided results incompatible with those of our actual science. In Chap. 3, I have pointed out that, apart from Hacking's recent comments (Hacking, 2014 chapter four), the discussions on the contingency issue have not involved the notion of style of reasoning. I want to fill this gap in the literature by drawing some important implications for the issue of contingency from the theory of styles of reasoning that I have developed so far. As far as the contingency issue is concerned, I shall address four fundamental questions. First of all, in the next section I shall discuss to which extent the emergence of styles of reasoning at a certain point of history is a contingent circumstance. In the following section I shall ask whether the endurance of styles of reasoning was inevitable by tackling the connected question as to why the styles are long lasting. Afterwards, I shall focus on the growth of knowledge by dealing with questions such as: if a certain style of reasoning continues to be employed and if Q is a 'live question', is science bound to converge on a single answer to Q? Finally, on the basis of my previous reflections I shall look into the question of the convergence of science in the long run. The answers to these questions will rely on some basic assumptions of the theory of styles of reasoning and imply a picture of the evolution of sciences in which *both* contingency and inevitability play a key role.

L. Sciortino, *History of Rationalities*,
https://doi.org/10.1007/978-3-031-24004-1_8

8.2 Styles of Reasoning, Contingency, Inevitability

8.2.1 *Inevitability and the Algorithmic Style of Reasoning*

The question I am going to examine in this section concerns the role played by contingency and inevitability in the genesis of styles of reasoning. A further development of the styles project can provide important insights into this question, although a comprehensive answer would require for each style of reasoning a study of cognitive history. I shall start by arguing that the emergence of the algorithmic style of reasoning was inevitable. In Chap. 5 I have maintained that the algorithmic style of reasoning emerged in ancient civilizations such as the Egyptian or the Babylonian: in order to solve problems such as Ahmes', the Egyptian mathematicians used a standard set of rules involving numbers that were put down on papyri and handed down from a scholar to another. I have also explained that the algorithmic style of reasoning is grounded in typical human capacities: for millennia human beings have unconsciously followed step-by-step procedures for fulfilling basic needs as animals do, e.g. spiders make silk for catching insects by following three basic steps (Devlin, 2005 chapter 5); furthermore, basic numerical competences are rooted in biological built-in resources. Importantly, I have also argued that when the first human societies started to make use of commodity money a concept of number *had to* come about. In this sense the emergence of the concept of number has been *inevitable.*

Although key elements such as the capacity of following step-by-step procedures of reasoning and the very concept of number have been a *sine qua non* for the emergence and development of the algorithmic style of reasoning, other factors have played a major role in its genesis. The introduction of zero, in addition to the positional system, allowed the representation of any number without ambiguities. The concept of zero was represented by the Babylonians by an empty space, it developed as a number in India, it was adopted by the Arabs and finally it wended its way across Europe when the mathematician Leonardo Fibonacci (1170–1250) promoted the Hindu-Arabic numeral system, which had appeared in Europe in the eleventh century through the Moors (see Seife, 2000). Interestingly, in around 400–300 BC also the Mayan people invented *independently* the concept of zero and a relative symbol. If one considers that this people did not come into contact with the earlier European civilizations, the emergence of zero appears an inevitable fact. However, it should be considered that in the ancient China zero was not used as a number but as an empty place indicator. Without the *number* zero there could not have been negative numbers and algebra would not have been possible. In other words, at least as compared with the *inevitable* invention of positive integers, that of negative numbers appears contingent.

The positional system made much simpler to perform calculations, as it can be noted by thinking of the difficulty of performing an addition by using Roman numbers. The Babylonians developed the sexagesimal system independently in about 2000 BC; in the Classical period (250–900 AD), the Mayan people introduced a base-20 positional system. Other number bases have been developed by different cultures but eventually the decimal system prevailed in large parts of mathematics. Since all the Indo-European languages share the base-10 system it is likely that the peoples who spoke *proto*-Indo-European used it. A plausible hypothesis is that the decimal system derived from the use of the hands for counting and that it spread around the world for its convenience as a basis for calculation (Barrow, 1992, p. 36). If this hypothesis is correct, given the basic features of humans, it seems inevitable that the base-10 system came to the fore and spread around the world. For sure, the use of positional systems in different ancient cultures, which did not come into contact, suggests that the emergence of a numeral system was an inevitable fact.

This brief account encourages the following reflections. The fact that the concept of number as well as the use of step-by-step procedures in mathematics emerged in different ancient societies suggests that, given any sufficiently developed civilization, it was inevitable that humans had to develop the algorithm style of reasoning. However, it is a merely historical fact that the fields of mathematics to which we apply the algorithmic style of reasoning are exactly what they are today: one might argue that even the coming to the fore of algebra is a fact contingent on the particular use of zero as a number in certain human civilizations. Some readers may feel that the extension of positive integers to negative ones was inevitable once the concept of zero came to the fore. This is a point in time in which it is possible, at least in principle, to imagine an algorithmic style of reasoning that deals with a very different mathematics, in this case an impoverished mathematics rather different from *ours*. One might consider that as the point in time in which contingency makes its way into the history of science. Of course, there is a sense in which a mathematics with negative and, above all, with complex numbers was inevitable. It has to do with the idea of success: ‘if mathematicians were to achieve the successes they have in fact achieved in proving results in number theory, they had to introduce complex numbers' (Hacking, 2014, p. 120). There is contingency as long as the success of number theory was not inevitable. But if the latter was inevitable, then so was the development of the algorithmic style to include negative and complex numbers.

8.2.2 Contingency and the Postulational Style of Reasoning

I shall now move on to the emergence of the postulational style of reasoning. According to some cognitive scientists, just as there is an ability to grasp the numerosity of sets of objects inherent in the human mind, so there is an innate sense of the basic conceptual principles of geometry. For example, a team of researchers has provided evidence for geometrical intuitions (e.g. symmetry, parallelism, congruence) in an Amazonian group who had no schooling and no experience with artefacts that employ geometrical concepts (Dehaene & al., 2006). In the mind of Netz and Hacking, to acknowledge the existence of such human resources is the starting point of the particular cognitive history of Greek geometry (Netz, 1999, p. 6) (Hacking, 2011) and, therefore, of the postulational style of reasoning.

Netz points out three crucial factors for the emergence of proof. The first one concerns the fact that the Greek society was ready to question almost everything: following Lloyd, Netz claims that the Greek social environment was fertile ground for forms of persuasion (Lloyd, 2000). The second factor concerns orality, i.e. the fact that in Athens 'the characteristic mode of political debate was oral' (Netz, 1999, pp. 292–293). Especially in Chap. 5 section 3, Netz shows how the structure of proofs, the use of formulae, the reference to the diagram, i.e. to an immediately present visual object, reflect the particular form of orality and the importance of persuasion in that *specific context*. In some sense, the demonstration was an oral argument written down next to a lettered diagram, argues Netz. The third factor concerns the existence of a literate environment: the invention of a lettered diagram would not have been possible in a preliterate society or a society which used pictograms (Netz, 1999, p. 58).

Netz's theses suggest that the emergence of the postulational style of reasoning was a fact contingent, especially if we compare it to the emergence of the algorithmic style of reasoning. Indeed, a lettered diagram is an abstraction conceptually similar to that of number; nevertheless, whereas the concept of number comes about independently in different societies, that of lettered diagram is a specific Greek invention. Its coming to the fore would have been impossible without the presence of a literate and argumentative society, according to Netz. In brief the emergence of the postulational style of reasoning has been contingent on social and historical factors. Another point concerns the way in which the innate intuitive comprehension of geometry was tapped. The postulational style of reasoning represents one of the many possible ways, surely not the only possible one, of tapping human cognitive resources and logical deduction: other ancient societies did use diagrams, but in radical different modalities. Nor is the postulational style of reasoning the only way for proving important geometrical truths: for example, Pythagoras theorem, which can be proven by using the postulational style of reasoning (i.e. relying on the similitudes of triangles), can also be proven by using algebraic methods or even by rearrangement (i.e. by making three copies of the original triangle and arranging them in particular ways). As Hacking put it, '[T]here could have been ample mathematics without human beings ever discovering demonstrative proof' (Hacking, 2014, p. 115). Ultimately, the specificity of the postulational style of reasoning is contingent on the features of the Greek society, e.g. factors such as literacy, orality and persuasion. However this specificity does not make the discovery of certain mathematical truths contingent since there might be other methods, different from the geometrical style of reasoning, for arriving at them.

8.2.3 *Contingency, Laboratory Style and 'the evolutionary tree of styles of reasoning'*

Let us turn to the laboratory style of reasoning. Referring to the experimental reasoning in the early times of civilization Hacking wrote that '[it] probably ha[d] no strict beginning. Human beings have been curious, looking, tinkering, exploring, even measuring, forever' (Hacking, 2009, p. 42). And, when he argued that the laboratory style of reasoning is a particular way of combining experimental reasoning and modelling, he added that in the earlier times 'the modelling was typically of the movements in the heavens' (Hacking, 2007, p. 3). It seems plausible that humans have always reasoned by constructing models. One could well make the hypothesis that modelling is rooted in ancestral human ritual practices of representing in order to make something happen in the future, e.g. the ancient Egyptians believed that by writing the name of their hostile people in pottery vessels and smashing them they could destroy their enemies (Van De Mieroop, 2011, p. 109). However, Hacking's thesis that experimental activity had no strict beginning is not plausible. In order to measure something it is necessary to take two steps, which are not reducible to counting: firstly, to adopt a unity of measurement; secondly, to possess a concept of number. Indeed, the value of a measurement typically represents how many times a predetermined length is contained in the length to be measured. In many cases, integer numbers are not sufficient because not all the lengths in geometry can be divided into an integer number of pieces of the same length. Therefore, the emergence of the concept of number and the algorithmic style of reasoning have been necessary for the coming to the fore of the earlier forms of experimental reasoning. This is to say that, as contingent as the emergence the laboratory style of reasoning may be, there is no contingency in the order in which the two styles of reasoning, the algorithmic and the laboratory one, came about. If one 'replied the tape' of our cognitive history, the event of the crystallization of the laboratory style of reasoning would be contingent on the emergence of the concept of number and, certainly, on the discovery of the irrational numbers, which was prompted, as far as we know, by the emergence of geometry and its postulational style of reasoning. In this sense the analogy with the Darwinian evolution is striking: as, say, the emergence of mammals is contingent on the emergence of primitive chordates, so the laboratory style of reasoning is contingent on the emergence and development of the algorithmic and the postulational style of reasoning. Algorithmic, postulational and laboratory style of reasoning are in relation of extreme connectedness as different varieties in the same species. However, although their order could not be reversed, each of them could have been different *in content*, e.g. it could not have achieved all the actual results.

To summarize it is useful to imagine a sort of evolutionary tree of the styles of reasoning: the algorithmic style of reasoning is in the tree 'trunk', as if it were an ancestor of all the other styles of reasoning. Several branches grow from the trunk but are separated from it by a knot to indicate a sharp discontinuity: the first branch that we encounter up along the trunk is the postulational style of reasoning, and in turn the laboratory style of reasoning may be viewed as a branch of the postulational style of reasoning, separated from it by a knot. Then, as I shall explain, the other styles of reasoning form different branches of the trunk. The birth of the tree has a strong character of inevitability, since the human cognitive resources and the universal human forms of socialization predetermined the emergence of the algorithmic style of reasoning. At the next stages both the postulational style of reasoning and the laboratory style of reasoning are the result of particular 'environmental pressures', e.g. the unique features of the Greek society. Note that trunk and branches, although separated by knots, are somehow connected in that they are part of one and only organism. No postulational style of reasoning could have been possible without the vast mess of problems about lengths, areas and volumes produced by reasoning in the algorithmic style of reasoning by the ancient civilizations anterior or coeval with the ancient Greeks; no laboratory style of reasoning could have been possible without the possibility of using real numbers to measure and a prior model of demonstration (Hacking, 2014, p. 140). There is no contingency in the order in which these styles of reasoning have emerged because there is no postulational style of reasoning without some of the achievements of the algorithmic style of reasoning and there is no laboratory style of reasoning without some of the achievements of the postulational and the algorithmic style of reasoning.

8.2.4 *Contingency and Pseudo-styles of Reasoning*

An idea of considerable significance for Paracelsus' way of thinking was the Pythagorean conviction that everything in the universe is intrinsically mathematical. Those who adhered to Pythagoras' school believed that numbers, to which they attributed mystical or divine significance, were associated to all sort of things that exist. This body of mystic tradition survived in the way of reasoning of Renaissance thinkers and of many naturalist philosophers still at Galileo's time. Rather than thinking of the universe as a system governed by laws to be discovered by performing experiments, these scholars directed their attention to single facts of observation that they interpreted as *signs*. Eventually, this conception of the universe was substituted by one in which single events have a cause and an effect, more than a meaning. However, it was not until this way of thinking underwent a mutation, whose consequence was the emergence of the statistical style of reasoning, that it became extinct. As I have explained in Chap. 3, according to Hacking this mutation consisted in a transformation of the concept of sign into a new concept of 'internal evidence', i.e. of evidence other than testimony. The statistical style was a new way of tapping this innate sense of evidence, a kind of evidence that was absent in the high sciences and that only in the seventeenth century became a standard of truth. In the evolutionary tree the style of the Renaissance way of thinking could be represented as a dead branch that, at a certain point, has a knot, which marks a point of discontinuity with the new branch of the statistical style of reasoning.

What is made clear by this brief survey is that it is contingent when, how and if certain human innate resources come to be exploited. Among these innate human resources I include not only the ability to grasp the numerosity of a grouping of objects or to imagine two parallel lines. I also include logic abilities such as deduction, which is the central method of the postulational style of reasoning, or the very innate tendency to use certain facts as 'evidence of something else', the evidence of things, which is the standard of evidence of the statistical style of reasoning. According to the theory of styles of reasoning I have developed, all these human innate resources are inborn features of human beings that at certain points of history are *co-opted* by a given style of reasoning. The historico-genetic style of reasoning is another case in point. Abductive reasoning has always been used by human beings: it was in Aristotle's *Prior Analytics* that Charles S. Peirce (1839, 1914) found the notion of abduction (Psillos, 2011, p. 119). When the seventeenth-century intellectuals directed their attention to fossils, minerals and geological layers abductive reasoning was co-opted to find out the best explanation of those *previously unnoticed* objects and its relevance for the study of nature became considerable.

The same considerations can be repeated in the case of the taxonomic style of reasoning. When Ray, Linnaeus and other naturalists set out to establish the relations between the different kinds of living beings, they found a way of tapping the human ability to classify—in fact, they put the folk classifications of plants and animals into a scientific footing. The co-option of these human innate resources, i.e. the abductive reasoning in the case of the historico-genetic style of reasoning and the common-sense view of natural relationships in the case of the taxonomic style of reasoning, has been a fact contingent on the emergence of a vast amount of evidence, an entire portion of the world previously unnoticed.

The laboratory style of reasoning played a role for the coming to the fore of the historico-genetic and the taxonomic style of reasoning, so that their emergence can be considered contingent on that of experimental science. Indeed, as Bowler maintains, what paved the way for the Darwinian revolution was the failure of the attempt to harmonize the new science inspired by Galileo and Newton with the contemporary understanding of the Bible (Bowler, 2003 p. 27). On the other hand it seems inevitable that the taxonomic and the historico-genetic style of reasoning had to rely on one another: e.g. without an unambiguous way of expressing the relationships between living beings, the historico-genetic style of reasoning would have been useless.

In the evolutionary tree of styles of reasoning both the taxonomic and the historico-genetic style of reasoning represent two different branches. Their evolution can be explained in terms of the concept of exaptation in biology. The latter refers to a character of an organism that comes to serve a new function different from the previous one: e.g. although the feathers might have evolved for regulating the body temperature at a certain point in time they were co-opted by natural selection in order to allow the flight. Similarly, when the historico-genetic style of reasoning and the taxonomic style of reasoning evolved, the abductive reasoning and the natural disposition for classifying took on new functions, perhaps not so distinct from the previous ones, but surely different in purpose.

To conclude, I have argued that the emergence of the algorithmic style of reasoning has been inevitable. However, the emergence of all the other styles of reasoning has been the result of both contingent and inevitable factors: on the one hand, particular historical, social, cultural and intellectual contexts have made it possible their emergence; on the other hand, there is a rationale in the order of their emergence, i.e. there is *no contingence* in the order in which they have appeared. This suggests that the emergence and the development of a certain style of reasoning may create the epistemological conditions for the emergence of other styles of reasoning; but *which* styles of reasoning come about, *whether* they come about and *when* they come about is *contingent*. There might have been different ways of tapping human resources and, above all, we could not have achieved the same results we have achieved today.

8.3 Why Do Styles of Reasoning Endure?

8.3.1 Explaining the World

In this section I shall argue that each style of reasoning is suitable for dealing with certain objective structures of our world just as certain methods of lock picking are suitable for unlocking a given type of lock. Since Hacking is anti-realist about theories, he has never put the accent on the ability of style of reasoning to represent the world. Consequently, the style of reasoning project makes no mention of what I consider an important fact: as I shall show, each style of reasoning is grounded in objective features of that portion of the world it studies. I suggest that this fact should be taken into consideration when addressing the question as to why a style of reasoning endures. Indeed, although the internal techniques of self-authentication are surely part of the explanation of the resilience of a given style of reasoning, they cannot account for its success in representing the world more or less well than another possible way of thinking. Since success must be part of the explanation of why we do not abandon a given style of reasoning, I claim that the ability of this very style of reasoning to account for certain objective features of the world is relevant to the question of why styles of reasoning are long lasting. Ultimately, in addition to the self-authenticating character of style of reasoning, it is possible to find *in the world* other reasons that justify the endurance of style of reasoning. I cannot investigate here how the interplay between these two very different causes (which concern the internal features of styles of reasoning and certain objective features of the world) can make a specific style of reasoning long lasting. I shall only present four examples that illustrate how each of the six styles of reasoning mirrors objective features of the portion of the world it deals with. They suggest that there is a deep connection between styles of reasoning and the features of those portions of the world they study.

8.3.2 *Laboratory Style, Historico-genetic Style and the Time's Arrow*

Many events are causally connected in time in a 'one-way direction'. In the example of Cleland (see Chap. 5), we may happen to see the shattering of a window by a baseball but we never see the ball going back along the same trajectory and the scattered pieces of glass suddenly jumping up and converging to the empty window frame (Cleland, 2002, pp. 487–494). Ever since Arthur Eddington's publication of *The Nature of the Physical World* (Eddington, 1928) this basic feature of our world is known as the 'time's arrow', whose consequence is the 'asymmetry of over-determination': to conclude that by throwing a ball the window glass gets broken one needs hypotheses on the ball, the resistance of the glass, the kind of throwing etc.; to conclude that the window glass got broken might be sufficient to find *one* of the many pieces of glass on the ground. Whereas causes are over-determined by their effects, the causal predecessors of an event under-determine it: there are much more traces than those needed for inferring that the window broke. Cleland notes that the undetermination of effects by their putative causes is both epistemic and causal: the shattering is epistemically and causally undetermined by the throwing of a baseball. On the other hand, the shards of glass epistemically over-determine the cause, but are not part of it.

Cleland's point is that the patterns of reasoning of the experimental *and* the historical sciences are grounded in this pervasive time asymmetry of nature (Cleland, 2002, p. 474). Her point can be developed: the time's arrow explains why the laboratory style of reasoning and the historico-genetic style of reasoning have just the features they have (e.g. their kinds of evidence, standard of truth and methods of reasoning) and why they succeed in, respectively, predicting phenomena and inferring causes. Indeed, in order to infer that an asteroid hit the earth about 65 millions of years ago scientists *must reason in the historico-genetic style of reasoning*. That is, they cannot perform an experiment in laboratory but they have to look for a 'smoking gun' (e.g. granules of shocked quartz), proliferate tentative explanations for their presence, use abductive reasoning and pick out the best explanation. The crux of the matter is that, *because of the over-determination of past events*, scientists do not need to find all the traces of the effects of the impact (and cannot look for all of them) in order to succeed in finding the cause. It is *a basic feature of the world that explains the success and the necessity of the historico-genetic style of reasoning*. Now, as the over-determination of past events by *few* present events explains the use of the historico-genetic style of reasoning, so the under-determination of future events by 'initial conditions' explains the use of the laboratory style of reasoning. What the physicists need is to test a hypothesized law or an entire theory, rather than choosing between different explanations. They can manipulate the initial conditions that *may* trigger the phenomenon, make models under different hypotheses, test them and reflect on the additional factors that may be relevant. In brief, *they must reason in the laboratory style of reasoning*—and they must do so because of the under-determination of future events from present initial conditions. Again, it is *a basic feature of the world that explains the success and the necessity of the laboratory style of reasoning*.

To conclude, the time's arrow is an objective feature of the world that explains why the laboratory and the historico-genetic style of reasoning are appropriate for finding out. It is reasonable to say that these two styles of reasoning are not enduring for contingent reasons but because they are suited for predicting the future from present events and inferring the causes from past events.

8.3.3 *Taxonomic Style of Reasoning and the Logic of Evolution*

Gould made a point (Gould, 2000) that can be pushed further in order to claim that the endurance of the taxonomic style of reasoning is ascribable to its capacity to mirror the treelike form of the interrelations between the living species, i.e. the causality of the evolutionary world. Linnaeus' system uses a binomial nomenclature: his taxonomic scheme establishes a nested hierarchy in which every living group is contained in a larger group, e.g. species within genera, genera within families and so forth. One could depict this taxonomy by drawing a sort of dichotomous schema in which each element bifurcates; the entire schema would give the impression of a progressively inclusive hierarchy. Gould notes that, although Linnaeus believed in the fixity of species, his nested hierarchy *implies* Darwin's branching tree, in which a main trunk ramifies into finer divisions. In other terms, although Linnaeus was a firm creationist, his scheme happened to capture an objective feature of the living world: the fact that the interrelationships among organisms follow a genealogical hierarchy built by evolutionary branching. Gould concluded that

> Linnaeus has endured because he combined the best observational skills of his time with a theoretical conception of organic relationships that happens to mirror—but not by pure accident—the topology of evolutionary systems, even though Linnaeus himself interpreted his organizing principle in creationist terms. (Gould, 2000, p. 4)

Gould also reminds us that the form of logic of Linnaeus' system was not 'invented' by him but was in use in other fields of knowledge since the dawn of Western story. For example, Aristotle adopted dichotomous maps for classifying the categories of reasoning (Gould, 2000, p. 7).

To sum up, a particular logic of classification, among the many others (e.g. those proposed by Linnaeus' contemporary scholars), happens to mirror a basic feature of the living world. We can draw two conclusions. Firstly, it was a *contingent* fact that Linnaeus proposed his system although, arguably, the old dichotomous logic of classification would have made its way into the study of living species. Secondly, provided that we continue to have interest in the living world, it is *inevitable* the taxonomic style of reasoning will continue to be used since it succeeds in mirroring the Darwinian causality, a *basic feature of the world.*

8.3.4 *Algorithmic Style of Reasoning, Postulational Style of Reasoning and the Phyllotaxis*

My third example concerns a very well-known problem posed by the mathematician Leonardo Fibonacci (1170–1250) in his book *Liber Abaci*:

> Suppose a newly-born pair of rabbits, one male, one female, are put in a field. Rabbits are able to mate at the age of one month so that at the end of its second month a female can produce another pair of rabbits. Suppose that our rabbits never die and that the female always produces one new pair (one male, one female) every month from the second month on. How many pairs will there be in one year? (in Boyer, 1968, p. 281)

His solution is an algorithm for calculating the number of rabbits after each month: starting with one couple, after n ($n>2$) months there will be a number of couples equal to the sum of the numbers of couples after $n-2$ months plus the number of couples after $n-1$ months:

1, 1, 2, 3, 5, 8, 13, 21, 34 ....

These are called Fibonacci's numbers. On many plants, the number of petals is a Fibonacci number: for example, lilies and iris have 3 petals; larkspur and aquilegia 5; delphiniums 8; cineraria 13; some asters and chicory 21. Then there are daisies with 34, 55 or even 89 petals and the list could go on. Furthermore, Fibonacci numbers are common in fruits' seeds, fruits' sprouts, flowerings and arrangements of leaves.

I shall focus on the *phyllotaxis* of plants, i.e. the arrangements of leaves along the stem. In a plant the leaves are arranged around their stems so that each leaf does not shadow the others. Suppose to go around the stem from leaf to leaf until reaching the leaf directly above the starting leaf (on the same vertical line); and then to go around for another number of turns in order to reach another leaf directly above the starting one and so on. What it is found is that both the number of turns necessary to reach the leaf directly above the starting leaf and the numbers of leaves met are Fibonacci numbers. Take the sunflower. Starting from a leaf, going leaf to leaf and turning clockwise we need one turn to reach the leaf directly above the starting leaf, meeting 3 leaves. Then 2 turns, 5 leaves; three turns 8 leaves; 5 turns, 8 leaves; 8 turns, 13 leaves. That is: exactly the Fibonacci series.

It is possible to say that by using the algorithmic style of reasoning we can know the number of leaves around the stem of a plant at certain fixed positions: the step-by-step procedure to follow is Fibonacci's recurrence relation. Whether the mechanism of phyllotaxis involves genetic factors (Pennybacker & Newell, 2013) or natural constraints (Douady & Couder, 1992) it is also possible to say that, as it were, some plants follow the Fibonacci recurrence relation. To conclude, by using the algorithmic style of reasoning we mirror a basic feature of the living world as if we adopted the same procedures adopted by plants, so to speak.

Let us take a further step. In mathematics two lengths A and B with A>B are in the golden ratio if their ratio A/B is equal to the ratio of their sum A+B to A; the value of the golden ratio is 1,61 …. Now, take a circle into golden proportions, i.e. a circle divided into two arc lengths whose ratio is the golden ratio. The value of the angle subtended by the minimum arc length is 137.5°. It is found that many types of plants have adjacent leaves positioned around the stem at 137.5°. Therefore, in order to find out the relative position of two successive leaves in many plants it is also possible to think in the postulational style of reasoning: project the points in which they grow into a plan and imagine them as two points in a circle, they form an arc that subtends an angle of 137.5°. Put in another way, by reasoning in terms of angles and circles in a diagram we happen to mirror a basic feature of the world: the way leaves are arranged by a plant. Incidentally, the ratio between two successive Fibonacci numbers tends to the golden ratio. This is not surprising since the number of leaves at each turn must grow under the constraint of being located at relative angles dictated by the golden ratio. This relationship between Fibonacci numbers and the golden ratio reminds us that there is consistency between what it is found *independently* by reasoning in the algorithmic style of reasoning and in the postulational style of reasoning.

8.3.5 *Statistical Style of Reasoning, Induction and Probabilistic World*

One of the results of Hacking's historical analysis in *The Emergence of Probability* (1975) is that the coming to the fore of the concept of probability opened the way to the *analytical problem of induction* and to the search for *stochastic laws of chance processes.* My first point concerns the analytical problem of induction. Inductive reasoning is grounded in an objective feature of the world, i.e. the unlimitedness of the space-time: we cannot search the whole world in order to reach certain conclusions; we have to infer them from *specific* facts. When the concept of internal evidence emerged, it was noticed that there are different degrees to which inferences are supported and the concept of probability allowed to measure their reliability. In this sense, the statistical style of reasoning is grounded in the existence of inductive reasoning, which in turn is grounded in the unlimitedness of space-time.

My second point concerns the existence in nature of stochastic processes. Ever since their discovery, many scientists have believed in the possibility of finding the causal mechanisms that could generate statistical laws. They conceived probability as an epistemic fact, i.e. a fact that simply reflects our ignorance of deeper mechanisms. However, the advent of quantum mechanics convinced the majority of scientists that statistical regularity is a brute and irreducible fact of nature. Indeed, the works of the physicists John Von Neumann, Simon Kochen, Ernst Specker and John Bell have showed that it is impossible to provide a 'classical' reformulation of quantum mechanics formalism. That meant the failure of the programme of hidden variable theorists who contended that the statistical laws are the manifestation of some unspecified deterministic laws.

If the world is only seemingly deterministic but at bottom *intrinsically probabilistic*, the use of the statistical style of reasoning is *necessary*. In other words, if stochastic processes are irreducible to deterministic processes the use of the statistical style of reasoning for representing this feature of our world is inevitable. That is, the statistical style of reasoning is grounded in the probabilistic nature of reality. As contingent as the emergence of the statistical style of reasoning may appear, *if stochastic processes are an objective feature of the world* and we continue to be interested in them, the statistical style of reasoning will endure.

8.4 Questions, Answers, Inevitability

In this section I shall address this question: supposed that a question is asked, is science bound to converge on a single answer? Hacking says:

> I am not a contingentist about the content of science, once the questions are intelligible and are asked. But I am inclined to contingentism about the questions themselves. (Hacking, 1999, p. 165)

Hacking maintains that, once certain questions are asked and the correct methods are used, the answers are free of the contingency of human history and represent a fingerpost for successive possible sciences (e.g. see Hacking, 1999). His contingentism about questions is well justified in the light of the fact that contingency plays a crucial role in the emergency of style of reasoning, which in turn bring into being new questions. However, as I have argued in Chap. 7, Hacking's inevitabilism about the answers appears to me correct but quite unjustified within his philosophy. When he suggests that there is only a *correct* answer to live question the adjective 'correct' might be, as in the case of entity realism, relative to the style of reasoning in which the answer is given. By the same token, his formulation of the contingency issue given in Chap. 1 ('If the results R of a scientific investigation are correct, would any investigation of roughly the same subject matter, if successful, at least implicitly contain or imply the same results?') is affected by the same issue: given the anti-foundationalism of the style of reasoning theory (see Sciortino, 2016b) how could we establish that a result R is correct? What makes the things worse is the fact that Hacking believes in the possibility of an alternative physics successful and progressive like ours but incommensurable with ours.

> Such imaginary stable sciences would not even be comparable, because they would be true to different and quite literally incommensurable classes of phenomena and instrumentation. (Hacking, 1992, p. 31)

As Léna Soler remarked, this is 'a new and fundamental type of incommensurability ignored by traditional philosophers of science' (Soler, 2011, p. 9), which is different from that I have been dealing with in Chap. 7. She called it 'machinic incommensurability' because it concerns a scientific practice whose instruments, laboratory procedures and measurement processes share no common measure with the *actual* scientific practice. Since these technical differences translate into different standards of truths, it is difficult to establish by which standards an answer to a live question is correct.

I suggest that, although within the framework of style of reasoning theory there are answers about which we do not have a universal justification, but only a style-dependent justification, there are several important cases in which we have good reasons to conclude that an answer is correct. A case in point is when a scientific result can be achieved by reasoning *independently* in different style of reasoning. I have already pointed out some of these cases. For example, I have already remarked that the Pythagorean theorem has numerous proofs, e.g. it can be proven by adopting the postulational style of reasoning, i.e. in terms of the similarity of triangles, or by rearrangement or, still, by using the differential calculus. Also, in the previous section I have shown how both the postulational style of reasoning and the algorithmic style of reasoning reach the same conclusions about the phyllotaxis of plants. Furthermore, in Chap. 4 I have explained that, although Aristarchus calculated the distances of the Sun and the Moon from the Earth by using the postulational style of reasoning, the same result can be obtained by reasoning in the laboratory style of reasoning. Note also that, as a matter of fact, instead of using demonstrations in order to achieve a geometrical result, one could prove a great number of particular cases by using computer programs, i.e. the algorithmic style of reasoning, and concluding by induction that the demonstration is correct.

This convergence of scientific results independently obtained by adopting different style of reasoning also concerns scientific theories. In *Flussi e riflussi* (Russo, 2003), Lucio Russo argued that although the theory of tides had already been developed in the ancient Greece later on it sank into oblivion and *different* parts of it were taken over by diverse intellectuals. After about 2000 years, Newton reunited these complementary parts by relying on distinct sources and reworked the theory. On the one hand Newton's theory of tides was not conceptually different from that developed in Antiquity, and on the other hand the theory literally rose from its ashes in the shape of one of the remarkable achievements of the laboratory style of reasoning. It is important to note that there are other cases in science in which there are no strong reasons for believing in the correctness of the ontology of a given theory. For example, as James Cushing noted, we could today have arrived at a *very* different worldview of microphenomena if, for contingent reasons, a causal quantum-theory program had been pursued (Cushing, 1992) (Cushing, 1994).

To conclude, although there are specific cases in which we are at a loss to know whether or not a scientific answer is correct, in other cases the fact that different style of reasoning achieves the same results suggests that those very style of reasoning and their methods are reliable. Then, if the style of reasoning is reliable, at least in certain cases it is plausible to believe that soon or later, by a series of fallible steps, science will obtain the *correct* answer to a 'live' question. In the light of these considerations, there are good reasons for being inevitabilist about the convergence of science on correct answers to single questions, but it is contingent that those questions have been asked.

8.5 The Long-Term Evolution of Science

I now want to examine the implications of the theory of style of reasoning concerning the long-term evolution of the sciences. I shall start by noting that, according to the style of reasoning project, over time there has been a growth of *many* epistemological items:

> Phenomena, [...] manipulative and technological skills [...] styles of thinking accumulate. (Hacking, 1983, p. 56)

Before looking into the consequences of this fact, I want to clear the ground from the common misconception that if science grows, in any possible sense, it means that it converges on truth:

> The phenomenon of growth is at most a monotonic increase of knowledge, not convergence. [...] 'Convergence' implies somewhat that there is *one* thing being converged on, but 'increase' has not such implication. (Hacking, 1983, pp. 55–56)

In this connection, it is useful to mention a point made by Howard Sankey: from no realist commitment does it follow that we will end up with a true complete theory of the world (Sankey, 2008, p. 261). For instance, the realist claim that scientific progress consists in advancing towards the truth does not imply that science converges on a *single* true theory: scientific inquiry might converge on different truths that do not add up to one whole truth. Similarly, to say that science leads to knowledge does not amount to saying that scientific inquiry will lead to the *whole* truth about the world. Sankey's point suggests that for a realist it is not *inevitable* that science will reach a final, coherent, single description of the world. But, of course, from the realist claim that science leads to knowledge it does follow that it is inevitable that science will converge on at least *some* truths, e.g. a (approximately) true description of a *portion* of the world.

Let us now keep realism aside. I want to investigate why one cannot expect that there is not *one* thing being converged on. Hacking says that there can be heapings up of knowledge without there being any unity of science to which they all add up. (Hacking, 1983, p. 55)

By spelling out this sentence it is possible to provide a picture of the evolution of scientific inquiry in the long run. The metaphor of 'heapings up' of knowledge suggests that some of the knowledge acquired by using a given style of reasoning is in some sense distinct from that acquired by using another style of reasoning[1]; the idea of no unity suggests that these supposed different parts of the our knowledge are not interconnected one with another. I shall clarify first what exactly accumulates and how; then I shall examine the question of the interconnectedness.

I recall that the discontinuous emergence of different style of reasoning implies that at different times of the history unrelated sets of questions have been considered important. For example, while after the emergence of the statistical style of reasoning problems in social mathematics such as the distribution of age at marriage has been a matter of study, they were not 'live' in the ancient Greece.

The metaphor of 'heapings up of knowledge' should not invite the thought that knowledge is simply something that is 'out there' and needs to be picked up. What a style of reasoning 'puts in a heap' is for example 'new objects', objects that did not exist before, such as the number zero in the case of the postulational style of reasoning or the concept of species in the case of taxonomic style of reasoning, or 'new phenomena', which did not exist before the emergence of the laboratory style of reasoning, e.g. the laser or the Compton effect.

Furthermore, as Stéphanie Ruphy points out, 'new objects' do not simply add further to all the objects of our knowledge but determine an 'ontological enrichment of the objects of science, to the extent that the use in scientific practice of different styles of reasoning widen and diversify the classes of propositions that can be true or false about them' (Ruphy, 2011, pp. 1220–1223). She gives the example of a forest fire: the statistical style adds statistical properties and the laboratory style adds controlled and purified versions of the natural phenomenon. The result is an object ontologically richer than the object 'forest fire' of the everyday life.

A first point we can make is that in as much as new questions, objects and phenomena are introduced by style of reasoning, their contingence is parasitic on that of the style of reasoning they belong to. Ultimately, the

[1] See also Sciortino (2016a).

contingency at the level of styles of reasoning implies a contingency at the level of the ontology. The next question is whether the knowledge acquired by all the style of reasoning converges on any unity. I think that one can give a straight answer in the negative for several reasons. Indeed, the concept of convergence implies that what it is converged on is something that for some reason is final and completed. But the process by which many items of knowledge emerge is open-ended. Firstly, within a style of reasoning new phenomena as well as new objects are in principle 'created' continuously. Secondly, new style of reasoning may emerge in the future, e.g. as consequence of the informational revolution, bringing into being new questions, instruments and previously unknown phenomena.

Furthermore, it might be impossible to connect all we have already found out by adopting the six style of reasoning, from the evolution of disc galaxies to how enzymes work (see also Hacking, 1996). Firstly, even though it were possible in principle to connect logically different parts of sciences there would not be certain that *actually*, soon or later, they would get connected. The history of physics offers many examples of different accounts that ended up being considered equivalent: e.g. the two different formulations of the differential calculus of Newton and Leibniz or the wave and matrix mechanics formulations in quantum mechanics. But, as Hacking points out, that was a *contingent* fact:

> [M]en like Laplace and Lagrange working around 1800, were in some sense obtaining consequences of Newton's laws of motion. [...] They had to invent the language in which the conclusions could be expressed. [...] They were not just joining up the dots to complete a picture. They had to put in the dots. (Hacking, 2002 [1982] -a, p. 175)

In addition, the theory of style of reasoning opposes two 'theses of unity', which Hacking has discussed in 'The disunity of science' (Hacking, 1999, p. 76). The first one is a thesis of *interconnectedness*: the idea that all the phenomena are related to each other. This is the conception of the world that urges scientists to unify all the fundamental forces in terms of a single field. The second is a *structural* thesis: the idea that the truths about the world are connected with each other and form a structure in which some of them imply all the others. In particular, the advocates of this thesis might maintain that there are certain causes, necessary or probabilistic, which imply all the other causes of phenomena; or they might claim that there are logical relations between all the laws of the universe. Some scientists are neither reductionists nor holists but they believe that all the explanations stand in logical relation (Hacking, 1996, pp. 46, 47).

If the laws of two classes of phenomena are not logically connected, it seems impossible that these two classes of phenomena are interconnected. So, to deny the second thesis means to deny the first one too. Now, the theory of style of reasoning denies the second thesis. Indeed, for instance, the kind of explanations provided by the historico-genetic style of reasoning are radically different from those provided by the laboratory style of reasoning. The historical explanations have the character of a story that cannot be tested in a laboratory.

The points I have made in this section make it clear that the knowledge that we acquire by adopting different style of reasoning has no unity and cannot converge on any kind of unitary description of the world. It is for this reason that Hacking's metaphor is appropriate: there are 'heapings up' of knowledge because each style of reasoning accumulates new laws, phenomena, facts and objects that cannot be connected by a single description of the world. The idea of convergence of science on a single true, coherent and complete description in the long run is at odds with the mechanism by which the style of reasoning produces knowledge: the growth of many different epistemological items is open-ended and never final and complete.

8.6 Conclusions

In this chapter, I have considered a number of connections between various ideas of the theory of style of reasoning and the issue of the inevitability of science. By addressing four fundamental questions I have discussed how *both* inevitability and contingency play a major role in the evolution of sciences. On the basis of my discussion, I advocate a position between the two extremes of contingentism and inevitabilism. Given the point of origin of the evolutionary tree of the style of reasoning, i.e. the birth of the algorithmic style of reasoning, the expansion of trunk and branches is a random growth away from simple beginnings that produce more and more complexity.

Bibliography

Barrow, J. (1992). *Perché il mondo è matematico?* Laterza.
Bowler, P. (2003). *Evolution: The History of an Idea*. University of California Press.

Boyer, C. B. (1968). *A History of Mathematics*. Wiley International Edition.
Cleland, C. (2002). Methodological and Epistemic Differences between Historical Science and Experimental Science. *Philosophy of Science, 69*, 474–496.
Cushing, J. (1992). Historical Contingency and Theory Selection in Science. *Proceedings of the Biennial Meeting of the Philosophy of Science Association, 1*, 446–457.
Cushing, J. (1994). *Quantum Mechanics: Historical Contingency and the Copenhagen Hegemony*. Chicago University of Chicago Press.
Dehaene, S., & al, e. (2006). Core Knowledge of Geometry in an Amazonian Indigene Group. *Science, 311*, 381–384.
Devlin, K. J. (2005). *The Math Instinct*. Basic Books.
Douady, S., & Couder, Y. (1992). Phyllotaxis as a Physical Self-Organized Growth. *Physical Review Letters, 68*(13), 1098–2101.
Eddington, A. (1928). *The Nature of the Physical World*. Cambridge University Press.
Gould, S. J. (2000). Linnaeus's Luck? *Natural History, 109*(7), 18–29.
Hacking, I. (1983). *Representing and Intervening*. Cambridge University Press.
Hacking, I. (1992). The Self-Vindication of the Laboratory Sciences. In A. Pickering (Ed.), *Science as Practice and Culture* (pp. 29–64). Harvard University Press.
Hacking, I. (1996). The Disunity of Science. In P. Galison & D. J. Stump (Eds.), *The Disunity of Science: Boundaries, Contexts, and Power* (pp. 37–74). Stanford University Press.
Hacking, I. (1999). *The Social Construction of What?* Harvard University Press.
Hacking, I. (2002 [1982] -a). Language, Truth and Reason. In *Historical Ontology* (pp. 159-177). , Massachusetts Harvard University Press.
Hacking, I. (2007). The Laboratory Style of Thinking and Doing. Retrieved from http://stm.ym.edu.tw/files/u1/HW_Ian_Hacking.pdf
Hacking, I. (2009). *Scientific Reason*. NTU Press.
Hacking, I. (2011). How did mathematics become possible? Retrieved from http://www.youtube.com/watch?v=PbKUsAR_8DY
Hacking, I. (2014). *Why is there a Philosophy of Mathematics at all?* Cambridge University Press.
Lloyd, G. R. (2000). Watching for the Eclipse. *The Times Literary Supplement*.
Netz, R. (1999). *The Shaping of Deduction in Greek Mathematics: A Study in Cognitive History*. Cambridge University Press.
Pennybacker, M., & Newell, A. C. (2013). Phyllotaxis, Pushed Pattern-Forming Fronts, and Optimal Packing. *Physical Review Letters, 110*(24), 1.
Psillos, S. (2011). An Explorer upon Untrodden Ground: Peirce on Abduction. In D. M. Gabbay, F. Pelletier, & J. Woods (Eds.), *Handbook of the History of Logic*. (Vol. 10).
Ruphy, S. (2011). From Hacking's Plurality of Styles of Scientific Reasoning to 'foliated' Pluralism, a Philosophically Robust Form of Ontologico-methodol ogical Pluralism. *Philosophy of Science, 78*(5), 1212–1223.

Russo, L. (2003). *Flussi e Riflussi*. Feltrinelli.
Sankey, H. (2008). Scientific Realism and the Inevitability of Science. *Studies In History and Philosophy of Science Part A, 39*(2), 259–264.
Sciortino, L. (2016a). E Pluribus Unum. Gli Stili del Pensiero Scientifico P. *?rometeo, 133*(34), 22–29.
Sciortino, L. (2016b). Styles of Reasoning, Human Forms of Life, and Relativism. *International Studies in the Philosophy of Science, 30*(2), 165–184. https://doi.org/10.1080/02698595.2016.1265868
Seife, C. (2000). *Zero. The Biography of a Dangerous idea*. Penguin.
Soler, L. (2011). The Incommensurability Problem: Evolution, Current Approaches and Recent Issues. *Philosophia Scientiæ, 8*(1), 1–38.
Van De Mieroop, M. (2011). *A History of Ancient Egypt*. Wiley.

CHAPTER 9

Epilogue

The idea that human beings have thought, known or found out in totally different ways in the course of history is widespread in the literature. However, philosophers and historians of science have given shape to this idea in different ways. For clarity's sake, I have expressed this point by saying that there is a concept, which I have called the concept of 'ways of thinking', of which, implicitly or not, different articulations or versions have been proposed. Of course, in place of 'way of thinking' other expressions, such as 'worldview', 'form of thought', way of knowing, style of thinking and many others would have been equally suitable for my purpose.

This being premised, we can say that this is a book on the concept of way of thinking and its different notions proposed in the context of historical epistemology. As I have argued in the first chapter, the thinkers in the wake of this latter tradition of thought reflected on the production of scientific knowledge by analysing scientific texts of the past and by making comparisons between different epochs and societies. What they realized is that in different times the production of scientific knowledge is based on different categories of thought, presuppositions and concepts. However, their ideas differed, among other things, about the role of these epistemological items, how they are related to each other in different times, how they have changed throughout history. The distinct notions of the concept of way of thinking that they put forward, such as 'mentality', 'a priori', 'thought style' and many others, reflect these differences.

L. Sciortino, *History of Rationalities*,
https://doi.org/10.1007/978-3-031-24004-1_9

I have presented these notions in Chap. 2 and in Chap. 3, while in Chap. 4 I have compared and contrasted them. My following point has been that these notions contain in embryonic form important philosophical issues concerning relativism, contingency and convergence to truth. In the case of some of them, such as Kuhn's or Feyerabend's, these issues have been well examined in the literature. But, in the case of Hacking's notion, the question has not yet received the attention it deserves although it represents the latest sophisticated version of a long tradition of notions of way of thinking that have been conceived in historical epistemology. As I have argued in the same chapter, the project of styles of reasoning has much to say about the relativism and the contingency issue. However, this issue cannot be tackled without developing Hacking's project. This has been my task, in particular, in Chaps. 5 and 6.

In Chap. 5, I have given substance to the notion of style of reasoning by characterizing six modes of thinking that, although different in nature, subject matter and historical trajectory share a set of common features. Styles of reasoning are not only specific ways of reasoning but also ways of doing and finding out; they are long lasting, have sharp beginnings, possess techniques of self-vindication and introduce new objects, standards of truth and candidates for truth-or-falsehood. The *internal* techniques of self-vindication—I have argued in Chap. 5—make styles of reasoning different from other ways of thinking, in particular those that are not in use in scientific research.

In Chap. 6, I have started to draw some major implications from the styles project. First of all, I have addressed the question as to whether there is an independent and atemporal justification for every claim made in a certain style of reasoning. My answer has been in the negative and my argument can be found in Chap. 7. I have argued that the claim of entity realism made by Hacking is a case of a claim in the laboratory style of reasoning that has no independent and atemporal justification outside it: whether the laboratory style of reasoning is a way of finding out nearly theory-free (as Hacking claims) or not, Hacking's characterization of styles of reasoning implies that entity realism is justified in the laboratory style of reasoning and unjustified outside of it. This is a thesis of incommensurability: two different communities that at different times adopt two different styles of reasoning may not have common research standards to assess the truth of a scientific claim. At a more general level, my point implies that Hacking's experimental realism—the claim that we *know* about unobservable entities—is not consistent with the epistemic relativistic implications of his styles of reasoning project.

Finally, I have examined the implications of the styles of reasoning project for the contingency issue. I have (1) given a picture of the evolution of sciences in terms of the emergence and stabilization of the styles of reasoning (2) discussed how *both* inevitability and contingency play a major role in the emergence and endurance of styles of reasoning (3) pointed out that each style of reasoning is grounded in certain objective features of the portion of the world it studies (4) argued that there are good reasons for being inevitabilist about the convergence of science on correct answers to single questions; although, as Hacking maintains, it is contingent that those questions have been asked. Point (3) and the self-authenticating character of styles of reasoning can explain why styles of reasoning are long lasting. This relevant point can be drawn from my discussion in Sects. 5.6.2 and 8.3. To finish, on the basis of the features of the styles of reasoning I have concluded that the idea of convergence of science in the long run on a single true, coherent and complete description does not jibe with the claim that the styles of reasoning continually enrich our present ontology.

Index[1]

[1] Note: Page numbers followed by 'n' refer to notes.

L. Sciortino, *History of Rationalities*,
https://doi.org/10.1007/978-3-031-24004-1

Printed in the USA
CPSIA information can be obtained
at www.ICGtesting.com
LVHW080614221123
764620LV00007B/413

9 783031 240034